ORIGINS OF THE MODERN MIND

ORIGINS OF THE MODERN MIND

Three Stages in the Evolution
of Culture and Cognition

MERLIN DONALD

Harvard University Press
Cambridge, Massachusetts
London, England   1991

Library of Congress Cataloging-in-Publication Data

Donald, Merlin, 1939—
  Origins of the modern mind: three stages in the
evolution of culture and cognition / Merlin Donald.
      p.   cm.
  Includes bibliographical references and index.
  ISBN 0-674-64483-2 (alk. paper)
  1. Cognition and culture. 2. Neuropsychology.
3. Cognition. 4. Intellect—History. I. Title.
BF311.D572 1991
153—dc20   91-9989
                                    CIP

Designed by Gwen Frankfeldt

In the beginning was the mounting fire
That set alight the weathers from a spark,
A three-eyed, red-eyed spark, blunt as a flower;
Life rose and spouted from the rolling seas,
Burst in the roots, pumped from the earth and rock
The secret oils that drive the grass.

In the beginning was the word, the word
That from the solid bases of the light
Abstracted all the letters from the void;
And from the cloudy bases of the breath
The word flowed up, translating to the heart
First characters of birth and death.

Dylan Thomas, "In the Beginning," 1934

CONTENTS

ORIGINS OF THE MODERN MIND

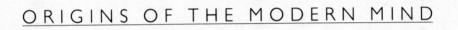

Prologue

In most areas of science, theories of origin are in the mainstream. The origins of the solar system, of the Earth, of its continents, mountains, and life forms, and of the human race itself are central issues in a variety of disciplines, from astronomy to zoology. Yet the question of human cognitive origins has not been in the forefront of cognitive science. In fact, cognitive science has built its edifice mostly on the study of two recent, and highly specialized, kinds of mind: literate English-speaking adults and computers.

In animal research, an important lesson was learned long ago: to do justice to the mental capabilities of animals, we must study a variety of species and place them in an evolutionary context. We cannot understand their cognitive capacities without accounting for their place in the biological order. For some reason, human cognitive research does not seem to have learned the same lesson. Cognitive science often carries on as though humans had no culture, no significant variability, and no history.

This neglect might reflect, in part, an aversion to theoretical speculation. But historically, the proscription of speculation in science has been unproductive, justified only by the strange illusion that no knowledge is better than some. Broad integrative theories of origin have traditionally predicted, synthesized, and inspired experimental work. Such theories have often *preceded* the strong confirmatory data upon which they would stand or fall; their function has been heuristic and imaginative, and their virtue has resided less in their correctness than in their ability to provide direction. Cognitive science needs more such theories, and more of that speculative wonder and respect for the whole that has raised other fields to such spectacular heights.

The objective of this book is to reflect upon the most likely scenario of human cognitive emergence. I consider data from many disciplines, and on this evidential base I propose a specific hypothesis on the origins of the modern human mind. Although archeological and anthropological data provide crucial insights, the main event is on the level of cognitive evolution; cognition is the mediator between brain and culture and therefore must have been the engine, as well as the locus, of change.

This is easily said, but human cognition is complex and multidimensional. How should its evolution be approached? There is a large literature comparing humans with other species on particular aspects of behavior, including reflexes, instincts, perceptual capacities, conditioned behavior, problem solving, insight, curiosity, communication, language, social intelligence, and memory. But unfortunately the continuum from reptile to mammal to primate to human has not always proven to be a smooth one; some animals appear better at one thing, some at another. For this reason, comparative studies have painted a somewhat confusing picture.

However, there is another possible approach, based on the underlying *structure* of mind. During the recent revolution in cognitive science, a new style of conceptualizing the mind has developed, one that is best described as structural. One major objective of structural models of mind is to describe modular architecture, that is, the overall configuration of components making up the totality of mind. Clinical neuropsychological evidence is central to such models, because it addresses the question of cognitive breakdown or dysfunction; components of mind sometimes break down independently of one another, providing clues as to their place in an overall structural scheme.

The essence of my hypothesis is that the modern human mind evolved from the primate mind through a series of major adaptations, each of which led to the emergence of a new representational system. Each successive new representational system has remained intact within our current mental architecture, so that the modern mind is a mosaic structure of cognitive vestiges from earlier stages of human emergence. Cognitive vestiges invoke the evolutionary principle of conservation of previous gains and are similar in principle to the many other vestigial behaviors we possess—for instance, baring the teeth in anger, or wailing in grief.

The modern representational structure of the human mind, I will argue, thus encompasses the gains of all our hominid ancestors, as well as those of certain apes. Far from being a diffuse *tabula rasa*, modern human cognitive architecture is highly differentiated and specialized. And despite our close genetic relationship to apes, the cognitive distance from apes to humans is extraordinarily great, much greater than might be imagined from comparative anatomy.

The key word here is *representation*. Humans did not simply evolve a larger brain, an expanded memory, a lexicon, or a special speech apparatus; we evolved new systems for representing reality. During this process, our representational apparatus somehow perceived the utility of symbols and invented them from whole cloth; no symbolic environment preceded them.

The problem of symbolic reference has always been the Achilles heel of computational approaches to language. The difficulty is this: to understand or use a symbol appropriately in context you must first understand what it represents, and this referential understanding is inherently nonsymbolic. For instance, to understand why it is amusing to name a reclusive hound dog "Raskolnikov," one must know something of Dostoevsky's novel, something of the brooding, asocial intelligence of hound dogs, and something of the style of contemporary humor. None of this knowledge can be obtained just by looking up other symbols. But that is all most computational algorithms can do. The programmer, or the user, must eventually provide meaning, since the computer has no knowledge of its own.

Moving to a nonrepresentational computer model (Brooks, 1989) won't solve the problem either; such systems must ultimately face the same limitations as an animal without symbolic intelligence. Just as traditional AI programs use nothing but symbols, most animals cannot use symbols at all, and no animal except humans has ever invented a symbolic device in its natural environment. The question is, how did humans, given their nonsymbolic mammalian heritage, come to represent their knowledge in symbolic form? Through what stages must this development have passed? How did humans bridge the tremendous gap between symbolic thought and the nonsymbolic forms of intelligence that still dominate the rest of the animal kingdom? During the relatively short time of human emergence, the structure of the primate mind was radically altered; or rather, it was gradually surrounded by new representational systems and absorbed into a larger cognitive apparatus.

One unusual aspect of this book is its incorporation of biological and technological factors into a single evolutionary continuum. I will argue that very recent changes in the organization of the human mind are just as fundamental as those that took place in earlier evolutionary transitions, yet they are mediated by new memory technology, rather than by genetically encoded changes in the brain. The effects of such technological changes are similar in kind to earlier biological changes, inasmuch as they can produce alterations to the architecture of human memory. The modern mind is thus a hybrid structure containing vestiges of earlier stages of human emergence, as well as new symbolic devices that have radically altered its organization. The structural relationship between individual human minds and external memory technology continues to change, and this book provides the beginnings of a conceptual framework within which our continuing mental evolution may be viewed.

The Need for a Theory of Cognitive Evolution

Mental Architecture as an Emergent Phenomenon

Neuropsychology and cognitive science are concerned largely with the fundamental structure of the modern human mind. Although some attention has been paid to the phylogenetic succession of changes that must have led to the modern mind (see, for instance, Anderson, 1983), emphasis has been placed mostly upon the modern structure of human mental capacities, without taking their evolution into account. It is not an exaggeration to say that theories of cognitive structure are built mostly upon studies of the human mind as manifest in literate, postindustrial society and upon studies of the capabilities of computers. The extraordinary range of theory that has resulted was constructed for the most part without the constraints that must be applied to evolutionary hypotheses: continuity with previous forms, consistency with selection pressures, parsimony with regard to the number and complexity of successive adaptations, and so on. The result is that structural, that is, modular, models of mind proliferate without regard to their biological feasibility, even within neuropsychology.

Within the traditions of biopsychology and comparative biology, on the other hand, humans have been viewed "from below," that is, in light of their mammalian predecessors. Phylogenesis has been a major concern to many thinkers in these fields, and efforts have been made to build the crucial conceptual bridge between the mental structures of apes and humans. A recent example is Lieberman's (1984) attempt to specify the adaptations leading to human language. Since apes, and particularly chimpanzees, are genetically so close to humans, the key to the unique power of the human intellect is sought in its contrast with that of apes. However, theories emanating from these fields have not been rich in detail about mental structure, particularly about higher cognitive functions. This is partly because there are few equivalences in the terms used to describe the cognitive capacities of animals and humans; and those that are used often fail to do justice to the complexities of human cognition.

But whether a given theory of cognitive evolution starts "from below" or "from above," it will wind up taking one of two broadly different approaches to higher function, which may be labeled "modular" and "unitary." Modular theories, sometimes called faculty theories, propose a number of quasi-independent cognitive "modules" that are responsible for each dissociable or isolable aspect of higher function. The specific arrangement of modules in each theory usually varies, but generally each mental module encapsulates a definable higher mental function, or a stage of such a function. There may be separate structures (and, by implication, separate evolutionary adaptations) for spatial reasoning, mathematical ability, musical talent, phonological skill, the oral lexicon, the written lexicon, visual imagery, nonverbal thought, and verbal thought, to name a few.

Modular theories of language evolution find their biological roots in the writings of Darwin (1871) and Wernicke (1874), who both supported the idea of a series of special human adaptations leading to speech. These notions have found a modern expression in the biological theories of language proposed by Lenneberg (1967) and Lieberman (1975, 1984). The most recent influential example of a modular approach is found in the neurolinguistic theorizing of Shallice (1988).

Unitary theories, by contrast, generally hold that higher function is achieved by a single cognitive structure, that is, a single adaptation, with an exception sometimes being made for the peripheral, or sensorimotor, mechanisms of language. Anderson (1983) made a persuasive case for this approach in his book *The Architecture of Cognition*. His defense of a unitary approach rests on three arguments: (1) human higher function has a very short evolutionary history, and there would not have been time to evolve special faculties, for instance for mathematics; (2) human intelligence is highly plastic, or flexible, and displays a variety of special skills that could not have been anticipated in evolution; (3) the various higher cognitive functions of humans have many features in common.

Anderson has argued that a single general framework, or functional architecture, is sufficient to support all forms of higher function, even language. His approach echoes the faith of an earlier generation in a general-purpose "learning" capacity as the crucial adaptation underlying higher function. In fact, it is no coincidence that his ACT⋆ framework is basically a diagram of memory structure.* What Anderson is saying is that the main "hardware" requirements of a computational model of human cognition (or of any complex mind) are general features of the system of storage and retrieval rather than separate specialized systems for various skills. It follows that the crucial human adaptation was some refinement of the memory system, especially of declarative and working memory, that enabled symbolic representation and language.

The unitary approach receives some support from primatology and comparative anatomy. A variety of studies over the past two decades have supported the *encephalization hypothesis*. Encephalization refers to the progressive increase in the relative size of the brain, particularly the cerebral cortex, during mammalian, and especially primate, evolution. Jerison (1973) and Passingham (1982), among others, have argued that the progressive growth in brain size characteristic of the great apes and hominids has followed a systematic trend, and that modern humans are simply an extension of that trend. Some particularly interesting data from Passingham show that the human brain has the proportions one would predict by extrapolating the trend of earlier primate brain expansions. Put simply, humans possess exactly the brains to be expected of large-brained primates. This proportionality extends even to the relative size of a variety of specific nuclei, ganglia, and substructures within the brain. In other words, there is no support in gross neuroanatomy for a completely novel language subsystem in the human brain, or for any other type of gross anatom-

---

*ACT⋆ is an acronym for Adaptive Control of Thought, a systematic computational approach to high-level cognition.

ical reorganization. This does not rule out some still-undetected molecular innovation related to higher function. But until such an innovation is uncovered, it appears that the most distinct property of the human brain is simply its extraordinary increase in relative size, which translates into increased numbers, and complexity, of memories, according to Jerison and Passingham.

The encephalization hypothesis in its strongest form holds that increased cognitive power is a direct result of increased encephalization. In principle, it argues for the kind of unitary structure of mind offered by Anderson. It implies that human cognitive power, including the higher aspects of language, is rooted in the superior computational capacity of the whole brain, and especially the cerebral cortex, rather than in any specific new structure. Lieberman (1984) walked the fine line between the unitary and modular approaches by assigning specifically linguistic adaptations, like control of the vocal tract and phonological processing, to specifically human neural speech processors, while leaving most higher functions to what he called a "central distributed neural computer." In other words, the front-end aspects of speech may depend on special neural processors controlling phonation, but verbal thought and higher cognition in general are made possible by a single biological adaptation: the expansion of the cerebral cortex. This aspect of Lieberman's theory is quite similar to an earlier proposal by Fodor (1983) (see chapter 3), which also placed the higher aspects of linguistic function in a general-purpose processor and the lower aspects in a self-contained language "module."

The form of any evolutionary proposal regarding language and thought will be greatly affected by the number, and types, of cognitive modules associated with human higher function. The simplest evolutionary proposal would be a unitary system supported by the growth of the entire cortex, such as the one proposed by Jerison (1973). In his theory, the history of language and thought becomes the history of cortical expansion. As the encephalization quotient goes, so goes cognitive architecture and intelligence.

A more complex evolutionary scenario might involve many individual modules, each serving a different cognitive purpose. For instance, there might be many specialized language and thought modules that possess some degree of anatomical localization. The evolution of such a complex mosaic structure, unique to humans, would imply that a series of separate adaptations led us from the cognitive architecture of apes to our present structure. A selection advantage driving the acquisition of each aspect of the mosaic would have to be specified. Moreover, the number of possible sequences of change would be greatly increased, and the empirical basis for making a choice between them would be very limited. Such a scenario would be difficult to construct, particularly since, on present evidence, the evolution of language occurred so quickly. By way of comparison, the neural mechanisms of mammalian vision evolved over millions of years, through a large number of species. For a neural structure of at least equal complexity to evolve and eventually (surely not in its initial manifestation) support language, we should presumably expect an equally long and arduous process of evolution.

Thus Anderson's principal objection to a modular theory stands: a complex modular solution to the problem of human higher function would not lend itself easily to an evolutionary model. Since we are not in a position to abandon evolution as an explanatory principle, we should be wary from the start of modular neuropsychological models which specify an architecture that depends on many built-in, specialized neural components. In any case, it is clear that any evolutionary hypothesis will be greatly affected by its creator's initial stance on cognitive structure. The reverse is also true. The credibility of any structural hypothesis will be affected by the evolutionary scenario it implies. Phylogenesis and structure are two aspects of the same thing—the one a motion picture, the other a single frame, both trying to represent the same reality. There are times when the single frame is clearer, and there are other times when a blur in a single frame is resolved by motion. Since our view of higher function is not very good to start with, we cannot afford not to try to use both approaches.

Culture as Evidence for Cognitive Structure

One element frequently left out of cognitive modeling is the element of culture, that is, shared patterns of acquired behavior characteristic of a species. But the cognitive capacities of animals directly affect the kinds of culture they produce, and in the case of humans, the opposite is also true: specific types of human culture have direct effects upon individual cognition. In fact, the uniqueness of humanity could be said to rest not so much in language as in our capacity for rapid cultural change.

If we wished to put the proposition even more strongly, we might assert that what humans evolved was primarily a generalized capacity for cultural innovation. Part of that capacity was linguistic communication; part of it was the ability to think and represent the environment. This is a very easily defended proposition from the viewpoint of evolutionary theory; the selection advantages accruing to a species capable of cultural invention would be tremendous. Archaeological and anthropological evidence strongly supports the idea that the rate of cultural change increased during hominid evolution, at first slowly (*Homo erectus* changed only slightly over a period of a million years), then more rapidly (early *Homo sapiens* achieved several major innovations over a period of 200,000 years), and finally at the continuously accelerating pace of our own species, *Homo sapiens sapiens.*

Dunbar (1990) has recently proposed that encephalization was driven not by the cognitive demands of toolmaking or spatial mapping of the environment but by growth in the size of social groups. In other words, it was not instrumental intelligence that drove brain expansion but rather social intelligence. Complex societies make great demands on memory: large numbers of relationships have to be analyzed, understood, stored, and serviced regularly in order to sustain a large group organization. With certain exceptions, the more advanced primates cluster together into larger and larger social groups, culminating in the human capacity for organizing and sustaining very large groups. It could well be the case that the intellectual abilities needed to sustain large groups are identical to those that enable cultural invention. The first adaptations in the hominid line might have been driven by the demands of social grouping, and cultural invention might have been its by-product.

The main difficulty with such a proposal is its vagueness about mechanism. What biological mechanism would enable continuous cultural invention, or maintain larger social groupings? Inevitably such proposals lead one to speculate about cognitive mechanisms operating on the level of the individual brain. Thus, the bridge from cultural to biological realms is necessarily cognitive, and a complete evolutionary proposal should address the cognitive level, even if its ultimate objective is to explain our capacity for cultural change.

In any evolutionary theory, cultural evidence must play an important role, since it is not reasonable to expect that every "module" of higher function is functionally present in every neurologically normal human mind. To provide an example, if Morton's (1980, 1981) well-known model of spelling* is taken at face value, the cognitive modules that constitute the heart of the visual path in the model— the visual input lexicon, and the graphemic output lexicon—are obviously absent in illiterate humans. Morton would not propose that illiterate humans are therefore missing certain essential biological modules. He has acknowledged that the modules do not in fact represent anatomical entities but only functional ones; and it is evident that the physical underpinnings of the visual path are normally present in the brain, since illiterate humans can usually be taught to read.

This, of course, would beg the question. The point is, some of his modules support species-universal traits that are not culturally bound (for instance, all the modules supporting speech); these probably represent distinctive biological adaptations. Other modules, like those supporting reading and writing, support behaviors that are obviously not species-universal and are largely or entirely bound by culture. The evolutionary basis for these, and thus the expectation of a distinct physical basis for them, may be very different. In other words, all modules, however closely reasoned on the cognitive or neuropsychological evidence, were not necessarily created equal, even though the neuropsychological criteria for their creation were similar.

If we accept the validity of the sorts of modules Morton and others routinely construct from neuropsychological evidence, it follows that the *actual cognitive structure* of an individual mind is heavily influenced by culture. Styles of reading and writing are culture-bound.

---

*Morton (1980) proposed a model of orthographic structure that contained four lexicons, or dictionaries. Two of these lexicons are primarily auditory-phonetic and are used in listening and speaking, respectively, while two are primarily visual and are used in reading and writing. The writing lexicon contains *graphemic* entries, that is, the central production systems for producing written script. The lexicons map onto one another through various functional paths, which can be selectively disrupted in patients with various kinds of reading disorders (dyslexias) and writing disorders (agraphias).

Thus, depending upon the specific skills demanded by literate culture, prospective model-builders have to introduce modules for different functions. For example, in the brain of a reader of modern English, as Morton and many others would agree, there must be a module that performs direct grapheme-to-phoneme translation. But in a largely ideographic writing system, such as ancient Chinese, there is no need for such a device; instead, there must be a device that performs very complex image-to-meaning-to-syllable translation, as well as direct image-to-syllable matching. This follows from the way the symbols are configured. There are other solutions to reading: hieroglyphs employ a rebus principle, rather than an alphabet. They may be read in various ways, depending upon the context, and on certain marker symbols that can be embedded in the text. Reading them involves a very different structural arrangement of cognitive modules than reading alphabetic material.

Neuropsychological dissociation of modules is taken as good evidence that the modules exist as functional units in the brain (Shallice, 1988). Presumably Shallice would scour the hospitals for speakers of ancient Egyptian who happened to read hieroglyphs, hoping to encounter various types of dyslexics; the academic purpose of the exercise would be to establish that ancient Egyptian reading did in fact break down in ways consistent with its putative internal modular structure. Thus, one might find dyslexics who have specifically lost the linkage between image, meaning, and sound at various points or in various combinations. The point of studying these pathologies would be to construct a functional model of the ancient reader's brain.

But why stop with reading? Every culturally bound skill deserves a similar neuropsychological dissection, and any dissociable, informationally encapsulated module that emerged would similarly receive the distinction of its own modular structure within the larger cognitive architecture. Reading attracts more attention because of its importance, but virtually any highly overpracticed skill is bound to have a distinctive modular structure that can break down in various predictable ways. This is not a frivolous notion: the brain of a professional tennis player is undoubtedly employing its resources in a very different way than it would have if, for cultural reasons, the same individual had grown up instead to become a very unathletic biblical scholar.

This idea receives some support from recent neurophysiology, especially in the study of cortical plasticity, or the malleability of the central nervous system. It has been known for some time that the immature brain is highly plastic; that is, it can grow connections, and lose connections, in many different ways, depending on early experience. Changeux (1985) has proposed an epigenetic theory of brain development, in which the young brain proliferates new connections fairly indiscriminately, that is, invents many possible routes of development, of which only a few will survive, due to the selective effects of experience. His ideas are a mirror-image of the neurological solution offered earlier by Hebb (1949) of selective synaptic growth.

Changeux was aware of the cultural corollary of this notion: namely, that culturally specific, highly redundant patterns of brain use would imply the existence of cultural traditions that have an indirect *neurological* instantiation—that is, the brains of many individuals in a particular culture are broadly programmed in a specific way, while in another culture they may develop differently, because use patterns are fundamentally different. This idea has gained credibility from recent studies of monkeys, using single-cell mapping of cortical regions. Merzenich (1987) and his colleagues have demonstrated that the cortical sensory region for the hand expands and contracts with demand, even in the adult. That is, the detailed topographic maps, or sensory fields, that project various skin sensations onto the somatosensory cortex are not static, fixed maps but dynamic computational resources. They recruit adjacent cortical columns when demand is heavy, that is, when the animal is required to make fine discriminations, and shrink back when demand is reduced. There are strict limits on plasticity; but the point is, the use of a given cortical area is decided by competition between its fixed input pathways, with the most active pathways dominating. The association cortex, with so many diverse input paths, may therefore be influenced in many directions.

Importantly, the latter is true in adults as well as infants. The implication of Merzenich's work is that when children or adults learn a novel skill like Braille-reading, they are literally reconfiguring their sensory cortex for the fingertips. There is a cost to this, and a limitation; in order to increase the cortical fingertip region they may have to trade off resources for some other aspect of somatic sensation, since cortical resources are limited. But the point is, the cortex is not cast in stone, not even in its sensory areas; its configuration is highly plastic, determined by use. And by disuse; failure to use a source of input will lead to shrinkage of its representation in the brain.

A corollary is that the evolution of cognitive structure *at the modular level* might have continued well beyond the point at which physical evolution had stopped. Cultures restructure the mind, not only in terms of its specific contents, which are obviously culture-bound, but also in terms of its fundamental neurological organization. Whether that organization is vested in a parallel set of specific brain adaptations or not (and obviously at times it is not), the brain sets fewer constraints than formerly thought on the process of cognitive evolution. Culture can literally reconfigure the use patterns of the brain; and it is probably a safe inference from our current knowledge of cerebral plasticity that those patterns of use determine much about how the exceptionally plastic human central nervous system is ultimately organized, in terms of cognitive structure.

The Organization of This Book

This book will propose two things: (1) an evolutionary thesis that is a best guess as to how we arrived at our present state, and a corresponding cognitive model of the transition from apes to humans; and (2) a structural proposal, an initial effort at working out the cognitive architecture suggested by the foregoing scenario. The point of the exercise, with its attendant risks, is to derive a model of higher cognitive structure that is compatible with available cognitive and neurolinguistic evidence, within conventional evolutionary constraints.

The approach adopted here calls for relatively few major biological adaptations, but is not as simple as a straightforward unitary model. It is difficult to reconcile a unitary approach with existing neurolinguistic data. However, the complex modular structure revealed by the data does not have to be attributed entirely to genetically bound adaptation. As explained above, some of the neuropsychological structure observed in the modern human can be seen as the product of biological evolution, and some as the imposition of cultural and technological constraints upon neuropsychological maturation and growth. But, even after considering how cultural factors might account for some cognitive structure, the biological residual still appears to require more than a simple, unitary adaptation, such as a larger brain.

The starting point for this venture is, not surprisingly, Charles Darwin. In 1871 Darwin proposed some ideas on human cognitive evolution that have since been largely forgotten. In Chapter 2, these ideas, and some of their forerunners, are reviewed. Darwin's proposals, although based on much less information than is now available, have the advantage that they were built upon a very broad review of the evidence; Darwin's approach was inherently interdisciplinary, and he drew from any source that was important to his argument, whether it was botany, geography, zoology, psychology, the social sciences, or medicine. It is equally true of the approach adopted here; the problem of cognitive evolution demands the widest possible range of information, but it now includes new fields that did not exist as separate disciplines in Darwin's time: neurolinguistics, anthropology, paleontology, neuroanatomy, and especially cognitive psychology.

The third chapter introduces the neuropsychological dimension of human cognitive evolution. It starts with Wernicke but moves rapidly into the modern era and reviews the available data on the underlying structure of what Churchland (1986) calls, aptly if awkwardly, the mind/brain. The fourth chapter briefly reviews some basic archeological data, tries to establish a rough chronology of "landmark" changes in hominids, and then discusses anthropological attempts at reconstructing the cultures of archaic hominids.

The central thesis begins to develop in Chapter 5, where the mind and culture of the great apes are analyzed and used as a starting point for building a cognitive–evolutionary bridge to the mind and cultures of modern humanity. The evolutionary proposal itself, consisting of three hypothetical transitions in the evolution of the human brain and mind, is described in Chapters 6, 7, and 8. In this proposal, two major biological adaptations are hypothesized, one that occurred with the evolution of *Homo erectus*, and a second with the evolution of *Homo sapiens*. A third adaptation is also proposed, as a logical necessity, for more modern humans; but the "hardware" supporting this adaptation is technological rather than biological. The third adaptation was nevertheless just as real, and just as revolutionary, in terms of cognitive architecture, as the first two.

There were therefore three major evolutionary transitions on the road from the mind of the ape to modern human cognition. The "transitions" were periods of rapid, radical change—the punctuations in the process of punctuated equilibrium. They involved much more than specific cognitive changes, or particular anatomical innovations: each transition involved a complete redesign of the form of human culture, as the direct result of a change in the architecture of human cognition. This does not rule out the possibility of additional graduated change between major transitions, especially within a single species—in fact, such change can be seen in several hominid species—but more emphasis will be placed on the major transitions in this scenario.

The first transition is described in Chapter 6 as a movement from the level of culture of apes and australopithecines to the level of culture of *Homo erectus*. Australopithecines showed no sign of a fundamental cognitive advance over their ape predecessors. Although Lovejoy (1980), Isaac and McCown (1979), and others have argued convincingly that there was a substantial change from apes in their social fabric and survival strategy, signs of innovation are lacking in the areas of tool manufacture and use, systematic cooperative hunting, migration, and other areas that indicate intellectual development. This is not the case with *Homo erectus;* there are many features of *erectus* culture that defy explanation in terms of simply expanding pre-existing primate cognitive structures, and this is the point at which the first qualitative break must be placed. The key innovation in *erectus* was the emergence of the most basic level of human representation, the ability to mime, or re-enact, events. This "mimetic" skill supported a culture that formed a necessary intervening layer between ape and modern human cultures. There are still cognitive vestiges of that long intervening period strongly in evidence in human cognitive architecture.

The second transition, discussed in Chapter 7, was the move from the culture of *erectus* to that of *Homo sapiens*. The biological evolution of modern humans was completed during this long transition. The key event during this transition was the emergence of the human speech system, including a completely new cognitive capacity for constructing and decoding narrative. There is a question of why the emergence of speech should have had such radical cognitive spinoffs, and part of the answer may lie in the circular interaction of speech with cultural change. The best level at which to describe most completely what probably went on is, once again, the cultural level.

The third transition was recent and largely nonbiological, but in purely cognitive terms it nevertheless led to a third stage of cognitive evolution, marked by the emergence of visual symbolism and external memory as major factors in cognitive architecture. External symbolic storage must be regarded as a *hardware* change in human cognitive structure, albeit a nonbiological hardware change. Its consequence for the cognitive architecture of humans was similar to the consequence of providing the CPU of a computer with an external storage device, or more accurately, with a link to a network. The limitations of an individual computer become an increasingly secondary factor as the knowledge of the network develops; computational "powers" are increasingly dictated by the network. In such a situation, the properties of the network may be more important for understanding what the machine can do than the properties of the machine itself. In a similar way, the structuring effects of culture and technology on the individual human mind need to be taken into account.

Just as neuropsychological considerations play an important role in constructing a credible evolutionary scenario, a modified neuropsychological model is an inevitable by-product of that scenario. Major adaptations in evolution are always built on pre-existing structures; this may be called the "if-it-ain't-broke-don't-fix-it" caveat (or, for the classically educated, the canon of parsimony). Thus, if modern apes are taken as prototypes of our immediate pongid ancestors (a reasonable position, provided it isn't taken too far), I would assume that (1) whatever cognitive capacities we can establish in apes persisted into the hominid line, albeit sometimes in vestigial or masked forms; and (2) any truly new biological adaptations—that is, those that were not simple expansions of existing capacities—emerged because they involved a qualitatively new cognitive function.

Thus, in proposing major transitions, I am proposing truly novel adaptations that could not be accounted for in terms of a simple diffuse expansion of primate skills. Enlargement of the brain, which continued throughout hominid evolution, does not in itself constitute or imply a major transition. The brain size of primates has been increasing since the advent of the great apes, but no major transition has occurred among apes in the cognitive realm comparable to that in humans. This is evident in the cultural similarities between different types of ape. They differ in many ways: in their social organization, in their diet; in their habitat; in their uses of found tools; and in their aggression and territoriality. But on the cognitive level, the differences between them are relatively minor. The chimpanzee has a cognitive edge over some other apes in manual skills, but the differences are sufficiently minor that chimpanzee culture is still basically only a variant on ape culture. Humans, on the other hand, are genetically closer to the chimpanzee than is any primate other than the gorilla; and yet our cognitive–cultural distance from the chimpanzee is enormous. There is no reason to think this distance can be attributed entirely to diffuse brain growth, since similar relative growth in the history of primate evolution has not had such major cognitive consequences. This in itself is a good reason to look for a chain of significant, specific cognitive adaptations that might bridge the gap from ape to human.

The structural consequences of the major human cognitive transitions are laid out alongside the evolutionary hypothesis in chapters 6, 7, 8, and 9. Each stage in development involved a structural change in cognitive organization, as well as a profound cultural change. A complex of new cognitive modules accompanied each adaptation. This implies at least two intervening biological "layers" of distinctly human cognitive structures. And since the later layers were built on what preceded, a vertical hierarchy is implied.

The most recent cognitive transition acknowledges the importance of external storage media on individual cognitive structure. It will be proposed that many features of modern human cognitive architecture, of which the most obvious is reading, originate at this level. They, like earlier innovations, rely on hardware changes, but in this case the hardware changes are external to the biological memory system and reside in the technology of the culture. The cognitive modules postulated at this level are perfectly real in the Fodorian sense, but not because they are physically bounded in the brain—rather, because they are physically bounded *somewhere*, often in an external memory locus. Without external loci, they literally could not exist. If the cognitive "modules" supposedly underlying reading and mathematics exist at all, they exist only as part of a larger organization imposed on the neural systems already in place at the time speech evolved.

But the external symbolic system imposes more than an interface structure on the brain. It imposes search strategies, new storage strategies, new memory access routes, new options in both the control of and analysis of one's own behavior. It enables new skill-complexes (like reading or programming) in which the locus of memory is partly or mostly external. The decipherment of the innate neuropsychological structure of mind thus depends on a careful mapping of the entire cognitive structure, including the external modules and memory loci. In subsequent chapters each of these points will be discussed, and the two major aspects of a theory, evolutionary and structural, will be laid out in detail.

Darwin's Thesis

Continuity and Discontinuity

There appears to be a great cognitive discontinuity between humans and other animals, despite overwhelming evidence in support of the physical continuity between humanity and antecedent life forms. Darwin was primarily concerned with the latter, that is, with the continuities in physical evolution. However, he tended toward consistency in his worldview, and consistency demands that, if the behavioral progress of animals unfolded alongside their physical evolution, as it evidently did, then the same should hold for humans. The Darwinian view thus favors, in principle, a continuity theory of language and thought.

With regard to the brain, the Darwinian view is no longer seriously questioned. Our brain, like our other organs, is essentially similar to its counterpart in all higher mammals. Its principal structures have the same functions, approximate locations, and names as they do in a variety of other species. In a very real sense the human brain is only one manifestation of what might be called the common mammalian nervous system. Our brain contains most of the same chemical neurotransmitters, is subject to many of the same diseases, and has the same types of neurons as those of other primates. Moreover, most higher mammals, including humans, have substantially similar cerebral organizations on a microscopic level; there is a typical topographic, laminar, and columnar organization according to which the sources of inputs and outputs, and the various interconnections between areas of the mammalian brain, are laid out. On the available evidence, the brains of apes and humans are so similar that one is left at a loss to explain the remarkable, and apparently discontinuous, nature of the mental capacities of humans in comparison with those of our primate cousins.

To complicate the picture, there is no direct information available on the brains or level of intelligence of the "missing links," the intermediate species that presumably bridged the gap from ape to human. All of our evidence is indirect: a few bones and artifacts and elaborate attempts at reconstructions of culture, based on remnants of seeds and pollen, patterns of tooth wear, geographical considerations, and our knowledge of modern apes and humans. The task of reconstructing the steps through which humans must have passed in their evolutionary transition is so difficult that many have chosen to ignore the problem.

One extreme approach, which some may take simply as a counsel of complete despair, is to proclaim a discontinuity in evolution when it comes to the human mind. This is the traditional position of Cartesian rationalists, a position that ultimately dates back at least to Aristotle. The human mind is qualitatively different, language is qualitatively different, the realm of symbolic representation is qualitatively different, from what went before. Therefore, the realm of language, meaning, thought, and intentionality somehow falls outside the reach of natural science. This viewpoint has its attractions, and in recent decades it has been the dominant opinion in philosophy and cognitive science.

This is changing. There is now a trend back toward synthesis, a rejoining of cognitive science and the Darwinian worldview. This is partly due to the limitations, now evident, of an experimental cognitive psychology that ignores the brain; and it is partly due to the presence of new research data. New biochemical and genetic data have radically altered our perceived age as a species. Twenty-five years ago the time-scale of human evolution was about five times longer than it is in the present version. We were thought to have descended from an ancient ancestor, *Ramapithecus*, who lived some 12–15 million years ago, who was thought to have deviated from the apes even further back in time, perhaps 40 million years ago. According to the modern genetic clock, that time can now be reduced: we are no more than about 5 million years from an ancestor we shared with the chimpanzee (Sarich, 1980). This estimate also agrees with improved dating of archeological data. Moreover, the oldest species within the genus *Homo* are now dated at less than 2 million years old, and the recency of fully modern humanity is startling: the oldest remains are only 50,000 to 100,000 years old.

At the time of the publication of Eric Lenneberg's classic work on the biology of language, 1967, the old time-scale was still in force, and the "golden decade" of 1967–1977, as Johanson and Edey (1981) referred to the great discoveries in archeology during that time, was only beginning. "Lucy," and the huge collection of very ancient australopithecine bones found at Hadar, Ethiopia, had not yet been discovered; skull 1470 had not been found; Jerison's and Holloway's important work on fossil crania had not yet been published; and modern cognitive science was in its infancy. Neurolinguistics did not yet exist as a discipline; Washoe and Sarah and many other apes trained to use symbols were still in the future. And neuropsychology was only beginning to break away from the confines of clinical neurological observation and develop its own methodologies. Transformational grammar was in its heyday; Chomsky's (1965) idea of an innate "language acquisition device" was accepted as a biological necessity. So complete was Chomsky's defeat of behaviorism that the doctrine of the uniqueness of human language was virtually unchallenged in psychology. Thus, when Lenneberg proposed his "discontinuity" approach to human language, he was in line with the times.

Lenneberg (1967) was one of the first, and the most successful, of the modern scholars who tried to draw a connection between neuropsychological, comparative, and evolutionary data on human language. Relying heavily on clinical studies of language abnormalities, on studies of twins, and on developmental data, he proposed a biological theory of language development. His language model was characterized by all the properties of a relatively self-contained biological system: the language device matures at its own rate, during a definable "critical period," and manifests unique properties in the perceptual, motor, and cognitive realms, which set it apart from the rest of higher cognition. The language device resides in definable brain structures, as shown by a variety of clinical studies. It is unique in evolutionary history, and although the properties of the system are not understood, it represents a break with the past. Lenneberg held that, although the capacity for language must be found in the brain, language has properties that require a special processor. Chomsky, who wrote an appendix to Lenneberg's book, went further than Lenneberg: he maintained that the brain contains specific metagrammatical information about the forms language can take. How such a unique language device could have evolved so rapidly might be puzzling to the biologist, but so be it: that was his problem.

The discontinuity approach was clearly at variance with the rest of behavioral science—and with anthropology, archeology, biology, and neuroscience—and it was not really so much a theory as a particular strategy for approaching the subject of language. To some extent, it was an attempt to postpone the inevitable. Both Chomsky and Lenneberg acknowledged the necessity of postulating a brain mechanism; in fact, both of them did, in their own ways, attempt to describe some of its properties. But, they argued, whatever that mechanism was, it operated on new principles. The principles of explanation available to explain other levels of behavior and mind simply did not apply here. Thus, it would be impossible to bridge the gap in evolution from below; the language mechanism had to be described on its own level first.

The fatal difficulty with a discontinuity approach is that it leaves language without a frame of reference in the external world. The symbols of language denote things and events that are known primarily in presymbolic or nonsymbolic ways. Without a knower, the symbols themselves are meaningless, truth has no means of verification, rules become endlessly (and, one suspects, needlessly) complicated. The magnitude of this problem is seen in artificial intelligence (AI) research, where the problem of context and meaning has overshadowed the impressive progress that has been made in other areas. To create the illusion that a computer "understands" a statement, when in fact it knows nothing of the world to which it refers, requires a fiendishly clever programmer. It is the ultimate Clever Hans illusion of all time.* By contrast, the innovative, incomplete, ungrammatical, intuitive language of the streets, the trademark of humanity, so effortless and subtle, seems to violate all of these rules in one way or another. The average human is the exact opposite of the AI algorithm: full of knowledge, short on rules and exact denotations, harnessing language to the pragmatic end of expressing and expanding underlying knowledge.

A continuity approach is also vulnerable to a whole range of criticisms (see Fodor and Pylyshyn, 1988), but it appears to be the only avenue open to us. It is not a new avenue; continuity theories have been proposed since the Enlightenment. Although Darwin is the touchstone of modern continuity theory, he had predecessors.

Pre-Darwinian Theories

Prior to the seventeenth century, the main concern of those who thought about the origin of language was whether language had a divine origin or was a human invention. The majority of Greek and Roman writers, and virtually all Christian and Moslem writers, assumed that language was a divine gift. Thus, God gave Adam a primordial tongue, which most Western writers assumed to have been Hebrew, while Arabs assumed it to have been Classical Arabic. Theories of divine intervention could be considered the original form of discontinuity theory.

In contrast, Plato believed that language had human origins, and he alluded to the gesturing of the deaf as evidence that the form of language was not fixed. Plato's position is compatible with that of some modern learning theorists, who hold that language is a by-product of a more general capacity for learning (for example, Skinner, 1957). The Stoics placed the origin of spoken language in onomato-

---

*Clever Hans was a famous performing horse who could supposedly count and carry out elementary mathematical operations such as addition (Pfungst, 1911, translated 1965). Hans indicated the solution to a problem by tapping his hoof on the ground the required number of times. It was eventually shown that the horse was giving the correct answers by watching his questioners for very subtle body-language cues that told him when to stop tapping. When a test was rigged so that no one who knew the answer was left in the room with the horse, he failed to give the right answers. For years, however, even his trainer had been fooled into thinking he could add.

poeia, or the mimicking of sounds associated with the thing being named. In fact, the word "onomatopoeia" means, literally, "the making of a word."

In the seventeenth century theories of language origin began to proliferate.* They generally held that language must have begun with manual gestures (Cresol, 1620; Dalgarno, 1661; Bulwer, 1644). Descartes (1637) asserted that it was the foundation of human reason and that it alone elevated man above his body, which was an automaton. Samuel Pepys (1661) observed that apes were quite humanlike and might be capable of learning sign language.

In the eighteenth century, fashionable theories of language origin had a biological slant, especially in France, and although evolution was not yet a catch-word, they all seemed to acknowledge that theories of language origin had to provide a bridge between an apelike prelinguistic state and modern human language. Our primal language was thought to be concrete—that is, lacking in abstract concepts— and based on facial expressions, gestures of the hand and body, and primitive vocalizations that had an imitative quality (Condillac, 1746; Rousseau, 1755). The earliest languagelike sounds were either monosyllabic interjections (Vico, 1750) or sing-song modulations (Blackwell, 1735). La Mettrie (1748) speculated that apes might be taught language. In England Adam Smith (1804) tried to construct a theory of language origins based on the laws of association and imitation. This explosion of eighteenth-century language theories came to an end with a final flurry of interest in the work of the Abbé Sicard (1790) with the deaf and of Itard (1801) with the Wild Boy of Aveyron. Interest in the subject waned for the next forty years and was not renewed until the theory of evolution was in the air.

Modern theories of our mental origins should really be dated from the publication of Charles Darwin's landmark book *The Descent of Man* (1871), although *On the Origin of Species* (1859), which did not itself speculate on language origins, triggered much debate. In fact, so great was the torrent of papers on the subject that in 1866 the overburdened Societé de Linguistique de Paris banned any further communications on the topic. The ban, alas, was unsuccessful, and

*I have relied heavily on Hewes (1977) in constructing this section. This brief section is not intended to be comprehensive; for more detail, the reader is referred to Hewes' extensive review.

since that date the subject of the origin of the modern mind has attracted views and speculations from virtually every branch of biology, anthropology, psychology, and linguistics.

With such a large literature at our disposal, we might expect a correspondingly large array of theories about our origins. But in fact there are relatively few specific proposals, and a great deal of redundancy between those that do exist. Prior to Darwin, gestural theories abounded. Since Darwin, each speciality has tended to focus almost exclusively on its own evidence. Thus, paleontologists have emphasized the importance of gross anatomical changes, particularly our erect posture and opposed thumb, in the development of intelligence. Comparative psychologists have emphasized our exceptional capacity for learning and concept formation. Linguists have emphasized our unique vocal tract anatomy and our capacity for grammar. Ethologists have focused on the similarities between human behavior and animal behavior. And animal behaviorists have set themselves the task of demonstrating how smart (and thus similar to humans) other species are.

The continuity/discontinuity question is theoretically independent of the question whether language is modular or unitary in its structure. Language could have come about either as the result of several distinct modular adaptations or as the result of a massive unitary adaptation that had an across-the-board effect on intelligence. In either case, there may or may not have been a qualitative break with the past. However, in reality, those who favor the continuity approach—learning theorists, primatologists, and some AI theorists—have tended to adopt a unitary theory, maintaining that only one cognitive adaptation was needed to "explain" human capacities. That adaptation is generally seen as a modification of existing primate intellectual structures. Those who favor a discontinuity approach, such as some neurolinguists and cognitive psychologists, tend strongly toward modular theories. Many of the language "modules" they propose represent uniquely human skills.

In reality, none of these subdisciplines alone can solve the problem of how language emerged; the larger form of the question is inherently interdisciplinary. Darwin, in characteristic fashion, took a very broad view in constructing his own theory, and in this century the task requires, if anything, an even broader, rather than narrower, reach. Although the question of mental origins, and the origins of language, is primarily a psychological one, the data of neuroscience, linguistics, and anthropology are also relevant to its answer.

Darwin on Animal Intelligence

One of Darwin's principal objectives in writing *The Origin of Species* (1859) was to establish natural selection as the mechanism underlying the continuous evolution of life, including human life. What had previously appeared as a divinely ordained, static hierarchy of Being, with human life at the top, became, in Darwin's view, a dynamic but mechanistic pattern of change; and while humanity still remained "at the top" in terms of its cognitive achievements, it could no longer be seen as a totally different order. Significantly, perhaps anticipating the magnitude of the controversy about to break, Darwin did not address the specific question of human mental origins until *The Descent of Man* (1871). In his introduction to that book, Darwin wrote that he had collected notes on this topic for many years "without any intention of publishing on the subject, but rather with a determination not to publish, as I thought that I should thus only add to the prejudices against my views."

Particularly controversial were his views on the mental powers of animals. Darwin argued that we tend to underestimate the cognitive abilities of animals, and he listed numerous anecdotal examples of learning in a variety of species. Darwin distinguished between stereotyped "instincts," which were inborn, and highly variable acquired or learned behaviors, which involved the use of experience and thus implied the presence of intelligence. In his view, animals, particularly mammals, possessed most of the higher faculties attributed to humans. Thus, their emotions and drives were so complex and variable that they could not be attributed to simple inborn programs. For instance, dogs manifest feelings of jealousy, possessiveness, resentment, grief, and loyalty and have social needs that imply a great deal of understanding of their social milieu. Darwin's feelings toward the emotional subtlety of dogs are shown in this anecdote:

> In the agony of death a dog had been known to caress his master, and every one has heard of the dog suffering under vivisection, who licked the hand of the operator; this man, unless the operation was fully jus-

tified by an increase in our knowledge, or unless he had a heart of stone, must have felt remorse to the last hour of his life. (p. 70)

He observed that monkeys are capable of elaborate deceit, even revenge, and in a story communicated to him by the zoologist Andrew Smith, he made his point:

> At the Cape of Good Hope an officer had often been plagued by a certain baboon, and the animal, seeing him approaching one Sunday for parade, poured water into a hole and hastily made some thick mud, which he skilfully dashed over the officer as he passed by, to the amusement of many bystanders. For long afterwards the baboon rejoiced and triumphed whenever he saw his victim. (p. 69)

Darwin was fascinated with the intellectual curiosity shown by monkeys, which was often so great that it exposed them to danger. Hearing of the instinctive dread they had of snakes, he once took a stuffed snake into the monkey house of the Kew Zoological Gardens, causing a commotion among the inhabitants. In his words,

> I then placed the stuffed specimen on the ground in one of the larger compartments. After a time all the monkeys collected round it in a large circle, and staring intently, presented a most ludicrous appearance. They became extremely nervous; so that when a wooden ball . . . was accidentally moved in the straw, they all instantly started away. These monkeys behaved very differently when a dead fish, a mouse, a living turtle, and other new objects were placed in their cages; for though at first frightened, they soon approached, handled and examined them. (p. 72)

Curiosity, however, overcame their dread:

> I then placed a live snake in a paper bag, with the mouth loosely closed . . . monkey after monkey, with head raised high and turned to one side, could not resist taking a momentary peep into the upright bag, at the dreadful object lying quietly at the bottom. (p. 72)

Darwin observed that the similarities between humans and higher mammals extend to the cognitive realm. Imitation is found in many species of animal. He gave the example of jackals and wolves, reared by dogs, who learned to bark; of birds imitating a variety of environmental sounds; of a dog, raised by cats, learning to lick her paws and thus wash her ears and face. He also observed that humans can be imitated only by primates, their closest relatives. This is consistent with most of his other observations, where imitation was restricted to species that were reasonably close together on the phylogenetic scale. The major exception appeared to be in birds, which could mimic not only other birds but also a variety of environmental sounds and the cries and calls of many animals, including humans.

Focused attention was, in Darwin's opinion, another important cognitive capability that animals share with humans. He gave the example of a cat patiently watching the entrance to a hole, waiting for its prey. In another example, he observed that circus monkeys were typically selected for training on the basis of their resistance to distractions; trainers quickly gave up on a highly distractible animal. Memory was another area in which he observed similarities between animals and humans, especially in their capacity to recognize other individuals after long delays, sometimes five or ten years or even more. Dreaming and imagination, as well as superstition, are also found in animals, wrote Darwin. He alluded to the movements and noises they make during the night and to their capacity for "imagining things" in the dark, just as humans do.

Reason was regarded by Darwin, as it was by Aristotle, as the peak of human intellectual power. However, even reasoning was to be found in animals, in some form. Darwin's evidence for this was behavioral: when confronted with a problem, animals pause, deliberate, and resolve. He gave the example of a bear making a current in some water with his paw, in order to draw a piece of floating bread within reach, and of an elephant blowing with his trunk on the ground behind an object, to drive it within reach. Monkeys and apes, as usual, were closest to humans in reasoning power. For instance, given boiled eggs to eat for the first time, monkeys smashed them and spilled most of their contents; afterwards, however, they carefully cracked the ends of the eggs and picked off the shells in small pieces, having understood the consequences of simply smashing them.

Another area in which animals resemble us might be called the wisdom of experience and age. Darwin related the well-known anecdote among trappers that young animals are more easily trapped than old ones and that it is impossible to catch many animals in the same place, with the same kind of trap. Caution in responding to the environment is a characteristic of older animals, as well as older humans.

Even though humans may remember more, attend longer, show more imagination, and solve more elaborate problems faster than other animals, the difference in intelligence is not qualitative but quantitative.

Darwin recognized that apes and other higher animals, like elephants, employ tools. Monkeys use stones to shatter palm nuts; orangutans use sticks as levers to get at food under heavy objects and use available objects as shields against stones and projectiles thrown by enemy orangs; elephants use sticks to drive away flies; baboons roll stones down mountainsides to drive away rivals. Darwin's evidence was gathered more informally than the careful observations of modern ethologists and animal behaviorists, but his conclusions were essentially similar.

With regard to abstract intelligence and self-consciousness, Darwin acknowledged the difficulty of judging the minds of animals. He did point out, however, that the presumption lies more with those who deny that animals can have abstract concepts than with those who accept that they do. Given our common mammalian heritage, he implied, the burden of proof is on anyone who would deny the continuity of mind. He was even willing to concede some small degree of self-consciousness to higher mammals. By self-consciousness he meant the ability to reflect on ultimate questions like death, one's origins, and the meaning of life. Animals may not possess this, but, he asked, is the nature of such questioning really any different from that of a dog reflecting, or rehearsing, the pleasures and pains of his own experience? And who can be sure that such reflection does not take place? Besides, as Darwin pointed out, culture and experience might be the source of self-consciousness; humans with very little education might also find it difficult to reflect, if the habit of such reflection had never been established. But in the case of humans, we do not doubt that, given the right exposure, self-consciousness could develop.

Darwin wrote at great length on the social and moral capacities of animals, which he saw as being closely interrelated. He spoke of the close bonding that occurs between members of a group, particularly among highly social creatures like dogs; of the altruistic behavior evident in close social groups; and of the importance of social reinforcers. He saw human morality largely as an extension of the social contract found in any reasonably advanced animal group. He gave as examples of social cooperation the pack-hunting of wolves and their sharing of food; the group-fishing of pelicans; and the cooperation of baboons in lifting heavy objects like stones, to expose insect nests. Mutual self-defense had been widely reported: bull bisons in America were said to herd the cows and calves into the center of the herd and defend the perimeter. Primates in particular were strong in mutual defense. Darwin related the following report about the behavior of baboons observed in Abyssinia: the pack had been attacked by dogs, and the baboons had managed to climb up on some high rocks, except for a young member of the group, who was surrounded by dogs. "Now one of the largest males, a true hero, came down again from the mountain, slowly went to the young one, coaxed him, and triumphantly led him away—the dogs being much too astonished to make an attack" (p. 101).

Social bonding between animals was a product of selection; Darwin related a report by Galton that most wild cattle in Africa could not tolerate even a brief separation from the herd and slavishly followed their leader. There could be only one self-reliant leader; if many were born they would be culled by lions waiting to prey upon any "individualists" who might wander from the group. Darwin did not speculate about the nature of social intelligence, but he was aware that it was a major part of what might be called the higher cognitive functions of all animals, including humans.

Darwin's Thesis on Language Origins

To Darwin, language was one of the most distinctive aspects of human mental life. However, he questioned whether it was qualitatively different from the communication systems of animals. Darwin first examined the question of the uniqueness of language in its perceptual and motoric aspects. He pointed out that many animals produce what we would call "tones of voice," which constitute modulations of their characteristic cries; for example, dogs bark in a variety of tones, which might express what we know as surprise, anger, fear, or joy. Thus, modulation of the "tone of voice" is not a uniquely human aspect of communication. Some creatures—for example, parrots—

can mimic and articulate quite complicated sequences of sounds almost as well as humans. And the capacity to imitate complex sounds is not uniquely human.

Moreover, Darwin noticed that animals seem capable of discriminating between different words and short sentences; thus, auditory discrimination per se, especially in the short time span of a few seconds, does not appear to be the basis of our unique ability. Humans use facial expressions and gestures that resemble those of animals, and we combine them with sounds, such as those used to express pain, fear, surprise, anger, and joy, to convey meanings that may also be conveyed through language.

Darwin knew that animal communication—for instance, bird song—is not always fixed; the particular notes of a bird song must be learned, just as the specifics of any human language must be learned. He reported the existence of bird-song dialects, which he found to be, in principle, like human dialects. Thus, the cultural variability of human language was not, for him, a qualitatively unique feature in the hierarchy of evolution. Darwin did not question the special nature of human language; but in all of these caveats he cautioned us to keep in mind the continuity of many features of our language with aspects of infrahuman communication.

Note that the manner of Darwin's thinking was psychological: he was thinking of the underlying structure of language, in a sense of its modular structure. If the presence in other species of voice modulation, vocal mimicry, complex auditory discrimination, communication by gestures and facial expressions, and variability of dialects is not sufficient evidence for inferring the presence of human-type language in other animals, then by extension he implied that the central question of what language is remains to be explained. Darwin wrote that the most distinctive trait of human communication must consist of some of the more abstract cognitive features of our language. In this aspect, his thinking was quite modern. And in looking for a "language center" in the brain, or some feature of neural organization that was crucial to the development of language, we should focus on these cognitive features, according to Darwin, rather than focusing exclusively on the auditory–vocal apparatus underlying speech.

Darwin was trying to move his audience gently away from their strong "humanist" prejudices and toward the realization that, even in the matter of language, we share a great deal with our mammalian cousins. This was important to his theory of gradualism and continuity. It was central to his gradualist argument to demonstrate that some of the building blocks of language were already in place in apes.

A century before Jane Goodall's work, and fifty years before Wolfgang Kohler, Darwin believed that apes possessed a capacity for abstract thought and tool use. His evidence was anecdotal and secondhand, but he was right, as he had a habit of being. He assumed that knowledge was not primarily linguistic and that language was therefore not a world unto itself; this view fits in well with the data of modern psycholinguistics, which have clearly established the effect of nonlinguistic context on the meanings of utterances. Ninety years before Chomsky's famous attack on radical behaviorism, Darwin had observed that human children learn language so incredibly fast that they must have special powers of language acquisition, but, unlike Chomsky, he did not see this special ability as a problem for the continuity approach.

Darwin could not have known that apes can be trained to understand and use symbols, as shown recently, but this knowledge would probably not have required him to make any major modifications in his position. Many things are missing from his discussion: he made no mention of grammar, he did not make specific proposals about the timing of putative evolutionary steps, and he had no specific model of language structure to propose as a result of his phylogenetic speculations. But considering its vintage, Darwin's discussion has survived the primary test of time: it remains relevant to the current debate, which began about 1975, about the emergence of language.

Buried in his chapters on human mental powers is a brief proposal, which I will call Darwin's thesis, on the transition from ape communication to human language. This thesis was an attempt to think through the steps that must have led to the sudden and dramatic acceleration of human cognitive ability, which appears to have accompanied the development of what he called "articulate language." Although lacking in detail, and largely ignored in our time, Darwin's thesis has some intriguing features, and provides a starting point from which to view later work. Darwin's proposal is presented in two chapters of *The Descent of Man* entitled a "Comparison of the Mental Powers of Man and the Lower Animals" and contains three key ele-

ments.* First, given the intellectual powers of humans without language (for example, deaf-mutes), there must have been a prelinguistic
change in primate cognition that raised the basic cognitive skills of
humans above those of apes and set the stage for language. Second,
the primal form of language was not sign language but rather speech.
Speech emerged in two stages, the first of which must have been a
rudimentary form of song, and the second, language itself. Third,
articulate speech, once established, interacted with the human capacity for thought, leading to the development of new forms of thinking.

Darwin believed that during the first stage of language evolution,
there was a prelinguistic expansion of primate intelligence, resulting
in a capacity to represent the world symbolically. Although he did not
explicitly spell out the implications of this statement, it is clear he
meant that human intentionality (our ability to represent, and make
propositions about, the world) must have emerged in some prelinguistic form. This implies the existence of archaic forms of hominid
cognition that rose above the intellectual level of the ape, without
involving what we know as language. Darwin did not specify the precise form these cognitive changes would have taken.

He proposed that during the second stage of language evolution the
"primal form" of language emerged. It was not some early version of
sign language; it was rather a *vocal* communication system. Darwin
acknowledged that our capacity for speech involved a major specialized adaptation, even though, in itself, a capacity to create a wide
variety of vocal sounds was not a sufficient condition for all the properties of modern human language. The vocal adaptation, in his view,
probably involved two aspects: (1) the direct imitation of various natural sounds; and (2) what might be called "rudimentary song," which
is a creative, prosodic, largely emotive use of the voice. Darwin believed that the constant use of the voice in this way would have led
directly to the strengthening and perfection of the modern vocal ap-

---

*In this review, I have relied on the second edition (1874), which contains a number of
corrections of and additions to the 1871 text. The second edition is somewhat rambling
and full of ideas which we would find questionable or simply wrong today, such as the
inheritance of acquired characteristics and the superiority of the "civilized" races. I have
taken the liberty of selecting what appear to be the essential elements in Darwin's argument and reassembling them in the form of evolutionary steps, or phases.

paratus, a Lamarckian position that does not meet with acceptance today.

Third, the existence of a primitive system of vocal communication drove a further general expansion of cognitive power, which in turn gradually led to complex articulate language. The result of this was our characteristic human capacity for what Darwin called "long trains of thought," which were, in his view, impossible without language. This increased cognitive power was not limited to vocalization and was sufficiently general to harness other modalities of communication where convenient or necessary—for instance, sign language. Here, again, he believed that constant use of language would have led directly to further development of the brain.

If we disregard for the moment Darwin's Lamarckian assumption that use of a structure leads to a permanent change which can then be directly inherited by one's offspring, the essential point of Darwin's thesis is that a prelinguistic cognitive revolution had to occur before any form of language could develop. This revolution was followed by the acquisition of rudimentary song, implicitly a form of speech without grammar, since grammar, for Darwin, seems to be the by-product of a general capacity for thought. Once some form of vocal communication was in place, a circular interaction between that system and the capacity for thought led to the development of an ability for longer and longer (and presumably more complex) trains of thought and their integral support system, speech.

Archaic Human Cognition

Interpreting Darwin's intent is not often easy, but here it is clear that a major prelinguistic improvement in the cognitive capabilities of apes is the step that led the emerging new species to its "hominid" status. Darwin does not specify the nature of the change, but it would, at the very least, have given hominids an ability to formulate an intention to communicate in a symbolic manner. If this were not the case, the second stage would not have taken the form that it did—rudimentary song.

To preserve his assumption that no qualitative change took place but only an improvement of existing capacities, we should recall what Darwin held to be the highest cognitive functions of apes. He considered that tool use was one of the highest human functions, as well as the closely related skills of problem solving and reasoning; another was the formation of abstract concepts. Although he knew apes possessed these abilities, he also knew they fall far short of the human standard; he pointed out that they might use available implements as tools in certain circumstances, but they do not appear to have the ability to design and manufacture tools.

Similarly, as Darwin wrote, their concept-formation skills are far more limited than those of untrained deaf-mute human children. From observing these children we know that human intelligence is superior to ape intelligence even without the use of language. These evolutionary advances in instrumental reasoning and the formation of abstract concepts must have preceded the arrival of language. Darwin might have meant that the general capacity that underlies our ability to manufacture tools was also the capacity underlying the use of intentional signals. Perhaps he meant that a general-purpose cognitive advance might have, in itself, started humans on the road to symbolic intelligence.

One highly specific feature of human communicative behavior that Darwin placed early in the line of hominid adaptations was the use of facial expressions. Darwin wrote that facial expression is greatly elaborated in humans, although basically resembling the range of facial expressions found in apes. Moreover, he noted that the use of gross body posture in humans is in principle similar to its use in apes, although more inhibited. Thus, we have to postulate a new level of expressive development in *Homo*, which probably occurred independently of vocal language. This development might have preceded symbolic language, since it would have lacked a biological rationale for emerging after language was already in place.

In the case of postural communication, Darwin observed that piloerection (raising the hair on the back) and gross postural display have obviously fallen off in modern humans, although vestiges remain, especially in our "hair-raising" reaction to situations of threat. Such vestigial features constitute a typical piece of Darwinian evidence; vestiges speak of our continuity with earlier life forms that found the adaptation more useful.

Great elaborations of facial expression and of accompanying emotional sounds, so characteristic of humans, are effective devices at very short range and reflect the intimacy of the social groups in which *Homo* was evidently living. Darwin held that in small groups facial expression is a very efficient device for the communication of emotion and that the continued heavy use of facial expressions by modern humans, even while possessing a tremendously powerful language, is a vestige of this early adaptation of *Homo*.

One of the most notable features of Darwin's thesis is that, despite his awareness of the importance of facial expression and postural signaling, he did not propose a gestural model for the first human language. He knew that gesturing occurs in apes, but it is emotional in content, stereotyped, and largely involuntary. Language is variable and voluntary, and thus gesturing must have changed fundamentally in nature to evolve into any form of language. Although he believed that gesturing may have evolved somewhat in early humans, Darwin did not believe in elaborate hand-signing as a precursor to spoken language, despite many earlier proposals to that effect. Perhaps this is because those hand signals that were known to him did not appear to possess the syntactic properties of language. American Sign Language (ASL) for the deaf, which is a fully developed symbolic medium (Klima and Bellugi, 1979), is a modern invention. In Darwin's time formal, standardized signing was still in its infancy. Darwin did not state explicitly why he had rejected a gestural model of primal language, but his own logic implies that if the primal form of language had been a formal system of hand signs, vestiges should remain and would be especially evident in purely oral cultures. The absence of such vestiges suggests that language found its origins elsewhere.

The archaic phase of human cognition thus seems to have involved, for Darwin, a general increase in intelligence, including problem-solving ability, toolmaking skill, and abstract concept formation that somehow allowed the formulation of shadowy, and presumably largely unfulfilled, communicative "intentions." Gestures, particularly facial expressions, remained tied to emotion. Humans still lacked true language, including sign language; the emergence of language would be tied to vocalization.

Vocalization: Rudimentary Song

Darwin considered sound as the primary medium of language. The extensive use of vocal communication, however, represented a major change from ape communication systems. Given the vocal shortcomings of our primate ancestors, how did the use of sound become so elaborated in *Homo*? What were the steps leading to speech? Darwin provided us with an explicit suggestion: some kind of primitive sound modulation, which we will call rudimentary song, came first, followed by more elaborate vocal imitations and finally by articulate language. The pressure to develop vocal skill had come from the preceding phase of general intellectual improvement. Once intentions could be formulated, the adaptive advantage of better vocal control became evident.

Darwin assumed that the voluntary vocalizations of early hominids were initially achieved with the standard mammalian vocal apparatus possessed by modern apes. He spoke of the refinement of the modern human vocal apparatus as a product of, rather than a precursor to, the earliest human vocalizations. This being the case, Darwin expected that the first use of the evolving new skill would have been in producing cadences, or modulations, that would have resembled singing more than speech. He pointed out that gibbons produce such modulations during courtship.

If the first hominids were at roughly the same vocal starting point as gibbons and chimps, feelings and emotions related to courtship, such as jealousy, love, and triumph, would have been the most likely stimuli for rudimentary song. The new sounds would have been accompanied by other signs and gestures, mostly manual and postural, as they are in apes. Given even a slightly greater amount of voluntary control than that of apes, hominids would be able to develop a range of cadences and modulations. Those cadences and modulations would have evolved into primitive forms of chanting and song.

In modern terms, Darwin was suggesting that the first aspect of voice control to evolve was prosody, not phonetics. Prosody is basically the background modulation of the voice during speech; it forms an "envelope" of emphasis and emotion around words, and its exaggeration is the basis of chanting and song. Song is difficult to define outside of a particular culture, but it is possible to broaden the definition to include almost any cadence or modulation of the vocalizations available to early hominids. Even in the surviving cultures of *Homo sapiens* there is such wide variation in how to define song that anthropologists have to adopt a very liberal definition. Indeed, this is exactly what ethnomusicologists have done: Herndon and McLeod (1980) give the example of Maori *haka*, which are not sung in the sense used by our culture, since there is no melodic line; rather, they are shouted rhythmically. Nevertheless, the Maori define it as song. A finer distinction would have to be made between song and prosody (the modulation of voice in speech) in modern human culture; however, such a distinction would be irrelevant in a prelinguistic context. In effect, rudimentary song would have been the precursor of both the prosodic envelope surrounding speech and the rhythmic, melodic, and harmonic aspects of song—that is, everything but the lyrics.

We know now that apes have very little voluntary control over their vocalizations (Skinner, 1957; Myers, 1976); they vocalize, but the patterns are highly stereotyped. An apelike creature that could purposefully modulate vocalization could produce quite a wide variety of sounds. In fact, strictly speaking, it could speak without alterations to its vocal apparatus if the required gray matter was in place. The anatomist Jan Wind pointed out in 1976 that although the complex human vocal apparatus increases the subtlety of speech, it is by no means a necessity and would not have been a necessity when speech first evolved. This is evident from the simple fact that human cancer patients can learn to speak through a simple throat tube, with a very limited repertoire of sounds; it is the brain, not the vocal cords, that matters most.

Darwin raised another important point about the order of succession; to put it simply, if speech had appeared before rudimentary song, what additional biological advantage would song convey? There are very few, if any, ideas or feelings that cannot be conveyed by means of symbolic language. Rudimentary song, even in modern society, appears in the context of emotions: love, praise, hate, pride, nationalism, rage, sadness, fear, bliss. But words can also express and produce emotions. It is not easy to find a good reason why song should have been adopted as a separate communication faculty if language had appeared on the scene first. The reverse, however, is not difficult to justify: language is a very clear improvement on rudimentary song as a system of communication. It vastly extends the range of things that may be communicated. Thus, if some form of rudimentary song came first, language would still have a good *raison d'être;* but if language appeared first, song would appear to be an unlikely sequel. This leaves only one other possibility, that language and song developed together; Darwin did not rule this out altogether, although he felt it unlikely.

One corroboration of Darwin's intuition comes from behavior genetics. The genetics of rudimentary musical skills appear to be different from verbal skill; in virtually all studies of the structure of intelligence, musical talent is isolated as a separate factor from verbal skill (Gardner, 1982; 1983). Moreover, language follows a completely different developmental course from musical ability: all normal children, across all human cultures, acquire verbal skills between ages two and three in a rapid succession of stages, while musical skills generally develop much later. Finally, aphasias (loss of language function) usually result from left-hemisphere injury, while aprosodias (impaired voice modulation) and amusias (selective loss of musical ability) more often follow injury to the right (Brain, 1965; Ross, 1981). All of these facts point to the same conclusion: language and rudimentary song are not aspects of a single system.

Another aspect of musical experience that Darwin mentioned is its link to dance, group ritual, and group experience. In all societies there are forms of ritual that involve dance and music. There is a very clear distinction between the social function of spoken language and the function served by rudimentary song; song bonds a congregation and moves emotion in a way that is hard for the verbal side of us to duplicate. Banal utterances can suddenly become moving and apparently profound when set to music. The bonding and emotive power of rudimentary song thus has an archaic, and profound, hold on human nature. The universality and robustness of the link between music and mass emotion speak to its deep roots in the past. Technologically primitive societies, which lack writing and complex organization, always have a highly developed set of rituals centered on some rudimentary form of music, whether of chanting, dance, or song.

Another way to put it is that the nonsymbolic and nontechnological aspects of music must have evolved separately from language. Of course, when language and related symbolic skills appeared on the scene, music developed to a new level that depended heavily on technical innovation and symbolic control. But the link between music and social ritual, and the evocative tie of song to emotion, are archaic vestiges and still in force.

Why did hominids first acquire rudimentary song and, only later, speech? This may seem a curious way to begin constructing a symbolic system. But Darwin realized where early humans were coming from—from a past that did not possess any form of symbolic communication. Darwin proposed that the earliest form of song might have simply been a series of variations on existing vocal cries, with a very important improvization: voluntary modulation. The advantage of the proposal is that it builds upon the available repertoire of apes, finding adaptive significance for every step of its acquisition.

Although he was vague on the order of succession, Darwin felt that a capacity for vocal imitation would be integral to this stage of vocal development. He gave the example, borrowed from Max Muller, of how the ability to imitate the growl of a beast of prey would communicate the specific nature of a threat and allow a specific response on the part of a tribal group. Imitation of human sounds would have served two purposes: reciprocal communication and cultural diffusion. Obviously, no form of language could have come into common use, even at this rudimentary level, unless participants in its development were able to reproduce the specific sounds that were being gradually assigned various meanings by the group. Thus, the earliest hominids progressed, if I interpret Darwin's suggestions correctly, from voluntary emotive modulations to a more skilled stage in which they were able to imitate the sounds of other animals and humans.

The Darwinian picture of early hominids that gradually emerges is of a tribal creature with a culture based on a rudimentary vocal communication system, one which perhaps served to coordinate group activity and to define individual and group emotions and intentions in a more flexible and extensive manner than is now found in apes.

Articulate Language

As we have seen, Darwin believed that constant use of the vocal tract in communication would have led to further development of the brain and vocal apparatus—a characteristically Lamarckian position. In addition, he believed that a continuous selection advantage in favor of those humans with more developed vocal skills would have facilitated the gradual improvement of vocalization. Darwin held that continuing conceptual development, contingent on continuing physical evolution of the brain, was the primary prerequisite for the emergence of language, rather than any specific change in vocal-tract anatomy. He noted that the aphasias, particularly those cases where the patient loses the use of what he called the "substantives" (nouns; this disorder is now known as anomia), illustrate how language is built upon underlying cognitive capacities. In the case of anomias, the patient clearly continues to understand the world but loses easy access to the linguistic labels that would normally apply. His point is that the underlying concepts are not linguistic constructs but rather are based on a level of understanding that is logically prior in evolution.

Darwin alluded to the signing skills of a famous nineteenth-century deaf-mute, Laura Bridgman, as further evidence that language was not in any way tied down to its vocal form, and that some form of generalized cognitive capacity underlay its superficial manifestations. Nevertheless, the generalized cognitive capacity of the human brain, in Darwin's version of events, must have been dormant until speech arrived on the scene. It was the vocal form of language, specifically phonological skill, that led to an accelerated evolution of linguistic ability. Thus, he presents us with an evolutionary dilemma: language is supposed to depend upon intellectual abilities that were useful before humans possessed speech and that nevertheless depended upon the speech for their full expression.

The final stage in this continuous chain of evolutionary development was a circular, iterative interaction between language and thought. Darwin believed that what he called "complex trains of thought" were totally dependent on language, just as mathematical thought depends on appropriate notation. The belief that advanced forms of thought, and higher forms of consciousness, were completely dependent on language was widely held in the late nineteenth century. In Darwin's theory, the appearances of adaptive cognitive skills would lead to competitive pressures toward their adoption by a species. Thus, any modifications of the brain that enabled more elaborate thought processes would become established through natural selection. Language was an integral part of that adaptation.

Conclusion: A Multidisciplinary Puzzle

Darwin wrote before either linguistics or psychology emerged as theoretical disciplines, and he had virtually no accurate chronological information on the evolutionary emergence of humans. Thus, we should not be surprised by large gaps in his evolutionary thesis, from our point of view. Nevertheless, several of his insights are worth holding onto as we approach the subject a century later.

First, there is the notion of a prelinguistic cognitive change that set the stage for language. Darwin realized that the utility of symbols would have depended upon a prior advance in thinking skills. Thus, the first stage in an evolutionary scenario should specify whether such an early, prelinguistic advance was necessary, and what it must have entailed. The question is precisely how early hominids could have become more advanced cognitively than apes, while still lacking language.

Darwin also suggested that the starting point for the construction of such a scenario would be the cognitive repertoire of apes. Thus, one task of any credible evolutionary theory of mind would be to bridge smoothly from the cognitive structure of apes to that of humans. Since cognitive structure cannot be directly observed, it must be inferred from a variety of indirect clues. One of those clues, following from Darwin's use of evidence, would be the functional organization of cognitive skills in the modern human and ape brains. Although he had very little to work with on this subject, a great deal of new information is now available and might shed light on the continuities and discontinuities of cognitive evolution.

Second, the vocal skills of humans constitute the main line of language evolution. Although intelligence, gesturing, and facial expression changed during the early course of human evolution, the specific emergence of language was linked to vocalization. Darwin distinguished between the modulatory aspects of vocal control, which we summarized under the label of "rudimentary song" and which correspond roughly to prosody and emotive vocalizations, and the purely linguistic aspects, which we know as phonetic. A capacity for vocal mimicry would be essential to the voluntary control of these new vocal skills. One implication of this is that a theory of language evolution should accommodate two successive vocal innovations, the first leading to rudimentary song, the second to rapid phonetic utterances, both with a built-in capacity for imitation.

Third, the emergence of language would have created new thought skills that depended upon language. The implication is that modern humans have at least two kinds of thought skills: those that do not depend upon language, and are thus archaic, and those that do. The meaning base of language, according to Darwin, was extralinguistic, and thus our nonlinguistic representations of the world must at least match our representation of the world in language. The relationship of thought to language is thus central to a theory of cognitive evolution. Precisely how does language enable more complex trains of thought?

What does this discussion imply for a neuropsychological, structural model of language? It suggests that our linguistic skill skates on the surface of a highly developed, distinctly human cognitive capacity; and that our search for the language areas of the brain may involve far more than the regions that evolved for articulate language alone. It also suggests an intervening layer of cognition between modern humans and apes. If such a vestigial layer of cognition existed, it should have left its mark on the observable structure of modern human cognition. There ought to be a nonlinguistic layer of cognitive skill in the brain, characteristically human and independent of language, since language came later.

Neuropsychological case histories might provide us with clues with which to reconsider, and reconstruct along more modern lines, Darwin's evolutionary sketch. Neurological breakdown allows us to observe how the mind functions without some of its component parts. Some functions always break down together and appear to be part of the same "processor," or cognitive module. Others may break down quite independently of one another, suggesting that their underlying neural organization might be independent as well. This line of thinking developed simultaneously alongside Darwin's theory of evolution, and Darwin was aware of it, although he did not attempt to integrate the new neuropsychology (at that time, perhaps, too new and chaotic to comprehend) into his ideas on the origin of language.

The discoveries of Dax (1836), Broca (1861), and Wernicke (1874) laid the foundations of our modern theories of neuropsychological structure and provided the first opportunity to investigate the underlying biological structure of language. This field has continued to develop in its own way and has led to its own kinds of evolutionary hypotheses, as we will see in the next chapter.

T H R E E

Wernicke's Machine

Modular and Unitary Models of Language

At the time Darwin wrote *The Descent of Man*, there was no consensus on the physical mechanisms of language, and he had no specific ideas about which aspects of primate brain function had undergone modification when language evolved. Shortly after *Origin of Species* was published in 1859, a number of nineteenth-century aphasiologists formulated the first brain-based theories of language organization. Their central notion was that human language was generated in specific neuroanatomical structures, each of which supported a linguistic subsystem. There was an ensuing search for specialized brain areas: one for decoding words, another for producing words, another for integrating visual and auditory symbols, and so on. Each neurologically dissociable function of language—each function that could be shown to vary independently of the other aspects of language—was thought to be localized in a specific region of the cortex.

The initial neurological evidence for a localizable physical mechanism underlying language lent support to the Darwinian approach. It was fortuitous for evolutionary theory that the brain regions identified with language were located in the neo-neocortex—the most recently evolved areas of the association cortex. It followed logically from Darwin's theory that the appearance of language on the behavioral and cultural level should coincide with the latest evolutionary developments in the primate neocortex.

The early discoveries of Broca (1861) and Wernicke (1874) served as the foundation stones of localizationist models of language. Broca (1861) described the symptoms of expressive aphasia and linked them to the third left inferior frontal convolution of the cerebral cortex, now called Broca's area. Although others had made the same claim, Broca brought the syndrome into the mainstream of neurology by suggesting that language was controlled from a specific brain region. Broca's initial model claimed that language was independent of other brain functions (since it could be damaged in isolation from other mental functions) and that it was localized in a particular part of the brain. From a Darwinian standpoint, this made good sense. The tertiary areas of the frontal cortex, of which the speech areas form a part, are a distinctly human evolutionary development. Broca's proposal thus implied (although he did not explicitly state this) that speech was a unitary skill, dependent upon a special feature of human brain anatomy that had evolved especially to support speech.

Broca's proposal ran into trouble almost immediately. The main problem was that language did not always break down in the same way. It was soon pointed out (Bastian, 1869) that visual language— namely, writing and reading—could be affected by brain damage quite independently of speech. This required that speech and visual language be treated separately in terms of underlying brain anatomy, but it did not completely undermine Broca's proposal. However, in 1874 Wernicke struck an apparently fatal blow to Broca's theory: he described a second, distinctly different kind of aphasia, fluent aphasia, that could be attributed largely to the destruction of part of the first temporal gyrus, now called Wernicke's area. The existence of at least two language areas concerned with speech led Wernicke to suggest a different kind of language model, one which treated speech as a complex of several underlying systems rather than as a unitary system. In his model the sound images of speech were stored in Wernicke's area and were sent along a neuronal pathway to Broca's area, which supposedly contained the detailed instructions for the articulation of speech sounds. Other fiber paths connected Wernicke's area not only to the frontal speech region but also to the various input modalities of language. This allowed Bastian's (1869) observations on dyslexia and agraphia to be accounted for, and it also led Wernicke to predict the effects of severing the fiber pathways between regions, the so-called disconnection syndromes. He predicted the existence of conduction aphasia, a disorder of repetition that results from severing the most direct path from Wernicke's area to Broca's area.

Wernicke's neurological model, imprecise as it may have been, served as the basis of the first modular models of language—the first models that postulated the existence of several independently variable linguistic subsystems. It was followed immediately by other similar models (for example, Kussmaul, 1877; Broadbent, 1879; Lichtheim, 1885), and the approach Wernicke took survives, unmodified in its essential features, to this day (Figure 3.1). Additional types of aphasia were described, including the "transcortical" aphasias—disorders of meaning and conceptual elaboration—in either the reception or expression of speech.

Although our intellectual debt to Wernicke and his successors can scarcely be questioned, the limitations of his approach are serious, particularly from an evolutionary standpoint. If Broca's original idea had been vindicated—that is, if speech had been localized to one brain subsystem, recently evolved—then Darwin's notion that humans had evolved speech through a linear unitary series of evolutionary innovations would have received a strong boost. But Wernicke had shown Broca's theory to be wrong soon after its publication. Subsequent

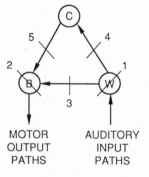

EFFECTS OF LESIONS:

1. Wernicke's aphasia
2. Broca's aphasia
3. Conduction aphasia
4. Transcortical sensory aphasia
5. Transcortical motor aphasia

MOTOR
OUTPUT
PATHS

AUDITORY
INPUT
PATHS

Figure 3.1 Wernicke's language machine: the standard model (after Lichtheim, 1885). W: Wernicke's region, responsible for language comprehension (first temporal gyrus); C: unspecified brain regions for conceptual elaboration; B: Broca's region, controlling language expression. The arrows indicate neuronal pathways between the language areas of the brain. Numbered cuts indicate how destruction of W, B, or their interconnections with other areas will result in one of the five major classes of aphasia.

models involved several language regions, each specialized for a different function. In these models, the speech circuit involved a speech motor region, an input decoding region, and a number of new pathways connecting these regions to various other structures within the brain, at the very least. Did each of these human "language centers" imply a special adaptation to the primate central nervous system? Or was the neural adaptation underlying speech spread out through preexisting structures? The evolutionary events preceding Wernicke's complex machine were not obvious.

In the last decade of the nineteenth century, the picture became even more improbable, as investigators tried to take other language disorders into account. Since reading and writing were affected by brain damage independently of speech, there had to be brain areas where the visual, or graphemic, form of words was encoded independently of their acoustic, or phonetic, form. Other features of language that complicated the system were the signing of the deaf, the verbal recognition of tactile objects, the connection of "knowledge" about objects to their verbal tags, and so on. Each of these could be independently affected by brain damage and thus "dissociated" from other speech deficits. Models became overly complex and unwieldy, particularly to someone looking at neurology from below, hoping for a fairly simple evolutionary modification that might help explain human language capacity.

Movements tend to produce a countermovement; and, in this case, the reaction came in a long series of critical papers starting with Jackson in 1874 and including Marie (1906) and Head (1926). Wernicke's most severe critic, Marie, denied the validity of Wernicke's clinical evidence. He argued that the cases presented by Wernicke were flawed both in important anatomical details and in the behavioral descriptions of patients' symptoms. Marie's principal anatomical criticisms were answered (Caplan, 1987), but the behavioral testing of patients was a more enduring problem. Jackson, Head, and later Goldstein (1948) believed that piecemeal testing of linguistic function could be misleading; aphasia was often embedded in a context of broader cognitive dysfunction and closely tied to disorders of the whole personality. As a consequence of their considerable influence, Wernicke's model came to be regarded by many neurologists as a gross oversimplification of clinical reality.

Particularly nagging was the problem of classifying aphasia. Marie had proposed that, in a sense, Broca's original idea was closer to the truth than Wernicke's. Marie believed that aphasia was a unitary disorder and that its apparent complexity was due to the simple fact that language had to operate through various sensory and motor pathways. Destruction of these sensory and motor pathways, and their corresponding cortical and thalamic projections, could masquerade as aphasia, since certain language functions would inevitably be damaged if the patient could not perceive accurately or modulate action properly. Marie considered that there was only one true language area, Wernicke's area. Broca's aphasia, in his opinion, could be seen simply as Wernicke's aphasia plus a specific motor problem with oral–articulatory coordination, called anarthria.

Marie's questions were never adequately answered, and even today some aphasiologists see virtue in his position (Lecours and Joanette, 1980). If aphasia was unitary, then the possibility that language depends upon a fairly simple evolutionary innovation could be considered once again. Early in the twentieth century the Gestalt movement and the Law of Mass Action (Lashley, 1929, 1949, 1950) heralded the return of the unitary hypothesis of higher function. Thought and language were once again seen as functions of the whole brain, and especially of the cerebral cortex. The influence of these ideas in neurology was evident in the work of Kurt Goldstein (1948), who proposed the existence of a global intellectual power he called "abstract attitude." Abstract attitude combined most of the properties of higher function: autonomous control of attention, conceptual integration, and independence from environmental stimulation. Many symptoms of brain damage could be seen as variants of a primary disturbance of abstract attitude. Although some aphasic symptoms, such as anarthria (impaired articulation) and word deafness, could not be explained away as impairments of abstract attitude, they were not regarded as high-level cognitive impairments. Rather, they involved sensory and motor interfaces with the higher cognitive system, and they could be attributed to damage to the perceptual and motor systems of the brain.

From a Darwinian standpoint, the demise of Wernicke's convoluted language model, and its even more convoluted descendants, must have been a welcome development. The evolution of a global intellectual capacity, broadly distributed throughout the cortex and fluid in its manifestations in the individual mind, made good evolutionary sense. It explained the plasticity of human cognition and the tremendous variety of human cognitive skills and cultures. It lent itself to the simplest (unitary) kind of evolutionary hypothesis.

Geschwind (1965), however, led a modern rediscovery of Wernicke's approach and provided a more solid foundation of neurological evidence to defend the modular model. His work has been greatly extended through the development of a new discipline, neurolinguistics. The modern neurolinguistic database is obtained primarily from experimental studies of amnesia, dyslexia, aphasia, agraphia, agnosia, and attention disorders. It also draws from follow-up studies of neurosurgical operations and from experimental studies of the normal population. On the basis of this body of evidence, neurolinguistic researchers are now, if anything, even more enamored of complex modular theories than were Wernicke and his contemporaries.

However, recent modular theorists are not strict localizationists, and many are indifferent to the functional neuroanatomy of the system they are studying. Their main interest is cognitive, rather than neural, architecture. But like Geschwind and other neuroanatomists, they believe that the essence of the unitary model—shared neural processing space for all higher function—is not compatible with the finding that language entails a set of highly specialized functions. Cognitive modules that form part of an innate, specific, human language adaptation would have more constraints on them than a general-purpose mechanism and would serve the special purposes of language more efficiently. Wernicke's hypothetical neuronal language machine was an attempt to solve this problem directly, by finding the physical mechanisms of language.

Fodor's Modules

Modular and unitary theories of cognition are finding some common ground: Fodor's book, *The Modularity of Mind* (1983), was the harbinger of this development. Fodor defined the properties of isolable cognitive modules, postulating a series of specialized "input" modules and one superordinate integrator at the peak of the cognitive pyramid, which he called the "central system." Fodor thus recognized the need for a unifying device much like the one Anderson specified, while conceding that certain aspects of cognition must be contained within specialized input modules.

Fodor proposed one very novel twist to entrenched cognitive dogma: he included language among the input modules, rather than making it a part of the central module. Utterances are themselves perceptual objects, he said, which have to be assigned a description, just like other perceptual objects. Input modules assign structural descriptions of their objects, and in this regard language falls into the category of an input system. Input systems also interpret the world and present their interpretations to the central system. They infer the properties of distal objects from their proximal manifestation at the transducer surface, like Gibson's (1950, 1979) earlier perceptual devices. Inasmuch as a linguistic input is a perceptual object, it requires this type of processing and specialized modules to achieve it.

Fodor's definition of input modules is a subtle and intricate one, which cannot be done easy justice. He did not identify perceptual systems with sensory modalities in the traditional manner: vision, hearing, balance, and so on. An input module, in his view, can be more narrowly defined than this (see Table 3.1). The brain may contain dozens of specialized perceptual devices for specific tasks, each enabling some special computation but simultaneously placing constraints upon how we can perceive the world. For instance, there

Table 3.1   Properties of vertically integrated input modules (after Fodor, 1983).

(1) Domain-specific (dedicated computational resource)

(2) Mandatory (automatic, obligatory processing)

(3) Inaccessible to consciousness

(4) Fast (near theoretical limits)

(5) Informationally encapsulated (impenetrable from outside module)

(6) Shallow outputs (processes only basic, highly salient features)

(7) Fixed neural architecture (privileged, dedicated paths)

(8) Specific breakdown patterns (as in agnosias, aphasias)

(9) Fixed ontogeny (standard pace and sequence of development)

might be a module for visual guidance of whole body movement, another for recognition of faces, another for the analysis of voices or for extraction of melodic information. Each module is domain-specific, that is, designed to perform the special, or idiosyncratic, types of computation demanded by a particular kind of input. The main purpose for having many input modules, instead of just a few, is that the computational demands of so many perceptions are very distinct, or eccentric. A general-purpose device would presumably perform such computations less efficiently.

Input modules, to Fodor, operate automatically; their functioning is mandatory. Humans cannot help seeing the world as a three-dimensional visual array, or feeling what their fingers run over as the surface of an object. Similarly, a sentence spoken in a known language must be heard as a sentence; it cannot be heard otherwise. Only highly trained individuals can, to some small extent, reverse this mandatory process: an artist, or a phonetician, might be able to analyze visual or spoken experience into elementary components. But their achievements only serve to emphasize the rarity of the event.

In addition to being domain-specific and mandatory, input modules are very fast-operating, their operations are generally not amenable to introspection or recall, and they depend upon fixed neural architecture that matures at a fixed pace and shows characteristic patterns of breakdown with brain damage. Finally, and most importantly, they are *informationally encapsulated*, that is, their perceptual machinery has access only to certain kinds of information. For instance, pressure on the eyeball causes apparent motion of the scene, despite the observer's knowledge that pressure is being applied. This is because the mechanisms that stabilize the visual world do not receive corollary discharges from the motor centers when the eyeball is moved passively. Knowledge that the eyeball is moving doesn't help, because the system is encapsulated, and the higher centers cannot influence our perception.

The idea that the brain (and mind) are made up of specialized modules has found some additional confirmation in recent neurophysiology. Specialized cortical and subcortical systems, with highly specific functions and connections, are the rule rather than the exception. The anatomical form of modules can vary. To a degree, cortical columns could be called self-contained computational devices that meet some of Fodor's requirements (Popper and Eccles, 1977). However, cortical columns are in themselves usually too specific to meet Fodor's functional definition. Clusters of cortical columns, organized into larger functional regions that serve specific computational purposes, are probably closer to meeting his criteria.

Discoveries of new functional regions, in which all of the cortical columns share some common function, have continued to proliferate after two decades of electrophysiological mapping of the cortex. There are, for example, as many as thirteen different visual cortical regions in some mammals and an impressive number of auditory and motor regions as well. These regions often seem to have the properties of Fodorian input modules: they are informationally encapsulated, function automatically and very fast, and seem to be dedicated to some specific computational function, albeit a very narrowly defined one.

The anatomical structure underlying a functional cortical region can vary, but there is some evidence that the large-scale organization of the cortex is gyral; that is, the internal architecture of a single gyrus suggests that it may grow, mature, and function as a unit (Goldman-Racik, 1982). Typically the columns contained within a gyrus, and the larger integrative pyramidal neurons within each column in the gyrus, share a common structure of input and output projections and are selectively sensitive to a specific domain of information. Moreover, the perimeters of the gyrus, in the troughs of its surrounding sulci, are less densely developed than the center; these appear to be boundary areas. Examples of gyral modules might be the postcentral gyri of primates, which are typically dedicated to topographic projections of the body senses; or the subregions of striate cortex, subserving various aspects of vision; or the first temporal gyrus in humans, which appears to be crucial to the processing of phonetic input.

Sometimes the domain of a cortical region can be startlingly specific: for example, within the somatosensory regions of the mouse brain there are whisker "barrels," functionally differentiated columns of cells that are really encapsulated miniregions, each dedicated solely to the computational demands of a single whisker on the mouse's snout. Removal of the whisker results in selective dying out of the cells in that specific cerebral barrel. Would such a device meet Fodor's definition of a module? In a sense it does; it is a fast, automatic, specialized device that is informationally encapsulated. Cognitive modules are typically defined on a broader basis, but this example does serve to emphasize the difficulties involved in deciding the level at which encapsulated modules can be specified.

The Unbounded Central Processor

The generality of the rule that brain structures are specific and dedicated is such that one might be tempted to ask whether there are any neural or cognitive functions that are not modular. If the rule is that brain devices are highly modular, where is the venerable "association cortex" of old, with its general-purpose learning capacity? And what room is there for the superordinate "central processor" so beloved of some AI theorists and cognitive psychologists? Where could it be located? What structures in the brain are large enough, and widely enough connected, that they could serve such a crucial general-purpose computational function?

Fodor defended the need for a central processor that was non-modular—that is, domain-general and unencapsulated—by arguing that many cognitive operations cut across cognitive domains. Perceptual learning, for example, cuts across various past representations of the environment, regardless of source. Beliefs and predictions, decisions and hypotheses, require a central system that can look at the contents of memory, or look at any of the outputs of input modules, to formulate its own integrated view of the world. Such a system must be unencapsulated, or unbounded. Language, in Fodor's view, is partly dependent upon the existence of an unbounded central processor, even though it also requires a special input module for its generation. This is because it must have access to the outputs of both perception and memory. The only way it could gain such access, given the internal constraints upon the "language module" itself, would be through access to a central processor of some sort.

A serious difficulty with this view is that there do not appear to be any good neuroanatomical candidates for an unconstrained central processor. All known cortical or subcortical regions have severe constraints upon their input and output pathways. The so-called tertiary cortical regions in the temporal, frontal, and parietal cortex, the newest and highest order of cortical organization, are broken up into numerous subregions and gyri, with distinguishable architectures, just like other cortical regions. Each subregion has its own distinct input and output paths, its own thalamic projections, its own relations to subcortical structures like the corpus striatum or the hippocampus or various other cortical areas.

Thus, in humans the insular and opercular regions, the dorsolateral frontal regions, and the areas around the angular gyrus and the supramarginal gyrus are all candidates for high-level integration; but each of them is architectonically distinct and subdividable. Even adjacent areas of the temporal or parietal "association cortex" often prove distinct in their internal neuronal wiring patterns. This fact in itself seems to argue against the existence of a central processor in any simple sense of the term. There seem to be a variety of high-level integrators, but on anatomical grounds each appears to have what Fodor calls a modular structure, that is, it has constraints regarding the information it can access.

Moreover, the functions of cortical association regions are not neatly definable. No single area seems to be truly the top of the cognitive pyramid, in the sense of being absolutely essential to all higher function. The effects of local injury to any one of these areas may be highly unpredictable from one patient to the next, and this unpredictability does not appear to be due to the limitations of psychological testing. The improved radiological data of the past decade has allowed more precise localization of the lesions causing aphasia, and our worst predictions have been confirmed. "Broca's region" and "Wernicke's region" are convenient fictions, the truth being that aphasia can be caused by a wide variety of lesions that spare these areas, while occasionally the complete loss of these areas will spare language function altogether, provided the adjacent white matter and basal ganglia are not damaged (Mohr, 1976; Stuss and Benson, 1986; Alexander, Naeser and Palumbo, 1987). The implication is that higher-level integration appears to be fluid and plastic in its underlying anatomy, and the anatomy itself looks modular throughout.

Does this matter? Could an unbounded central processor be supported by a somewhat arbitrary, largely software-configured (learned) device that had no fixed instantiation in predetermined anatomy? This would fly in the face of everything we now know about brain function and about efficiency of function. If simpler cognitive functions are largely dependent upon the inherent efficiency of anatomically fixed modular devices, then it is not unreasonable to expect that more complex functions, like discourse construction, would also require a specialized device, albeit one that had the required interconnections to read the outputs of lower-level systems. Taking Goldman-Racik's gyral-clustering model to its logical extreme, we might argue that if a tertiary region were to serve as a common pool for general-purpose processing, it should have been designed for precisely that purpose. Ideally, it should consist of a single, presumably rather large gyrus, with a relatively homogeneous internal architecture, so that it could be configured and reconfigured for various functions on demand. It should have extensive connections to every perceptual and motor module, it should mature late, as a unit, and it should continue to mature long after other cortical structures have finished maturing. Its loss should be devastating for all higher function. On the present evidence, no such cortical region exists.

Nevertheless, there are cortical structures that fulfill some of the requirements of a central processor. Eidelberg and Galaburda (1984) proposed that a particular subregion of the inferior parietal lobule (IPL) might serve a very broad computational function, allowing horizontal integration of information from various input modules. Their starting premises were somewhat different from Fodor's, and they did not propose that the IPL was truly a general-purpose device. Nevertheless, they attributed functions to the IPL that were very general. Various other authors, starting with Goldstein (1948), have attributed aspects of broad cognitive integration to the frontal lobes, especially the dorsolateral frontal region, which is still sometimes loosely referred to as the "supervisory system" (Shallice, 1988). However, the local neuroanatomy of the dorsolateral frontal region is complex enough to have justified at least six large Brodmann subregions; this does not support the idea of a unitary functional organization for this entire region.

Thus, in the typical neuropsychological model there is no Fodorian central processor; there are usually several central functional regions that contrive to produce the illusion of universal access to memory and perceptual inputs. But each aspect of thought is the output of some special device. Thought and language operate despite the absence of a central device, much in the manner suggested by Minsky

(1985) in his speculations about how the computational mind might be configured. Although he was not concerned with hardware, his proposal that autonomous "agents" worked together to produce various apparently integrated forms of computation amounts to the same kind of solution. (The medieval Scholastic concept most like a central processor, the so-called agent intellect, seems uncomfortably close to this idea.)

An evolutionary scenario is implicit in Fodor's proposal, just as in most structural models, whether or not the scenario is made explicit. The scenario implied by Fodor's modular structure would involve at least two major evolutionary adaptations, both uniquely human. One adaptation would be a special vertically integrated language module controlling certain linguistic functions. The other would be a more general cognitive change, involving the development, or at least enlargement, of a superordinate, nonmodular cognitive processor that would play a supervisory role over conscious processing.

Although Fodor's proposal could be seen as a prototype of global model construction in cognitive science, many variations on his structural theme have been proposed. Most neuropsychological solutions would be compatible with his postulation of input modules for language and various other special kinds of computation, while disagreeing with the idea of a single central module. For instance, Shallice (1988) has proposed a slightly different modular structure, based on his work with Norman (Norman and Shallice, 1986). Instead of a single central processor, Shallice's conceptual brain contains several superordinate central devices that enable higher cognitive function. Shallice returns to the old problem of central control, or attention. If the brain contains many parallel modules and relatively autonomous subsystems, how can it avoid the conflicts and discoordination typical of decentralized systems? Whereas Fodor proposed an unbounded central system to handle any form of processing that cuts across modalities and input modules in general, Norman and Shallice proposed that, in place of a single central system, there is a hierarchy of control systems of increasing generality. Local control is handled by specialized subsystems, such as the one for visual attention, while complex routine activity is handled by a second-tier system, which they call "contention scheduling." There is, however, a supervisory system at the top of the pyramid, and even the supervisory system is probably broken up into various subsystems, although they are not clear what these are.

Norman and Shallice's approach presents the prospective evolutionist with difficulties. For one thing, very little about their central attentional hierarchy is uniquely human. Apes, and other mammals, presumably require the same sort of hierarchical control of attention and thus the same trio of local schemata, contention schedules, and central supervision as humans. Nothing in their proposal specifies precisely what must have changed in the course of evolving our unique cognitive skills. Unlike Corballis's (1989) theory, for instance, their theory does not spell out a special cognitive feature that might lie at the heart of humanity's remarkable achievements.

On the other hand, to be fair to Norman and Shallice, it may well be that the supervisory hierarchy did not change in qualitative terms in humans. Perhaps it only grew larger in capacity. Perhaps all of the qualitative changes that underlie symbolic thought took place elsewhere. If this was the case, then the evolutionary scenario leading up to modern humans might be similar both for Fodor's modular structure and for the one proposed by Norman and Shallice. That is to say, the cerebral cortex, including the supervisory hierarchy, might have grown larger in a fairly nonspecific way to accommodate cognitive growth in hominids. In addition, at least one totally new type of input module evolved, the one for language.

Neuropsychological Aspects of Evolution

The above evolutionary scenario is implicit in Fodor's idea of the modular mind; but there have been several explicit attempts to create a neuropsychological model for the evolution of both language and higher cognitive functions. The majority of these attempts have focused on the three most salient features of the human brain: encephalization (brain size), the localization of language, and the lateralization (sidedness) of cerebral function.

The most salient distinct physical feature of the human brain is its size relative to the rest of the body. Within the primate line, the jump from ape to human involved roughly a tripling of relative brain size and a doubling of the number of neurons (Jerison, 1973; Passingham,

1982). As relative brain size increased, neural density decreased, accompanied by an increase in dendritic growth and in the number of synapses. The growth in synapses is thought to reflect the increasing number and complexity of interconnections between neurons in larger nervous systems. This increase in size and complexity is especially prominent in the cortex, cerebellum, and hippocampus; and within the cortex, it is most evident in association areas. Since the idea that human language and knowledge depend largely upon increases in processor size, rather than upon changes in structure, is uncharacteristic of Wernicke and his neurolinguistic descendants, discussion of encephalization will be postponed until the next chapter.

The issues that are most important in any neuropsychological approach to evolution are the localization of language and the lateralization of human cognitive function. Language is a unique human trait, and thus an attempt to localize language amounts to an attempt to discover what is special about the human brain. Cerebral lateralization is also a distinctly human feature, at least in degree, and language appears to be the most highly lateralized of human cognitive systems. Thus, it is not surprising that the evolution of language and lateralization have been approached as two sides of the same coin. The evolution of cerebral laterality was already an important matter to Jackson (1868), and many neuropsychological thinkers have speculated on this subject, culminating in the modern era in a number of articles by Geschwind (1965, 1984) and his colleagues.

Although laterality provides a vehicle for considering the evolutionary scenario preceding language, it is not always the primary phenomenon of investigation. Aphasiologists have been interested in the fine-grained analysis of language breakdown, a modular approach that is to some extent independent of the laterality issue. An important related question asks what the human mind can do without speech and language: can language alone account for all that is distinctly human in cognition? The answer to this question has implications for any prospective evolutionary scenario. The integrity of the language system is also an issue. Some theorists have argued for different evolutionary courses for speech and language capacity in the more general sense; Lyons (1988) sees the two as having evolved separately, with a general language capacity having preceded the more specific adaptation for vocal language.

Neuropsychological theories must also account for the available evidence on language ontogenesis, which can provide clues about the underlying structure of language. It was the ease and rapidity with which children acquire language under all but the most adverse conditions that led Chomsky (1965) to argue for an innate language acquisition device, as we have seen. Although he never tried to specify what physical form such a mechanism might take, he implied that a special neural device must have evolved at some point in human evolution. In making this proposal he relied upon the neuropsychological data pointing to a critical period for language acquisition, particularly speech acquisition, compiled by Lenneberg (1967). Chomsky thought that the innate speech device was capable of computational functions that are both distinctly human and discontinuous, in the Cartesian sense, with all of earlier evolution. Chomsky's proposal was remarkable in that he attributed innate, specific grammatical knowledge to the brain of the human child, implying the evolution of a specific language device that contained what are normally thought of as both hardware and software features.

Other ontogenetic hypotheses have de-emphasized innate mechanisms and relied on learning to allow for acquisition of specific semantic and syntactic features. Parker and Gibson (1979) have modeled the course of language evolution on cognitive and linguistic ontogenesis, placing more emphasis on improved human learning capacity than on innately stored knowledge. Thus, where Chomsky emphasized innate modular mechanisms, Parker and Gibson used essentially the same neuropsychological database to propose a more diffuse, unstructured cognitive expansion, whose by-product was a general capacity for symbolic representation. These two positions are at the extremes of the nativism-empiricism continuum; most neuropsychological theories are located more comfortably in the middle of the road.

Language Evolution and Cerebral Laterality

Hemispheric lateralization of function is the most global, and inclusive, level at which higher cognition appears to be modular. The initial division of function between hemispheres early in the hominid line probably did not include language. Lateral differences appeared earlier in mammalian evolution; some lateralization of cortical function appears in lower mammals. In rats, the right hemisphere assumes the role of controlling orientation in space (Denenberg, 1984), while in nonhuman primates the left might control unimanual skills (though the behavioral evidence for this hypothesis is thin). Anatomical asymmetries in primates and other mammals are more solidly established: the right hemisphere tends to be heavier, and the left Sylvian fissure tends to have a lower, more posterior projection in cats, monkeys, and apes (Kolb and Whishaw, 1990), just as in humans. Some earlier division of labor between the hemispheres might have served as the basis for a later, uniquely hominid, pattern of specialization. For instance, the left hemisphere might have gradually evolved the basis for an extended system of communication, while the right evolved an extended representation of the external environment.

Early theories of left and right cerebral organization run the gamut from master–slave to more balanced theories. Master–slave theories usually attribute some form of global dominance to the left hemisphere and relegate the right to the role of a slave, holding tank, or echo chamber for the left, at least in the majority of humans. Jackson (1868) initially called the left the "leading," and the right the "automatic," hemisphere in right-handed persons; thus, in both perception and language, the left was dominant. The right possessed word representations, he held, but these could not be used in forming *propositions*, which is the essence of language; they could only be used automatically, in straightforward associations.

In 1874 Jackson reconsidered the role of the right side and suggested that both hemispheres might have their own unique specializations. He proposed that the right parieto-occipital region was the leading side in conscious visual imagery (the "voluntary revival" of images), with the left serving an automatic role. The left frontal region was seen to be the leading side in expressive language, with the right serving as an automatic linguistic slave system. Jackson had moved from a position of total left-dominance to a balanced, or reciprocal, dominance view, attributing the leading role in some functions to the right, so that neither hemisphere was globally dominant over the other. Thus, early in the history of neuropsychological theory, both hemispheres were thought to be involved in both imagery and

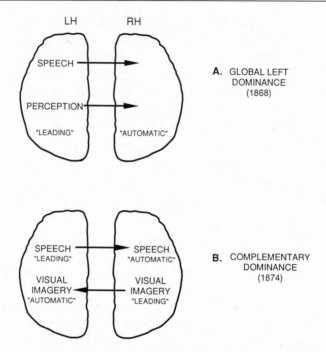

Figure 3.2 Jackson's first (1868) and second (1874) proposals regarding the dominance of the right and left cerebral hemispheres. The "leading" side in speech is able to construct propositions, while the "automatic" side is only capable of involuntary associative language. The "leading" side in vision can voluntarily generate images, while the "automatic" side can only retrieve them involuntarily. LH: left hemisphere; RH: right hemisphere.

language; laterality did not imply an all-or-none division of function. Rather, it was a matter of which hemisphere was in the lead, that is, of where control finally resided.

Early theories of dominance linked speech to handedness, and it was widely held that speech in the left-handed was localized on the right (Broca, 1861; Jackson, 1868). This theory survived in some form until Penfield and Roberts (1959) presented conclusive evidence that the left hemisphere was dominant for speech in most left-handers as well as right-handers. The notion of a dominant hemisphere that controls language, skilled motor activity, and consciousness is not entirely dead. Bogen (1969, 1985) and Sperry (1966, 1968) have attributed consciousness to the left hemisphere; Gazzaniga (1983) has claimed the right hemisphere is mute and primitive; and Kimura (1976, 1979) has argued that not only language but skilled serial motor control in general may be vested in the left hemisphere. Corballis (1989) similarly attributes most of what we recognize as uniquely human in cognition to the left hemisphere. All of these authors, however, concede some special, mostly visual, functions to the right side of the brain.

Jackson (1874) was one of the earliest to support this idea. He believed that both sides of the brain were normally involved in "verbalizing," that is, formulating and comprehending the content of language. Since speech depends on perception, especially visual imagery, he believed damage to the right hemisphere affects how language is used. Many others have come to essentially the same conclusion, holding the left to be verbal–logical, or propositional, and the right to be visual–synthetic or appositional (Humphrey and Zangwill, 1951; Hécaen and Angelergues, 1963; Gazzaniga, 1970; Sperry, 1966). Bogen (1969) has proposed a strong form of this theory, proposing that the neurologically normal mind can alternate between two modes of thought, which are essentially right- and left-hemisphere modes.

In some form, this hypothesis is very widely accepted. It could be called, without exaggeration, the established theory of cerebral laterality of function. But the assumptions underlying this theory are rather curious. Specific memories and thoughts are not generally believed to be localized; in fact they are held to be distributed throughout the system bilaterally. No one is willing to claim that aphasic left-sided stroke patients cannot possess and comprehend the semantic contents of events, or that they are incapable of thought. But the "computational style" of right-hemisphere processing apparently differs. Why? What could there possibly be in the left hemisphere that predisposes it to analytic thought? Or in the right to appositional thinking? Is language secondary to the left hemisphere's more fundamental capacity for analytic thought?

By extension, the computational style of the right might "force" anyone unfortunate enough to have to read with their right hemisphere (perhaps because of early brain damage) into using a graphemic, or visual, reading strategy. Kolb and Whishaw (1990) have made a proposal to this effect. The assumptions underlying the proposal are left unstated, for the very good reason that they are quite radical: the implication is that the left hemisphere has recently evolved not so much an exclusive capacity for language as a different computational style from the right. That computational style is generalized to a variety of functions: planning, thinking, perceiving, memory, and representation. Because the computational style of the right hemisphere is vestigial, it cannot think or communicate analytically, employ syntax, or reason symbolically. The left hemisphere, enjoying the benefits of its more recent evolutionary acquisitions, has a virtual monopoly on these functions; thus, when the right hemisphere is forced to read, it falls back on a different reading strategy.

Corballis and Beale (1976) wrote that cerebral cognitive duality might have been driven solely by the lateralization of language. In their view, more concrete, or sensorimotor, forms of cognition depend on bilateral symmetry. This is because the environment is bilaterally symmetrical, and perceptual representations and motor activity have to be able to respond with equal efficiency from either the right or the left. Manual skills, on the other hand, could be lateralized, and might actually benefit in neural efficiency from being represented in only one hemisphere. Thus right-handedness, and left dominance for the hands, emerged.

Hand control involved improved sequential motor skill, that is, motor programming. In Fodor's terms, such programming skill would probably not be informationally encapsulated, given the interaction of manual skill with other perceptual and motor events. Given such skills, the left hemisphere was better equipped to handle the expressive, or motor, aspects of speech when it first evolved, and thus left-dominance for language began. Then, given a bias toward the left hemisphere for the *production* of language, speech *perception* might also have become lateralized to the left. This would have occurred for reasons of efficiency, since speech perception is so closely tied to vocalization, and speech would be better served by a single hemisphere.

Thus, Corballis and Beale (1976) proposed that speech effectively "invaded" the left hemisphere, leaving the right to perform preverbal styles of thinking that preceded speech in evolution. The reason the right has some capacity for speech perception is that some referents of speech are nonsymbolic, and the right hemisphere might have forms of knowledge that would be useful in language comprehension.

Moreover, some forms of thinking (for instance, in the fine arts) lend themselves to nonverbal modes of expression, and the right hemisphere might be better equipped in this regard, since it is not heavily invested in analytic or linguistic modes of thought.

Corballis and Beale thus defined the capacities of the right hemisphere by default, or negation; it does not have language, except in a very limited way, so it manifests a form of cognition that is, by implication, more primitive. The default hypothesis is in many ways easier to handle in biological terms. There is no need to postulate two scenarios, one for each hemisphere. The presence of manual skill and language are the major determining factors in the difference between hemispheres, and they share certain properties. Language was our major evolutionary innovation, and since the right hemisphere has less of it, it retains our primitive modes of cognition. By implication, if we can describe the functions of the right hemisphere, we might catch a glimpse of what human thought was like, at least in general qualitative terms, before language came along.

The default hypothesis is attractive because it is simple. Corballis and Beale's underlying assumption appears to have been that, prior to language, higher cognitive skills were more or less symmetrically represented. After reviewing the literature on right-hemisphere function, they concluded "not that the right hemisphere is intrinsically specialized for these functions, but that the left hemisphere is deficient in them, presumably because of its specialization for language and verbal processes . . . If this is so, then there is no need to assume that right-hemisphere specialization evolved separately from left-hemisphere specialization. Rather, it is acquired by default, a by-product of the lateralization of language representation" (1976, p. 103).

Recent comparative data on animal lateralization, although still very limited, suggests that this early version of the default hypothesis was wrong. The right hemisphere might have become specialized for spatial functions *before* the left specialized for the control of certain motor functions. Thus, each hemisphere's special style may have its own evolutionary origins. Corballis has acknowledged this and revised his view slightly in a more recent article on the evolution of laterality (1989).

Kinsbourne (1974) provided a somewhat similar perspective on the biological reasons why lateral asymmetries may have developed. He reviewed the functions of the brain, more or less in the order in which they have appeared in evolution. On the most fundamental level, that of arousal and orientation, he could see no reason for asymmetries, since the environment was symmetrical and behavioral demands equiprobable from any direction. Thus, a worm has to be able to turn with equal facility in either direction and has symmetrical sensory and motor systems; the same principle applies even to higher animals. As long as behavior remains on a rudimentary, stimulus–response level, the need for bilateral symmetry would remain. However, abstract information processing and other tasks that do not require a symmetrical interaction with the environment (for instance, unimanual skills practiced by monkeys) might become represented on only one side of the brain. Asymmetry, Kinsbourne argued, is the default mode; it appears when the biological need for symmetry relaxes. Thus, Kinsbourne also falls into the default hypothesis camp.

Steklis and Harnad (1976) pointed out that it could not have been a coincidence that both speech and hand dominance are usually in the same (left) hemisphere. They used the label "serial motor praxis" to describe the common element between the two types of skill. They argued that although apes show cerebral dominance as individuals, the distribution of dominance in the group is about evenly split between right and left. Steklis and Harnad observed that, from gross anatomical clues, the first deviation from symmetry in the left–right distribution seems to have occurred in australopithecines, simultaneously with the appearance of erect posture and humanlike hands. They thought it likely that the first language was manual–gestural; and since manual control was already left-dominant, the speech regions became established on the left.

Kimura (1976, 1979) pursued this question on a more empirical level, linking language loss (aphasia) to a loss of complex motor performance. According to Kimura's proposal, laterality was primarily driven by the need for an efficient sequential motor device for the hands, and was linked to speech only secondarily. Kimura's theory was built largely on clinical evidence of factorial clustering. Problems with complex motor performance, clinically referred to as the apraxias, often accompany aphasia and result from lesions to the left association cortex. Geschwind (1965) believed that apraxias were largely the consequence of disconnecting the motor cortex from the speech areas and that the speech regions were necessary to the typical tasks demanded of apraxics. Kimura turned this idea around and suggested that apraxia is actually a direct disability of serial motor control, not merely another aspect of language disturbance. The reason for the association of aphasia and apraxia, Kimura argued, is that the left hemisphere is primarily specialized for serial motor control, whether manual or vocal. In other words, apraxias are the result of direct damage to those parts of the cortex involved in complex motor skills, which include speech. Kimura (1976) also observed the frequency with which free hand movements occurred; she found that people tended to produce such movements only during speaking. Moreover, right-handers, who are overwhelmingly left-lateralized for speech, move their right hands; while left-handers, who tend more toward mixed dominance, move both hands. Thus, hand movements appear to be involved with speech areas of the brain, both in clinical breakdown and in normal activity. Finally, when Kimura reviewed the few (at that time) existing case studies of signing disorders in the deaf, she found evidence of a clinical breakdown of signing linked to destruction of the speech areas on the left side of the brain.

Kimura therefore suggested that the initial evolutionary development of manual dexterity for tool use led to the use of the hands in gestural communication. Vocal speech probably came later and, as another complex motor function, it was left-lateralized as well. She concluded that the left hemisphere "is particularly well adapted, not for symbolic function per se, but for the execution of some categories of motor activity which happened to lend themselves readily to communication." Curiously, Konrad Lorenz (1965) once proposed a closely related idea: that the unique human capacity for manual skill was dependent upon the left hemisphere. Kimura's theory amounts to a strong form of the default hypothesis.

The default hypothesis has its attractions, but it leaves unanswered (and, in fact, largely unasked) the question of precisely what special neurophysiological features of the left hemisphere give it such an advantage over the right. There are tremendous gaps in the reasoning by which left-dominance for manual skill is coupled to speech. Why should a cognitive module dedicated to manual skill—that is, to the sequential programming of hand action—generalize to language? The evolution of language, as will become clear, involved a great deal more than serial motor control; it involved representation, intentionality, syntax, the invention of lexicons, and so on, none of which could be explained simply by extrapolating a capacity for serial motor control. Just because manual skill was represented on the left, it does not follow that higher levels of linguistic control would necessarily reside on the left.

Geschwind's Search for the Speech Module

Geschwind (1965, 1970; Geschwind and Galaburda, 1984) and his colleagues have also examined the functional and anatomical asymmetries of the brain in an evolutionary context. However, Geschwind approached the problem from a very different perspective. Geschwind was a localizationist. He argued in favor of a localized language area in the human left hemisphere; and in a seminal article published in 1965, he tried to explain most of the symptom clusters of aphasia as the effects of disconnecting the speech region from sensory and motor areas of the brain. Thus, "agnosias" could be seen as modality-specific naming deficits, associated with isolation of the relevant sensory area from the speech regions. "Apraxias" could result from disconnecting the speech areas from the relevant areas of premotor cortex. The key unifying element was a left-sided cortical area specifically devoted to language. Disconnection of that area from other regions would leave the patient without access to the language system.

Geschwind's "language regions" bore some resemblance to Fodor's central unbounded processor. In his version, the language regions not only produced language; they allowed cross-modal perceptual comparisons, assigned labels to inputs, and enabled certain kinds of thought. Above all, they were seen to be at the peak, or at least near the top, of a hierarchy of cognitive modules, each of which was carrying out its assigned function. Confirming in principle many of the basic tenets of the early aphasiologists, Geschwind emerged as the modern apologist for Wernicke's approach.

The functional geography of the brain explored by Geschwind was based on neuroanatomy and rooted more in fundamental neurophysiology than in neuropsychology. The best-explored part of that geography, the sensory and motor cortex, is not only well-defined in function but its anatomy shows a continuous evolutionary history, in which the human brain can be located with some precision. In the same tradition, Geschwind explored the anatomy of cortical association areas involved with speech, looking especially for asymmetries of the human brain as clues to forming hypotheses about precisely what functions were to be found in them.

Early in his research (1965), Geschwind singled out the inferior parietal lobule (IPL) as a crucial structure in language. It is situated at the junction of the visual, somatosensory, and auditory cortical areas. Geschwind surmised that cross-modal associations were the basis for speech development in humans; his clinical observations had shown that many aphasic symptoms were due to isolation of the IPL from association regions devoted to the higher processing of vision, audition, and touch. In order to make comparisons between the perceptual outputs of completely different sense modalities, those outputs would have to be encoded at quite an abstract level and fed into a central comparator system, which could read the perceptual outputs at that level. Thus, cross-modal comparisons require abstraction, if not symbolic representation; assuming it supports cross-modal comparisons, the IPL thus loomed large in the perceptual hierarchy.

Later work by Geschwind's colleagues (Eidelberg and Galaburda, 1984) showed that the IPL could be divided into six subregions, of which three were in the angular gyrus. These regions have their cytoarchitectonic counterparts in monkey and ape brains; but as the authors observed, "The human parietal lobe is distinguished . . . in its massive expansion in size and in the elaboration of subregions." And whereas right and left parietal subregions tend to be symmetrical in monkeys and apes, they are strikingly asymmetrical in humans. In particular, two subregions of the angular gyrus appear to be lateralized. The PEG subregion constitutes about one quarter of the angular gyrus and tends to be larger on the right side; its damage leads to unilateral neglect, suggesting its involvement in spatial attention.

The PG subregion, which corresponds to part of Brodmann's area 39, forms the larger part of the angular gyrus and is usually larger on the left. It apparently descends from a supramodal association area in the monkey, which receives inputs from several polymodal sources in the Sylvian region. PG is heavily interconnected with an area of the temporal lobe, the planum temporale (PT), which corresponds roughly to Wernicke's area. In fact, they observed a significant correlation between the anatomical asymmetries they could measure in those two regions. Eidelberg and Galaburda concluded that the leftward lateralization of language to the angular gyrus is closely related to the size of the subregion PG. The authors concluded that "the posterior Sylvian-angular gyrus relationship is simply one link of a chain of interconnected regions of the brain that are specialized for language function, lateralized behaviorally for the most part to the left hemisphere, and bear a characteristic profile of anatomic left–right asymmetries."

Geschwind, and Eidelberg and Galaburda, were attempting a higher degree of precision in attributing cognitive functions to specific right- and left-hemisphere regions. They were looking for the kind of exact localization that has usually been found in the physiology of sensory and motor systems. In contrast, most neuropsychologists, like Kimura and Corballis and Beale, eschewed this ambition and settled for much less exact localization, usually broadly assigning function either to right or left.

One complaint that might be registered about Geschwind's approach is that he did not try to deal with certain of the major features of language. For the most part, he dealt with its perceptual and motor aspects: where phonemes and graphemes might be represented, for example, or where control of articulation resides, or which paths allow visual and auditory representations to be compared. He did not address the problem of syntax, or semantic representation, or the lexicon. Moreover, he and his colleagues did not articulate in detail the kinds of linguistic functions to be attributed to specific structures like PT or PG. He would have replied, perhaps, that such questions are premature; but it is equally true that, in the absence of a larger cognitive model, the anatomical data are hard to read. The history of anatomical asymmetry in primates and humans remains a major source of information, but in itself it is not sufficient for the construction of an evolutionary scenario. What is needed is a functional evolutionary theory, rooted primarily at the cognitive level, and only secondarily at the anatomical level.

Corballis: Generativity and Left Dominance

Corballis (1989) has recently proposed a theory of cerebral laterality that tries to deal with some of the limitations of earlier versions of the default hypothesis. A major objection to the earlier theories was the apparent lack of common cognitive elements between the serial motor skills thought to be controlled from the left hemisphere and language, which evolved later. Although language is a serial motor act, its essential features are surely in its unique syntactic and phonological rules, which seem fundamentally different from motor skill. Why should specialization of the left hemisphere for serial motor control, and especially manual skill, result in a left-bias for the rule-governed aspects of speech and language, unless they depend on some common feature, for which the left hemisphere is specialized?

There are two aspects to Corballis's proposal. The first part is a persuasive reiteration of what has come to be the shell of the standard theory: the human pattern of right-handedness and left dominance for language originated with the refinement of manual skill, particularly toolmaking, in the genus *Homo*, about 2 million years ago. High-speed articulate speech evolved more recently, within the past 100,000 years. The chronological evidence for this sequence comes mostly from archeological data. The second part of Corballis's proposal postulates a novel reason why articulate speech had to be built upon the unique cognitive style of the human left hemisphere, which evolved earlier for the control of motor skill, or praxis. The reason is that there was, and is, a common cognitive adaptation underlying human praxis and language: "generativity"—that is, the "general ability to form multipart representations from elementary canonical parts." Generativity is thus both an analytic and a combinatorial skill.

In language, this ability manifests itself in the combination and recombination of phonemes and words into unlimited sets of utterances. This combinatorial skill is also important in the comprehension of speech; to perceive the speech stream accurately, the listener must segment and parse the input into its salient features. A related point was made some time ago by Warren (1976), in a discussion of the evolution of auditory perception in humans. In Warren's terminology, all mammals, including humans, share the ability to perform "holistic pattern recognition" (HPR), but only humans are able to perform a kind of auditory analysis called "identification of components and their order" (ICO). The latter involves categorical perception and retrieval of the elementary components of language.

In visual perception, image generativity involves "parsing" images into components, which can be recombined and separately manipulated; this can be seen in the breakdown of visual imagery into elementary geometric components, or "geons" (Kosslyn, 1987). As observed by Kosslyn (1988) and Farah (1984), the left hemisphere is able to break down a stimulus into categorical elements and therefore better able to generate novel images of its own than the right. Although it is not proven that the left monopolizes imagery, it appears to be the dominant hemisphere for actively generated imagery.

As in his earlier theory coauthored with Beale, Corballis has stayed with the default hypothesis: the right hemisphere has its own cognitive "style," but its style is not distinctly human. Rather, it is the default form of knowledge, a vestige of earlier adaptations, a style of cognition resembling that of other mammals. The essential breakthrough in human cognitive evolution was the emergence of one truly novel adaptation that set us apart: our capacity for generativity, which is typically lateralized to the left hemisphere. Thus, in Corballis's view, the right hemisphere (in isolation) should illustrate the cognitive armamentarium of archaic hominids and the left that of modern humans.

In his evolutionary scenario, Corballis argued that generativity first emerged in the left hemisphere for praxic skill and was essential to the development of toolmaking, where categorical analysis of the fashioned object was essential to the ability to reproduce it, that is, generate new ones. This generative ability expanded its range and eventually enabled the development of, among other things, language. The evolution of language was secondary, somehow dependent upon the pre-existence of the underlying generative capacity. Since generativity is the distinguishing feature of left-hemisphere cognition, laterality of function must have been a key neuropsychological aspect of human cognitive evolution, perhaps because it allowed humans to retain their traditional cognitive skills in the right hemisphere, while developing new ones in the left.

Unlike Geschwind, Corballis is not concerned with the finer details of anatomy. He does not speculate about the possible role of particular parts of the left hemisphere, such as the angular gyrus or the frontal operculum, in enabling a capacity like generativity. He is more concerned with describing the general cognitive role of the left hemisphere, as a central cognitive system with a characteristic cognitive style. Corballis's hypothesis is thus more in the Jacksonian tradition than in Wernicke's. He does not try to specify particular aspects of thought or language and localize them in particular parts of the left hemisphere. Rather, he assigns each entire hemisphere, presumably organized as a whole, a larger cognitive role. The generativity theory somewhat resembles Jackson's earliest views on lateral asymmetry, in that both visual and linguistic thought are portrayed as left-dominant skills.

Generativity involves a process of combinatorial–sequential analysis that is not unlike the concept of analytic thought except that it is not language-bound. By definition, a generative device must have access to a wide variety of high-level inputs, presumably originating in various specialized input modules. Thus, Corballis's generative left hemisphere would require internal wiring arranged so that it could select and apply rules to its combinatorial procedures. There would be no inherent reason why such a combinatorial device, once invented, could not be used on a variety of qualitatively different inputs: for instance, in the visual analysis of manmade objects and tools, or the auditory analysis of phonemes, or the combination of words into sentences.

But, given that Corballis has not placed any constraints on how this skill might be distributed in the left hemisphere, there is also no reason why there could not be several subsystems within the left hemisphere, each with its own generative capacity. For instance, visual analysis might be independent of vocalization in its neural representation. There could be a number of other such devices, all generative in their own right, serving various perceptual and linguistic subsystems. The theory in its present form holds only that, no matter how many devices have generativity as a property, they are all lateralized to the left and, to a degree, are interdependent on an overriding left-hemisphere computational style.

Corballis's inclusion of visual generativity among the skills of the left hemisphere is based on Kosslyn's (1987) theory that the visual imagery of the left hemisphere is categorical (that is, broken down into identifiable parts) whereas that of the right is holistic (seen as a unity or Gestalt). Generative operations act upon categorical elements; thus, they may be independently manipulated in imagery and ultimately recombined. The process of visual invention depends on such recombinations of elements. For example, if an observer notes that a chair is made of a seat, legs, and a back, each forming a distinct category, then it is possible to envisage endless possibilities for new chairs: ones with more or fewer, larger or smaller, more or less ornate, components. If a chair is perceived only as a whole, however, it cannot be decomposed, and there are no elements to recombine. To fabricate a chair, the component parts must be decomposed and recombined; thus generative ability may have been essential to toolmaking.

However, it is a very long leap from the visual analysis of a machine or manufactured tool to the verbal analysis of sentences, and it is not at all clear that the latter capacity can be "explained" by postulating a single generative processor. Strong psychometric and neuropsychological evidence suggests that verbal and visual–spatial skills are orthogonal to each other and should be regarded as different intellectual factors (Gardner, 1983). A test such as the Mechanical Aptitude subtest of the Differential Aptitude Test epitomizes an analytic, recombinatorial visual test. In the subtest the testee is asked to calculate relationships between the parts of various mechanical devices, drawn in two dimensions. But competence on this subtest has very little correlation with verbal intelligence. In fact, integrating across many factor-analytic studies of the intelligence distribution, researchers have found that visual and verbal analysis constitute two of the most robust and independent intellectual factors underlying human intellectual skill. This argues against a single equipotential generative capacity available for any use and suggests at least two relatively independent generative systems. Of course, this does not rule out the possibility that both systems are on the left; but it does weaken the bond between them, as well as the case for linking them anatomically in evolution.

Is generativity just another label for rule-governed action? Some linguists (such as Jackendoff, 1983) have pointed out the presence of common rule-governed properties underlying both language and action. Corballis is claiming (1) that the presence of this common element, which he calls generativity, implies a common origin and mechanism for all forms of (apparently) rule-governed behavior (inasmuch as it is rule-governed, if not in all respects); and (2) that the absence of generativity in animals, including apes, has been demonstrated. The first claim is closely tied to the second; if generativity is a single adaptation, it must be unique to humans, since the behaviors it supposedly explains—toolmaking and language—are uniquely human.

If generativity, whether visual or linguistic, can be demonstrated in any aspect of ape behavior, Corballis's evolutionary hypothesis would disintegrate. This question is addressed in Chapter 5.

Moreover, if generativity is a broad, generalizable skill, as it must be if it applies to actions as diverse as speaking a sentence and visually parsing a three-dimensional object, then there should be no areas of human skill that are not potentially generative. The generative system can be seen as Corballis's substitute for Fodor's unbounded central system: it receives inputs from all lower-level encapsulated modules and applies a rule-governed, analytic intelligence to them. If it can be shown that independent skill systems do not interpenetrate in their generativity—that generativity occurs at a lower level, that of the input module—then Corballis's thesis would again be at risk. Similarly, his claim that generativity is linked exclusively to the left hemisphere is vulnerable to any demonstration that the cognitive style of the human right hemisphere includes a generative element. These questions will come up again as we extend our discussion in subsequent chapters.

One final remark about the generativity hypothesis. Like Geschwind's search for the anatomical modules of language, it leaves out semantic reference altogether. But semantic reference is the quintessential core of human intellect and language. Visual generativity, of the type described by Kosslyn, does not depend on semantic reference, but generativity in language does. Surely this alone suggests that generative language is sufficiently different from generative visual analysis that the two do not necessarily have to be assigned to the same hemisphere. Granting that Corballis claims not to be attempting a global theory of higher intellect, he has in fact proposed a theory that goes well beyond the question of laterality, and to the heart of the nature of uniquely human intelligence. But the question that must be asked is, given the reach of his ambition, did he go far enough?

Despite these reservations and open questions, the notion of a left-hemisphere mode of computation that is generative is much more powerful than the earlier idea of left-lateralized serial motor control. There was no compelling rationale for Kimura's theory, which placed linguistic functions in the same hemisphere that specialized at an earlier stage of evolution for serial movement control. The motor subroutines for hand movement or vocalization could just as easily be in a different hemisphere from language. However, generativity is basic to both praxis and language, and both might have been its evolutionary by-products.

Although some conceptual progress has been made, the question of which function was lateralized first, or what the primary function of the left hemisphere might be, has not been resolved with any certainty. The question may or may not have any direct relevance to the evolution of language. One could argue that the evolution and structure of cognition are indifferent to anatomy; the same cognitive architecture could, in theory, be implemented in a variety of different physical arrangements. But in the real world the physical and mental aspects of evolution are so closely interconnected that determining the structure of modern human cognition may prove impossible without a careful consideration of the underlying functional neuroanatomy.

The Right Hemisphere and Semantic Reference

Corballis assigned the left hemisphere a leading role in both visual thinking and language, but some writers have gone even further in asserting the superiority of the left hemisphere. Gazzaniga (1983) reviewed twenty years of experience with testing split-brain patients and concluded that "the cognitive skill of the right hemisphere without language is vastly inferior to that of the chimpanzee"; and also that "the price of lateral specialization for language on the left is a state of rudimentary cognition for the right hemisphere" (1983, p. 536).

This idea is reinforced to some extent by the findings of Poizner, Klima, and Bellugi (1987), who found that deaf subjects can develop aphasias in sign language with left-hemisphere lesions similar to those that interfere with speech. In a carefully documented series of case studies, they showed that right-hemisphere lesions do not lead to aphasia-like symptoms in the same population. Even when language is channeled through visuospatial input modules, as it is in sign language, the left hemisphere appears to be dominant. Taken at face value, this seems to indicate that the physical mechanism of language is localized on the left, in the normal case, for all aspects of language: speaking, reading, writing, signing, recognizing, recalling, and listen-

ing. It also confirms the linkage, in terms of gross localization, between categorical visual event perception, which underlies the comprehension of hand signs, and speech.

But there are reasons to doubt whether the dichotomy between the
cerebral hemispheres is so absolute. Zaidel (1983) has questioned the
validity of most of the split-brain cases for the study of right-
hemisphere function. The same question has been raised by others
(Myers, 1984; Patterson and Besner, 1984). Myers made a particularly strong case against accepting Gazzaniga's denial of any linguistic
competence to the right hemisphere. Of the 28 split-brain patients in
the East Coast series, 17 had major neurological deficits in addition to
a severed callosum, such as right cortical atrophy, mental retardation,
or left hemiparesis. In 9 patients the corpus callosum was not totally
split, and interhemispheric communication was therefore possible. In
addition, a number of the 28 patients either had not been tested for
right-hemisphere language or their results had never been published.
The 3 patients in the series who were thoroughly tested, who were
free of serious ancillary brain damage, dementia, or mental retardation, and who had a complete callosal transection all showed some
evidence of right-hemisphere language comprehension.

Based on his work with 5 carefully studied split-brain cases, Zaidel
claimed that "the disconnected or isolated right hemisphere has access
to a substantial auditory vocabulary and to fairly complex pictorial
representations of meanings of words in terms of common experimental situations" (Zaidel, 1976, p. 202). Zaidel's cases were better documented than most previous split-brain case reports. He used standardized tests, established better controls on cross-cueing between
hemispheres, and carried out a one-year follow-up of his results.

Zaidel used the Peabody Picture Vocabulary Test to test the right
hemisphere's comprehension of language. The Peabody requires the
patient to choose an appropriate pictorial representation of a single
verbal idea (for example, to point out the picture that matches a concrete noun like "cow" or a more abstract idea like "justice"). The right
hemisphere was tested independently of the left, by restricting the
visual stimuli to the left visual field and requiring a response from the
left hand only. In his selective sample of split-brain patients, the right
hemisphere's level of comprehension had a mean-age equivalent of

almost twelve years. Zaidel concluded that the right hemisphere possessed a great deal more proficiency in the comprehension of words
than Gazzaniga was willing to concede.

Zaidel then analyzed the vocabulary test results by further breaking down, separately for each hemisphere, the verbal responses to
concrete nouns, action names, and verbs, separately. The right hemisphere was inferior to the left in all categories, as might have been
expected, since the performance of the left hemisphere was close to
perfect. But the mean right hemisphere results of the split-brain patients were well above chance, and fairly close to those of groups of
aphasics, who typically show considerable retention of language comprehension, especially of single words.

The right hemispheres of the same five split-brain patients per-
formed badly on the Token Test (De Renzi, Faglioni, and Scotti, 1970),
which requires the patient to remember longer, nonredundant auditory phrases, such as "touch the large red square and the small yellow
circle." Zaidel concluded from this that the right hemisphere was lacking in auditory short-term memory but possessed a significant degree
of auditory comprehension of words, as well as the ability to map
words on to a pictorial representation of their meaning.

The key finding here is the duplication of function across hemispheres. As far as can be determined, the left hemisphere, isolated
from the right, possesses normal linguistic competence. That is, no
parts of language seemed to be missing when the right hemisphere
was removed from the picture by disconnection and the left hemisphere was tested. Therefore, the existence of any competence at all
in the right hemisphere implies duplication of that function. In Zaidel's studies, the duplication was selective. Duplication appears to
have been limited to three aspects of language: the ability to perceive
the phonetic components of the auditory message, to find the appropriate lexical items, and to match the lexical items to certain aspects
of their meaning, visually represented.

One might be tempted to try a maturational explanation for the
duplication of elementary auditory comprehension of words. According to Lenneberg's venerable theory, cerebral lateralization for language occurs gradually during the critical period for language acquisition (Lenneberg, 1967). Therefore, one might expect some right-
sided vestiges of the earliest stages of acquisition, when language is

still supposedly bilateral in its representation. In other words, the right hemisphere in adults should have a childlike language skill, a remnant of the early bilateral language phase. The problem with this explanation is that normal young children do not have all of the specific skills found by Zaidel in the right hemispheres of split-brain adults. The vocabulary used in the Peabody test is not an infant's vocabulary; and even the pictures are fairly sophisticated. Moreover, many of the young child's abilities are missing in the isolated right hemisphere of the adult. The young child is able to comprehend strings of words as long as those in the Token Test quite early in his development. This aspect of the child's skill is lacking in the right hemisphere, as are the expressive skills that normally accompany the development of comprehension. Most importantly, action names and verbs, some of which are acquired very early in childhood, are also missing in the right hemispheres of split-brain adults.

Thus, the isolated right hemisphere does not look anything like a vestige of child language. This should not be entirely surprising, considering that Lenneberg did not believe lateralization was complete until puberty. If vestiges continued to build in the right hemisphere at a decreasing rate throughout the gradual lateralization of the cortex, the right hemisphere would be expected to possess a great deal more language than Zaidel found. (It is interesting that Zaidel reported that the average isolated right hemisphere's vocabulary is about equal to that of a twelve-year-old, granting that it possessed only a very restricted part of a twelve-year-old's linguistic competence.)

Zaidel's findings are compatible with the observation that stroke patients with severe aphasia often retain the same aspects of language as those found in the right hemispheres of split-brain patients. In fact, when he compared his five cases with the performance of Broca's and Wernicke's aphasics on the same test, he found a substantially similar pattern of results (Zaidel, 1983). This suggests that some severely aphasic stroke patients are probably comprehending language not with the spared areas of the left hemisphere but with the right hemisphere.

The evidence for the linguistic monopoly of the left is derived from the very robust clinical fact that aphasia rarely develops following right-hemisphere damage unless the patient has suffered from early left brain injury, and from the fact that unilateral cerebral anesthesia of the right hemisphere will not cause loss of speech, while the same procedure on the left temporarily blocks language function (Milner, Branch, and Rasmussen, 1964). Both of these observations, however, need to be regarded with caution. Unilateral anesthesia, which involves injecting sodium amytal into one carotid artery, has a very transient effect, and testing of the patient is necessarily limited to a few simple and brief procedures. There is no opportunity to test the higher linguistic aspects of language, and the temporary aphasia of these patients could be restricted to a fairly low-level impairment of the language system, involving input and output modules rather than the central aspects of language.

Moreover, the typical right-hemisphere stroke or trauma patient does not manifest obvious aphasic symptoms, and therefore is not usually thoroughly tested for language functions. There is some evidence, however, that the left hemisphere does not enjoy quite the monopoly that some would give it. A long-standing claim in the neurological literature that the speech of right-damaged patients is flat and lacking in voice modulation, or prosody (Brain, 1965), has been recently confirmed, and a classification of these symptoms has been proposed (Ross and Mesulam, 1979; Ross, 1981). Aprosodias, as disabilities of this type are called, do not usually follow from lesions of the left hemisphere; rather, they result from damage to right-hemisphere regions homologous to left "language" regions.

Prosody is nonphonetic and is considered by some to be nonlinguistic. However, as mentioned in the discussion of Darwin's thesis, prosody is a uniquely human vocal feature, involving voluntary voice modulation that cannot be found in other primates. It is closely tied to the timed meaning of utterances and is often used to convey meaning. Whereas the production of the phonetic elements of language depends upon an intact left hemisphere, their modulation in prosody depends upon the right. At the very least this suggests that the actual production of human vocal speech is, to some extent, controlled from both hemispheres.

In addition, the isolated right hemisphere of the split-brain patient has the ability to decode spoken language at the level of phonemes and morphemes and to understand individual words, particularly frequently used words (Gazzaniga, 1970), although it has difficulty with syntactic features. Phonemic and morphemic analysis of input, search and location of lexical addresses, and some degree of semantic references are thus within the purview of the right hemisphere. Gainotti, Caltagirone, and Miceli (1979) examined right-sided stroke and tumor cases and found them to be unusually vulnerable to the presence of semantic distractors in stimuli, suggesting that they had a significantly reduced critical capacity for evaluating the meaning of words. This would reinforce the results with split-brain patients and suggest a role for the right hemisphere in the semantic analysis of words and sentences.

This points to the conclusion that the right hemisphere has some language competence and some comprehension skills that seem to be dependent upon generative capacities—at least upon the analytic, categorical skills that lie at the heart of generative capacities. This fact alone spells trouble for Corballis's theory. But another finding spells even greater trouble. A study by Gardner, Brownell, Wapner, and Michelow (1983) explored the role of the right hemisphere in complex discourse comprehension. Right-damaged patients were tested on the recall, understanding, and integration of narrative and on their comprehension of humor. They had no difficulties with the phonology of single words or the syntax of single sentences but considerable difficulty with longer narratives. Compared with age-matched controls, and with aphasics, they had greater difficulty recalling the main events of a narrative, remembering the sequence of events, and attributing actions to appropriate agents in the narrative. They also had difficulty understanding emotional or incongruous features in the narrative and showed some disturbance in their understanding of humor. In a word, they could not get the gist of the stories; they seemed to lack a framework, or schema, for their understanding.

Note that these patients had intact left hemispheres and no aphasia in the normal clinical sense of that term. The errors they made were high-level errors; the lexical and syntactic elements of the narrative were understood, but meaning was disrupted at the level of discourse comprehension. Keeping in mind that the input elements of a story must be encoded linguistically, it is hard not to argue for a role for the right hemisphere in the construction of schemas for discourse comprehension. Discourse comprehension involves a uniquely human high-level analysis of linguistic input, a level of analysis sometimes called metalinguistic (see Kintsch, 1988). Therefore, the right hemisphere seems to be implicated in the highest and most characteristically human of cognitive skills.

If decoding words and sentences taps generative skill, then surely the decoding of longer discourses made up of words and sentences demands even more powerful decompositional and recombinatory capacities. It follows that the right hemisphere, although perhaps only partly implicated in the special encapsulated linguistic speech module in Fodor's sense, is involved in a generative mode of processing of the outputs of that module. Thus, the exclusive attribution of generative skills to the left is questionable, and the difference between right and left may be only one of degree. It may also prove to be the case that language generativity itself breaks down into a variety of subsystems, distributed in a complementary way between both hemispheres.

The Case of Brother John

Gardner and colleagues' (1983) study of discourse comprehension brought out the importance of looking at large-scale discourse in assessing deficiencies of language. It also raised the question of the relationship of the "language machine"—the latter-day descendent of Wernicke's machine—to knowledge and semantic representation. Gardner and colleagues would admit that their patients were free from aphasia in the normal definition; thus, their "language machines," or speech modules, were presumably left-lateralized. Nevertheless, their right-damaged patients suffered a considerable loss in the ability to understand straightforward discourse. Was this loss truly linguistic, or part of a broader intellectual deterioration?

This raises some difficult but absolutely central questions. If a patient can understand a sentence perfectly but cannot organize and paraphrase a short story, there has been a breakdown in semantic representation. Note that semantic breakdown in this case does not appear disguised as general dementia or as selective anterograde amnesia, as it does in Alzheimer's disease. In fact, some of the elements of the story are remembered but distorted, and some of the meanings are misunderstood—never the meanings of individual words and phrases but the schematic assignment of actions to agents, the interpretation of events, and the accurate extraction of themes from narrative. It might be tempting to attribute these misunderstandings to a disturbance in the semantic representation of events, independently of language per se. But can these events be represented without language? What is the human mind capable of without an operative language system?

This question was considered by Lecours and Joanette (1980) in a case study of a paroxysmal aphasic, whom they refer to simply as Brother John. Brother John was a 50-year-old man who worked as an editor of letters for his religious order. He had suffered from epileptic seizures for 25 years. Although he had a few grand mal seizures when he had his first epileptic attacks as a younger man, they had been brought under control by medication. When tested he suffered largely from focal seizures and paroxysms, which were classified as short spells (1–5 minutes, occurring up to five times a day) or long spells (1–11 hours, occurring about once a month). Sleep electroencephalography revealed typically epileptic spike-and-wave activity at the left frontotemporal electrodes, and during a seizure he showed slow waves over most of the left side of his scalp, with a peak in the mid-temporal area.

The important fact about Brother John's spells was that they selectively shut down language processing, while he remained conscious and able to remember what he was experiencing. During his long spells, his cognitive abilities returned gradually, progressing through several stages of recovery. The authors recorded his responses and comments in detail and thus had an unusual opportunity to observe what a person can do in the absence of speech (including internal speech) and written language.

During a long spell Brother John would reliably pass through a series of stages which successively resembled most of the various clinical subtypes of aphasia. He would initially manifest the symptoms of global aphasia, and after about half an hour his symptoms gradually changed, going through a period of jargonaphasia until he began to look more like a case of Wernicke's aphasia, characterized by anomias and paraphasias. As his recovery progressed, his symptoms gradually changed again, and he looked more like a conduction aphasic, while in the final few hours he was left mostly with amnestic symptoms.

The authors held up this case in support of Marie's doctrine of the "oneness of aphasia," a doctrine based on the idea that the various types of aphasia are really only variations in severity of a single syndrome, which may or may not be combined with anarthria. More importantly for the present review, they also examined nonlinguistic cognitive function in their patient and concluded that, even while his language was very severely impaired, the patient was able to record and remember complex events in the environment and talk about them accurately after his seizure.

The case is too rich in detail to review completely, but the conclusions are worthy of detailed consideration. Even while he was globally aphasic, dysarthric, or capable only of producing neologistic jargon, Brother John was very much aware of his disability and able to cope systematically with it. Often, he was able to anticipate a spell by a few minutes and arrange with a friend to cover for him at the office. He would arrange things so that, during the worst part of his spell, he would not inconvenience himself or others. He would consciously fight a tendency to become depressed and sleepy, and he kept a portable radio handy, in order to periodically test his ability to comprehend what he heard on it. In other words, he was fully conscious and able to cope with the situation, despite the absence of language. Moreover, he had full recall of the episode later.

The extent to which he retained the ability to cope with practical challenges was quite remarkable. One episode, while he was traveling in Switzerland, was particularly striking. He found himself at the peak of one of his seizures as he arrived at his destination, a town he had never seen before. He took his baggage and managed to disembark. Although he could not read or speak, he managed to find a hotel and show his medic-alert bracelet to the concierge, only to be sent away. He then found another hotel, received a more sympathetic reception, communicated by mime, and was given a room. He was able to execute various procedures which formed a framework for linguistic operations; for example, he was able to point out to the desk clerk where in his passport to find the information required to fill out his registration slip, while not being able to read it himself. Finding himself too hungry and miserable to sleep, he went to the hotel restaurant. He could not read the menu, but he pointed to a line which he thought might be the hors d'oeuvres and randomly chose an item, hoping he would like it. In fact, it was a dish he detested, but he ate it, returned to his room, and slept for the remainder of his paroxysmal attack. When he awoke, he went to the hotel desk and explained the episode in detail.

A number of important conclusions can be drawn from this case. First, despite the complete absence of language, internal or external, Brother John was able to cope in ways that are uniquely human. He was capable of coherent thought, able to recognize music, voices, and faces, and the uses of objects and places. His spatial orientation and basic mechanical intelligence were intact. His episodic memory for the events of his seizures was accurate (this was objectively verified by the authors) and organized. During a seizure, he was able to remember and execute a request given to him weeks earlier by the authors— for example, taping his spontaneous speech as it recovered during the spell.

Second, visuographic language appeared to recover somewhat independently of the auditory–oral system. Early in the recovery process, he usually went through a period when his ability to communicate through reading and writing was much better than his auditory comprehension or his ability to speak. During this phase he would resort to paper and pencil for communication, and he was able to solve arithmetical problems, such as correctly multiplying two- and three-digit numbers, while remaining aphasic.

Finally, both gestural ability and practical knowledge were intact. He could imitate or reproduce on demand a wide variety of gestures. He could tune a radio, operate an elevator, and, as the Switzerland episode shows, assess and respond appropriately to a social situation of some complexity. All of this was achieved in the absence of visual or oral language, and in the absence of internal speech as well. In his own introspective account, he claimed that he could not "find the words" for things and events. Nevertheless, he could think about them coherently, deal with them appropriately, and remember them later.

This case suggests strongly that human intellectual skill is uniquely powerful, even in the absence of language. Most of the behaviors demonstrated by this man during his seizures were well beyond the abilities of any other primate or mammal. Language was obviously not the vehicle by which he assessed events, formulated plans, and evaluated his own responses. Nor was it important in his vivid and perceptive episodic registration of events or in the execution of gestures and mime. Moreover, his semantic representations of the world were accessible and useful to him, quite independently of language. This, of course, did not apply to his work, which inherently involved the use of written symbols, or to any knowledge encoded primarily in symbolic form. But the boundaries between the linguistic and nonlinguistic worlds have seldom been so clearly seen, and the independence of the one from the other so nicely demonstrated.

The Brain without Language

Neuropsychological studies are our best, and in some cases our only, source of knowledge of the way the human mind can function in the absence of one of its "modules." Wernicke's model was the start of a way of thinking about the structure of mind, and modern neuropsychological models have extended and enriched that approach. The lateralization of human higher cognitive function is, in itself, less interesting than the demonstration that higher function is organized in dissociable functional modules, some of which are left-lateralized.

The case for a single, unbounded central system, whose state is roughly equivalent to what we experience as consciousness, appears to be strengthened by the case of Brother John. Brother John's loss of language was roughly like losing (temporarily) a sense modality; but there was no loss of awareness. Language inputs and outputs were simply not available, so he muddled through without them, just as deaf people do without sound, and blind people do without vision. Gnosis and praxis were intact, as were episodic storage, self-representation, working memory, social intelligence, and scripts and schemas for action. Despite the loss of internal speech, he was able to cope with a variety of complex, uniquely human situations and problems.

Had language been an inherent part of the central system, or even if the central system had become dependent upon language for its functioning, the integrity of the central system would surely have been shattered. At least, he might have appeared somewhat demented, that is, confused, disoriented, and unable to cope with novel or complex situations. But this did not happen; he was fully aware, with no breakdown of basic conscious functioning.

What was lost, then, from knowledge and memory, by excising all forms of language and internal speech? In particular, what happened to his semantic memory store, and what of propositional knowledge? The verbal forms of this material were not available to him, and any cognitive operations that were dependent upon symbolic representations could not be executed. But he evidently knew many of the things that he normally would have expressed, and probably learned, via the language system. Therefore, a great deal of his knowledge must have been nonverbal, or at least stored outside of the language system in nonsymbolic form.

Nonverbal knowledge is often described as a right-hemisphere property, as if the memories of the left were usually encoded in verbal form. But the knowledge base that allowed Brother John to cope was not restricted to the "holistic" or Gestalt-like perceptual processing identified with the right. His intact abilities included planning ahead, carrying out social interactions, evaluating complex events, and dealing with a variety of uniquely human devices such as telephones, automobiles, elevators, hotels, and, especially, people. He apparently needed language for none of these things.

The coping ability of Brother John is a surprise to many people. But why should it be, when we consider the coping skills of people who are raised without language, that is, the congenitally deaf who receive no training in vocalization, sign language, or reading? Although the majority of deaf people in modern society receive help and training in communication, this was not the case for most of our past history, and it is still not the case in many societies. The illiterate deaf-mutes of the past, some of whose lives are recorded, were effectively without a system for symbolic communication. They lived without any access to the oral–auditory language system. Most of them had no formal sign language and had to rely on spontaneous gesture and mime. Since they did not form a community, and were a small, virtually invisible, minority in society, there was no opportunity to develop a formal sign language for their use. And prior to the last century, *most* people in the world were illiterate, including the deaf. Therefore, they could not use written or printed media for communication. Thus, the congenitally deaf person, until very recently, was in a situation quite similar to that of Brother John.

Of course the deaf-mutes were neurologically normal, where he was not. Moreover, their condition was permanent, and his wasn't. And he possessed language most of the time, and they didn't. Nevertheless, during one of his spells, he found himself in a comparable position with regard to external linguistic communication. And from the evidence, his position with regard to internal speech was similar to theirs: just as they could not harness an internal phonetic, written, or sign language that they had never learned, he could not access the one he had learned.

From the extensive available historical evidence on the education of the deaf (Lane, 1984), it is clear that in the past their intellectual development was greatly held back by the absence of a common sign language. In isolation, deaf people were capable only of gesture and mime and were employed, for the most part, in menial occupations, usually manual labor. Once deaf people began to be placed in schools, they evolved their own signing systems and gradually gained access to literate culture. But prior to widespread literacy and the invention and standardization of sign language, they were isolated from symbolic culture and the knowledge skills it made available.

However, despite the constraints placed until recently upon their linguistic development, deaf-mute people have always been capable of many of the more complex, and uniquely human, cognitive skills. Darwin (1871) was aware of this and used this fact to emphasize the differences between ape and human intelligence, even without language. Just like Brother John during one of his spells, Darwin's examples were able to cope with the basic social framework of behavior. They were quite able to carry out the type of adaptive behaviors Brother John showed during the Switzerland episode, for example, when he located a hotel and managed to convey what he wanted to the hotel clerk. They had normal praxic skills and were usually employed as domestics or in the trades, jobs that for the most part demanded a level of social and manual skill that was uniquely human. They were able to solve problems and learn new facts about the environment—new places, new faces, new skills—when they moved from place to place.

There is no reason to believe their episodic memory or their conscious awareness was in any way impaired by the absence of symbolic language. Deaf people who learn some form of language typically recall episodes from their youth, despite the absence of normal language development. Thus, the human brain, without language, can still record the episodes of a life, assess events, assign meanings and thematic roles to agents in various situations, acquire and execute complex skills, learn and remember how to behave in a variety of settings. Brother John was an exceptional neurological case: usually brain injury does not "excise" language so precisely as this, and usually the patient does not have the opportunity to recover completely and reflect on the experience. But his behavior without language is entirely consistent with what we know of the deaf-mute person.

Why does the typical isolated right hemisphere not display the skills of the deaf-mute or Brother John? One possible explanation is the presence, even in the split-brain cases, of extensive permanent brain damage that impairs cerebral function in a nonlocalized way. Adults who have received either hemispherectomies or massive unilateral brain injury often remain impaired not only by large-scale brain damage on the lesioned side but by a disturbance of function on the "good" side as well, due to destruction of millions of cross-hemispheric and thalamocortical circuits.

A better comparison with Brother John would be with the typical aphasic stroke patient. Severe aphasia sometimes occurs simultaneously with a general cognitive loss, but often severe aphasics are still able to cope with many practical and social demands, despite permanent brain damage, a crippling language loss, and accompanying depression. Aphasics typically still use gestures and mime, and many do not suffer agnosia and apraxia, or any evidence of amnesia or disorientation. In fact, these symptoms are sufficiently independent in occurrence from the aphasia per se that they are described separately from the aphasic symptoms themselves.

Conclusion: Wernicke's Machine in Evolution

What do the above data mean? What is the device that is destroyed when Wernicke's left-hemisphere machine, made up of left-sided brain regions supposedly dedicated to language function, is damaged or eliminated, or, in the case of the deaf-mutes of history, prevented from developing? If humans can function at a high level without it, if their awareness—or rather the integrity of their person—is not destroyed in its absence, and if objects, events, people, and social situations can be encoded, assessed, remembered and reacted to in the absence of this device, where does it fit into the scheme of human cognitive structure?

Fodor's notion that the language system is, at least in part, a vertically integrated, somewhat self-contained input module, outside the mainstream of thought and awareness, continues to be strongly supported by neuropsychological data. The language module supposedly contains the encoding and decoding devices for speech and graphic representation (of which signing and writing are two examples), and the lexicon, or lexicons, that serve the system. Whether the same module includes grammar is debatable; but it does not seem to contain some of the basic human semantic reference systems, which often remain intact during clinical language loss. On the other hand, in the absence of language, it is impossible to test many aspects of meaning, especially those that are inherently linguistic; and thus it remains possible that certain aspects of semantic representations are unique to the language module proper.

Corballis's attempt to place generativity at the center of the left hemisphere's unique capabilities is an attractive idea but falls short of accounting for the apparent dissociations between different forms of generative skill. Various aspects of perceptual generativity, as manifest in visual decomposition, and in the decoding of signs and phonetic utterances, may have elements in common, but this does not necessarily imply their inclusion in a single evolving system, and it is not yet clear that all these skills are uniquely human. There are other problems: it is not proven that generative ability is restricted to the left hemisphere, nor is it known whether or how a generative device might be harnessed to a semantic reference system.

Syntax appears to operate at a different level in the system and does not break down in a clear-cut manner in aphasics. Apparent "agrammatisms" may be better described as specific types of semantic disorders (Shallice, 1988). In fact, pure agrammatism is a suspect symptom in aphasiology inasmuch as its careful analysis has raised more questions than it has answered; Badecker and Caramazza (1985) have suggested that apparent agrammatisms are often only anomias for verbs and closed-class words.* The evidence for an independent "grammar module," or a specialized device that applies grammatical rules to roughed-out prelinguistic constructs, is not good. Semantic reference is another problem area: semantic reference is retained to a degree in deaf-mute people, as it was in the case of Brother John and in people with related neurological syndromes. This is not to say that there is not a form of semantic reference that might be unique to language; but there are obviously types of semantic reference that do not depend on language. The whole question of reference and representation looms large as the central issue of cognitive structure and will receive extensive treatment in later chapters.

Where would a general-purpose adaptation like generativity fit into the system? Inasmuch as it applies to the production and decoding of speech, it would have to be built into the speech-input module itself. Quite independently (since these things break down independently) it would have to be built into the input module for visual symbolic representation (which might or might not be identical with the module for visual image generation hypothesized by Kosslyn). This suggests separate evolution of two or more independent input modules, each of which is generative.

But the central system or systems also have a generative aspect. Fodor places syntax at least partially in his hypothetical central system. Discourse and narrative are undoubtedly analyzed and produced at this level as well. Plans, scripts, schemas, and conscious awareness are all typically allocated to the unbounded central system. This would appear to have been yet another evolutionary adaptation featuring generativity; and yet, it is not clear that the central system is

---

*Function words are sometimes called closed-class words because they constitute a fixed number of items at the core of a language. There are only 363 function words in English, in contrast with hundreds of thousands of content words, such as nouns, adjectives, and verbs, which make up most of the words in the dictionary and are continuously growing in number. Function words carry meaning only in relation to other words and serve to define relationships by connecting, substituting for, and modifying words and phrases. They include determiners (the, those, neither), pronouns (he, she), conjunctions (and, but, or, nor), and auxiliary verbs (do, be, have, will, may). Anomia (the inability to retrieve the words needed for an intended utterance) for certain important closed-class words could thus prevent the proper construction of sentences and look like agrammatism, whether or not grammar per se had been affected.

completely lateralized to the left, or that it is a unitary system in its own right.

Would generativity have emerged independently in so many ways, or is it perhaps just an inadequate word trying to account for too many separate cognitive features? Visual decomposition, perceptual recombination, phonetic segmentation, verbal invention, toolmaking, parsing, and grammar are but a few of the complex acts that are supposedly explained and enabled by this remarkable adaptation. Yet for reasons that have already been cited, some of those activities are central while others are probably attributable to a lower level in the system, that is, to an input module. This does not lend itself to a simple or convincing evolutionary scenario.

One way to resolve this dilemma is to avoid any direct linkage between the different generative systems that humans evolved—that is, not to postulate "generativity" as an adaptation in itself. In fact, one might extend this to the entire enterprise of proposing separate "cognitive styles" for the right and left hemispheres. The difficulty with proposing that the left is generative, analytic, propositional, logical, or sequential, while the right is nongenerative, holistic, appositional, intuitive, or spatial, is that we have to find a common element among quite disparate skills and then construct an evolutionary scenario to explain why these disparate skills had to cluster together in one hemisphere (presumably because they all tap some common processor, or some mysterious "processing style"). But there is no evidence for any such common processor, and the conceptual linkages between the clusters of skill attributed to each hemisphere are loose and poorly documented.

However, if we abandon this strategy of clustering all the distinctly human adaptations in the left hemisphere, under the aegis of one all-encompassing cognitive adaptation, what alternatives are left? The strong form of the default hypothesis—the idea that only the left hemisphere advanced, leaving the right with vestigial forms of cognition—has to be abandoned; the right hemisphere has its own uniquely human features as well. The modular approach is still credible—this much is clear from the neuropsychological evidence—but only a modular approach that de-emphasizes laterality as the dominant issue. Laterality could still emerge as an issue in the cerebral representation of particular aspects of cognition, but there would be no need to postulate some common property to tie together the skills apparently clustered on one side of the cerebrum.

Where then does laterality fit in? Clearly, some aspects of speech and language are usually left-lateralized. The perceptual and motor skills supporting speech, writing, and formal sign language are largely on the left, and the lexical addresses of words appear to be largely on the left. Although most humans are right-handed, the lateralization of praxis is not so dramatic as is the lateralization of speech, and the pattern of lateralization of perceptual skill is even less well established. The right hemisphere is capable of prosodic expression and some phonetic decoding, at least at the level of words and simple phrases. And the right hemisphere is probably capable of some degree of componential visual analysis, both of which could be considered both categorical and generative. Although apraxias are typically left-hemisphere symptoms, they may still be attributable to the disconnection of language from action, as Geschwind proposed; thus we cannot necessarily place the control of praxis exclusively in the left hemisphere. Praxis might be bilaterally represented and selectively impaired when its control by language is damaged.

In summary, the case for clustering all uniquely human cognitive skills together under one umbrella, and placing them in a single line of phylogenetic adaptation, lateralized exclusively to the left hemisphere, is not strong. It is not established that all aspects of linguistic function are left-lateralized; in fact, it seems more likely that both cerebral hemispheres are involved in distinctively human forms of cognition. But perhaps the most important conclusion to be drawn from the neuropsychological literature is that human intelligence without language has properties that set it apart from ape intelligence, just as Darwin predicted. Among the uniquely human capacities found in the complete absence of language are a capacity for spontaneous gesture and mime, which can be retained after language loss; toolmaking and praxis in general; emotional expression and social intelligence, including an ability to comprehend complex events and remember roles, customs, and appropriate behavior. These fundamental abilities, robust and so important to human survival, might have emerged early in the human line, before language evolved. Their neuropsychological dissociability from language suggests a distinctly human, but prelinguistic, level of cognitive development and a possible basis for an early hominid adaptation that set the scene for the later arrival of language. This possibility is addressed in detail in Chapters 5 and 6.

Neuropsychological theories of brain evolution are a useful exercise, but one general criticism of this approach might be that it is too narrow in its functional view of language and thought. Wernicke's machine is not the whole story. A broader view of the evidence—especially from anthropology and biology—might help place the neuropsychological aspects of human cognitive evolution in perspective. In addition, the neglect of representational systems and symbolic reference needs to be addressed if we are to deal realistically with the applications for which human language has apparently evolved. Lateralization and language localization can be placed in perspective as part of a pattern of human cognitive evolution that can best be seen as part of a larger adaptation with social and cultural dimensions.

The Chronology of Anatomical and Cultural Change

Markers in the Chronology of Change

The reconstruction of human prehistory is a necessary component in tracing the origins of the human mind, but much of the evidence available on primates and early *Homo* is tangential to the present inquiry, and I shall try to focus selectively on findings that are central to the emergence of thought and language. It might be argued from the outset that gross morphology is irrelevant. Bipedalism, the opposed thumb, an elaborate vocal apparatus, and even a large brain are not in themselves necessary or sufficient for language. Each of these features may be found to some degree in other species that possess nothing resembling human language, and humans have been shown to have language capacity in the absence (owing to genetic abnormalities or disease) of some or all of them. Even trying to find a simple cause-and-effect relationship between the modern human supralaryngeal vocal tract and speech would be a futile exercise. Wind (1976) has made this point convincingly.

However, in evolution a major anatomical change inevitably signals a concomitant functional change; thus, the study of morphology can help establish how many major evolutionary steps were taken on the way to modern humanity, and when those steps were taken. Anatomy has been, above all else, the evidence used to establish the succession of species in the line of descent. It gives us a time-tag to place on major change. A gross anatomical change such as the emergence of the human vocal tract is part of a larger biological picture, and a product of interactions with continuing cultural change. The vocal tract would not appear in isolation from a radical change in hominid communication; this in turn would not occur without the existence of appropriate cognitive skills.

The modern ape is the logical starting point of our speculations. Our closest genetic relatives, chimpanzees and gorillas, differ from modern humans along a number of physical dimensions, but these differences are mostly of quantity rather than quality. The physical resemblances between humans and apes are striking, and the anatomical transition from ape to human is primarily manifest as a series of geometric transformations. The cranium of the ape is smaller, with a lower dome, while the face protrudes much more ahead of the orbital plane, producing a prognathous appearance. The intervening species in the hominid line fall fairly smoothly on a line of gradual transformations, until the modern human skull and face shape is reached.

The same applies to other aspects of anatomy. In most of its aspects, the human body seems to have emerged as a transformation of the ape body. The great apes are shorter and more heavily built, with longer, more powerful arms and a pelvic girdle that does not permit them to walk fully erect. Because they do not walk erect, their rib cage and lung structure are different from our own, as is their vestibular and motor-control apparatus. Also as a consequence of posture, their heads do not sit squarely on top of their spinal cords but rather lean forward, although somewhat less than most four-legged mammals. Thus, the hole at the base of their skulls (the foramen magnum) through which the spinal cord enters the brain is at the back of the cranium, and their necks are much more muscular than ours. They possess hands which, although not human in form, are remarkably similar to our own when compared with their presimian forebears.

These transformations could have accumulated gradually over millions of years, or they might have occurred in relatively discrete steps, that is, over thousands of years. The latter view has come to dominate in recent decades. The theory of punctuated equilibria holds that major evolutionary changes take place in sudden, radical steps that

lead to the emergence of new species (Gould and Eldredge, 1977).*
The chronology of evolutionary changes leading to *Homo sapiens
sapiens*, established primarily in terms of gross anatomical differences, appears to support this notion. This does not rule out very
gradual change as well; this also occurs. But there are periods when
evolution seems to accelerate, and radical change is more likely.

The changes from Miocene ape to human appear to have involved
a number of successive but discrete adaptations, each adaptation moving a little further along the continuum toward human form. Table
4.1 shows a working interpretation of this succession and the estimated time at which each intermediate form appeared. Note that,
although these species appeared successively and are sometimes presented as though one blended into the next, the exact time and order
of transitions to new species are not known. The process of speciation
is thought to occur rapidly, in a small, in-breeding founder population
that is geographically isolated from the larger ancestral species. The
result, in terms of the archaeological record, is a series of document-
able new species, with little or no fossil evidence on transitory forms
during speciation.

The fossil record is better than average on the question of human
emergence, perhaps because we are of such recent origin. On the basis
of this record, as well as on genetic and biochemical evidence (Sarich,
1980), the hominid line of evolution deviated from the line that led
to the chimpanzee about 5 million years ago. Australopithecines, the
earliest documented hominids, existed from approximately 4 million
to 1.5 million years ago. *Homo* came into existence about 2 million
years ago. The earliest species regarded as *Homo*, the habilines, existed from about 2 to 1.5 million years ago; but the habilines were in
many ways more similar to australopithecines than to humans. The
first distinctly human species was *Homo erectus*, who lived from 1.5
to about 0.3 million years ago. Archaic *Homo sapiens*, our immediate
ancestor, came into being about 250,000 years ago (Stringer and Andrews 1988).

---

*Discrete steps in evolution are not defined as instantaneous changes, or saltations, in
single individuals. Even the sudden, radical punctuations in the process of punctuated
equilibria would require hundreds of generations. Such steps are discrete and sudden only
in a relative sense, when contrasted with much longer periods involving millions of years
of relative stasis that have been observed in the fossil remains of many species.

Table 4.1   Approximate time-line for the succession of hominids, in years before present.

5 million years: Hominid line and chimpanzee line split from a common ancestor

4 million years: Oldest known australopithecines
  • erect posture
  • shared food
  • division of labor
  • nuclear family structure
  • larger number of children
  • longer weaning period

2 million years: Oldest known habilines
  • as above, with crude stone-cutting tools
  • variable but larger brain size

1.5 million years: *Homo erectus*
  • much larger brain
  • more elaborate tools
  • migration out of Africa
  • seasonal base camps
  • use of fire, shelters

0.3 million years: archaic sapient humans
  • second major increase in brain size
  • anatomy of vocal tract starts to assume modern form

0.05 million years: Fully modern humans

Jerison (1973) and others have developed the concept of an encephalization quotient, or EQ, a ratio between the expected brain size of a species, based on regression lines calculated from samples of hundreds of different species, and the actual size. Jerison reports that the regression lines of brain size on body size are different for fishes and reptiles than for birds and mammals, with the latter having, on average, larger brains for the same body mass. Of all the mammalian species measured, humans had by far the largest EQ, possessing about 6.9 times as much brain as an average mammal with a similar body weight. The nonhuman species with the closest EQ to humans was the chimpanzee, with an EQ of 2.6.

Taking his analysis one important step further, Jerison examined the hominid endocast data, obtained by casting brain molds from fossil hominid cranial reconstructions (see Holloway, 1974) and estimating body mass from skeletal reconstructions. His analysis revealed the time course of hominid brain expansion over the past 5 million years. This brain expansion, relative to the brains of apes and prosimians, is illustrated in Figure 4.1, based on data of Passingham (1982) and Jerison (1973).

The most primitive primates, the prosimians (P), have average mammalian EQs of 1. Great apes (GA) have about 2.5 times more brain for the same body mass, while hominids have increasingly more. The earliest hominids, australopithecines (A), had larger average EQs than apes, but they still overlapped the ape range, as did the habilines (H). *Erectus* (E), however, moved outside of the ape range; thus, the first great demarcation line in EQ had been crossed with *Homo erectus*, whose EQ was roughly double the ape EQ. The latter ·was averaged across early and late *erectus*, but in fact the later *erectus* skulls had a significantly larger cranial capacity than the earliest ones. The cranial expansion culminated in modern humans (S), who have almost 3 times as much brain, per gram of body mass, as the average

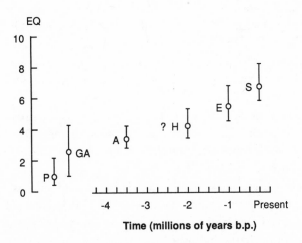

Figure 4.1   Time course of hominid encephalization. Means and approximate ranges of the EQs of each species are shown by the open circles and vertical bars, respectively. P: prosimians; GA: great apes; A: australopithecines; H: *Homo habilis*; E: *Homo erectus*; S: *Homo sapiens*.

brains at various times in our evolution. The major points of interest
ape. Thus, within the primate line, from prosimians to apes to humans, there was a sevenfold increase in relative brain size.

The succession from ape to human involved two particularly large
changes in EQ, both of which occurred rapidly over a period of a few
hundred thousand years. The first occurred with the emergence of
Homo erectus, who reached an EQ double the average of the great
apes and roughly five times the mammalian average. The second major change occurred with *Homo sapiens*, whose increased cranial capacity is roughly three times that of the great apes, and seven times
the mammalian average. Using estimated EQ alone as a time marker,
this suggests there were very major cognitive changes in the human
line about 2 million, and 200,000, years ago.

Another useful marker is any evidence of radical cultural change;
and in hominids evidence of radical cultural change has tended to
correlate with increases in brain size. *Homo erectus* was characterized
by major cultural innovation as well as a larger brain: *erectus* produced tools of great refinement, eventually domesticated fire, and migrated over much of the globe. *Homo sapiens*, with the largest hominid brain, produced the exponential rate of cultural change that
characterizes modern human society.

But cranial capacity is not the only relevant anatomical time
marker. Erect posture is another important one. Approximately one
million years after the hominid line deviated from the chimpanzee
line, the australopithecines adopted bipedal locomotion, with concomitant changes in their method of respiration, in cranial and spinal
anatomy, and in the anatomy of the hands. Although their cranial
capacity was slightly larger on average than that of great apes, the
range of their EQs greatly overlaps the range of ape EQs; thus, erect
posture evolved before the first major increase in the relative brain
size of hominids.

There are other markers, derived from a finer analysis of fossil
crania. Exterior brain molds have been obtained from some of the
better-preserved crania of hominids by Holloway (1974). As the skull
grows in concert with the brain, its interior surface contains a crude
imprint of the major surface characteristics of the brain. This has
allowed the recreation of the gross surface appearance of hominid

are the surface correlates of cerebral laterality. LeMay (1984) has shown that *Homo erectus* already possessed the characteristic asymmetry in the angle of the Sylvian fissure (higher on the right, longer on the left) that is found both in later fossil species and in most modern humans. This superficial asymmetry, assessed in hundreds of modern humans with radiography, is reliably correlated with the relative volume of a specific language region of the cortex (the planum temporale), on the left side, as well as with degree of handedness and speech laterality. Another related asymmetry, the slight clockwise protrusions of the frontal and occipital lobes, called fronto- and occipito-petalia, can also be found in fossil species (LeMay, 1976, 1984). Both of these findings indicate the presence of gross morphological right–left asymmetries in early hominids.

Unfortunately, at least for those who are looking for a direct link between laterality and the appearance of characteristically human culture, the Sylvian markers are also present in australopithecines and apes (LeMay, 1984). Thus, gross surface asymmetries in the cerebral cortex must be more archaic than the cognitive changes most characteristic of humans. However, Holloway and de la Coste-Lareymondie (1982) have argued that the other measurable markers of laterality, the fronto- and occipito-petalias, were found first in archaic *Homo* and were absent in australopithecines. Thus, some markers of laterality may have changed during the course of emergence of the hominids, but the significance of this change is not clear.

Another cranial measure that has received considerable attention in the literature on human language evolution is the flexure of the basicranium, that is, the curvature of the basal surface of the skull. Lieberman (1984), who has summarized the relation of this measure to the shape of the modern vocal tract, reports that the angle of the major articulatory muscles of the pharynx is a direct function of this curvature. The human vocal tract is characterized by an extreme descent of the larynx and a high pharynx that rests at right angles to the oral cavity. The length of the soft palate and the descent of the tongue and larynx are characteristic only of modern humans.

Lieberman's analysis is summarized in Figure 4.2, which shows the mean basicranial flexures of gorillas, chimpanzees, australopithecines, Neanderthalers, and modern humans. The primate mouth is long and

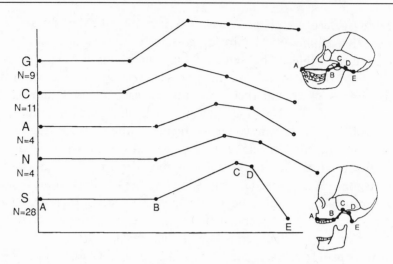

Figure 4.2   The curvature of the base of the skull (basicranial flexure) in selected primates and hominids, modified from Lieberman (1984). The curvatures are outlined against profiles of a chimpanzee skull and a modern human skull (not drawn to scale), to show approximately where the five points on each curve were measured. G: gorilla; C: chimpanzee; A: australopithecine; N: Neanderthalers; S: sapient humans. The number (N) of skulls in each sample is indicated on the left. Some of the critical constrictor muscles of speech are anchored at right angles to the line connecting point D to point E, which is much more steeply sloped in modern humans.

narrow and largely filled by the tongue. The adaptation central to human vocalization is the descended larynx, which leaves room for an elaborate supralaryngeal vocal apparatus that can create a greatly increased variety of sounds. The figure shows the contour of the basicranium—that is, the base of the skull—which reflects the shape of the upper vocal tract: points A through E are bony landmarks that trace the shape of the cranial area where many of the vocal muscles are inserted. In the gorilla and chimpanzee, the basicranium is relatively flat, reflecting the lack of an enlarged supralaryngeal and pharyngeal area. The australopithecines are also relatively flat in this regard but had started to move toward more flexure in the basicranial line. The Neanderthalers, who are very recent, had much less basicranial flexure than modern humans. Assuming for the moment that Lieberman is right—that the existence of the human supralaryngeal vocal tract is closely tied to basicranial flexure—this places a very recent date on the human vocal apparatus.

These markers can be used together to draw a time-line on which major cognitive landmarks might be located. The critical dates are 4 million, 2 million, and 200,000, years ago. Although the precise lineal relation of the australopithecines to the hominids is a disputed topic, it is a reasonable assumption that australopithecines in general are either directly ancestral to *Homo* or share a common ancestor with us, and that various species of *Homo* are ancestral to modern humans (Johanson and White, 1979; Passingham, 1982). On this assumption, humans must contain within their makeup some vestiges of their major transitions from the ape, and it is likely that these changes would continue to be in evidence both anatomically and culturally. Thus, a guiding principle in reconstructing the cognitive succession should be to look for evidence of earlier cognitive adaptations within the cognitive structure of modern humans.

Thus, the time course of human morphological evolution shows periods of very rapid, and fundamental, change—for instance, in the transitions from ape to australopithecine about 4 million years ago, then to *Homo erectus* 2 million years ago, and finally to *Homo sapiens* 200,000 years ago. In between these rapid changes, there were periods of less dramatic change, such as the well-documented variations in the specific anatomy of australopithecines, from early *afarensis* to the later gracile and robust subspecies, found over their collective lifetime of about 2.5 million years. There was also an apparent gradual increase in brain size of *Homo erectus* over a period of 1 million years, as seen in the increased average EQ of the later *erectus* skulls, specifically in Peking Man.

From the evidence summarized in this section, the demarcation lines between species are the large markers in our evolutionary scenario: bipedalism (− 4M years), followed by cranial enlargement and some changes in the gross morphological asymmetry of the hemispheres (− 2M years), then a second cranial enlargement and the modern vocal apparatus (− 0.2M years). The significance of some of these markers, particularly of the last, is controversial. They can only be interpreted in terms of available reconstructions of the cultures of successive species of hominid.

Bipedalism: The Australopithecines

In the chain of evolutionary change that led from the ancestor we shared with our fellow primates to *Homo sapiens sapiens*, the first major event was the adoption of bipedalism. Earlier in this century the "brain-first" theories had dominated; that is, it was held that the brain expanded in size before bipedalism emerged in the evolutionary sequence. However, those theories depended heavily on the Piltdown skull, and when the finding was exposed by Weiner and Oakley (1954) as a fraud, brain-first theories were largely discredited. Dart's (1925) evidence for the existence of a very ancient bipedal species, the australopithecines, had questioned the "brain-first" theories thirty years earlier, before the exposure of the Piltdown hoax; Dart had shown that australopithecines possessed small brains, about the same size as an ape, while possessing erect posture and fairly modern hands. Through the work of LeGros Clark (1967) and the use of radiocarbon dating, the australopithecines were shown to be much older than any existing hominid remains, as much as 2 million years old. More recently, Johanson and Coppens (1976) produced evidence of early australopithecines that were almost 4 million years old. The era of australopithecines thus extends from about 4 million years, or more, before present, to about 1.5 million years.

There were several species of australopithecine, the most ancient being the *afarensis* species discovered in Ethiopia (Johanson and Taieb, 1976) and the most modern the *boisei* species found in Olduvai Gorge, in Tanzania (Leakey and Leakey, 1959). The various species of australopithecines had essentially the same anatomical features, EQ, and survival strategies. Thus, they will be grouped together in this discussion.

Australopithecine bipedalism involved a great many structural changes in primate anatomy, not only to the pelvis and thigh bones but also to the foot, hand, rib cage, and respiratory apparatus. Gould (1980) has called it the greatest single adaptation in the line of human evolution, and from the viewpoint of gross physiological and anatomical change, he is right. Bipedalism involved the abandonment of the safety of the trees and the loss of prehensile feet, the adoption of a much less energy-efficient means of locomotion, and much greater risk from terrestrial predators. What would have been the immediate advantage of adopting such an innovation? Freeing the hand for extensive tool use could not have been the *initial* advantage; tool manufacture, even in the crudest form, did not appear until well over a million years after bipedal locomotion. Nor could the initial reason have been to achieve greater height for better distance vision in the savannah; without such an ability already in place, primates would have perished soon after leaving the forest. Lovejoy (1980) argued that bipedalism could not have emerged without a complete change in the social structure and survival strategies of australopithecines; the risks of bipedalism were simply too great.

Social stability among humans is a function of, among other things, pair-bonding between males and females to reduce male aggression, food sharing, a stable home base, and sharing of infant care. This family structure is not typically found in ape society, although some elements of the pattern, such as occasional food sharing, do occur in chimpanzees. In effect, Lovejoy was suggesting that the presence of bipedal locomotion in australopithecines signaled the emergence of a characteristically human family structure, with an increase in birth rate and group size, social stability, and social cooperation in raising young and obtaining food. Australopithecines may not have produced complex cultural artifacts or ranged across the planet to various new habitats as *erectus* did, but they probably achieved a social structure that had many human characteristics.

Aside from social change, there is no evidence of a profound change in intellectual capacity in the reconstructed culture of the australopithecines. They did not manufacture tools, although they undoubtedly used them, perhaps in a more flexible way than apes. Their modified hands allowed them a precision grip, and their erect posture freed their hands for carrying things, digging, and more accurately throwing projectiles. But their habitat did not change; they stayed in sub-Saharan Africa, within the same narrow geographical zone, with approximately the same range of action. Australopithecines might be regarded as bipedal, terrestrial primates, with more sophisticated manual skills and a social structure that had some essentially human features but a mind that was still essentially apelike in its cognitive aspects. A critical step had been taken toward human anatomical form and social structure, but the cognitive restructuring that led to human mental powers was yet to come. Bipedalism, in itself, did not lead to that restructuring.

Encephalization: The Hominids

The aspect of physical evolution most often linked with the emergence of human cognitive capacity is relative brain size. The strong form of the encephalization hypothesis holds that primates may eventually have acquired a "critical mass" of cortex, which gradually increased overall cognitive capacity and eventually led to the human capacity for language (Jerison, 1973). The notion of a critical cortical mass is, as we have seen, integral to the unitary hypothesis that a single adaptation—brain size—would suffice to trigger the novel cognitive capacities of humans.

But how should encephalization be measured? Is overall brain size, even the EQ, a useful metric? Hebb had proposed (1949) that the ratio between the association cortex and the rest of the brain would be a more useful measurement, since the association cortex at that time was thought to be the locus of memory and learning capacity. Hebb suggested that it was the increased capacity to form associations or memories, implied by the expansion of association cortex, that served as the basis for humanity's great intellectual skills. His position amounted to a belief in the regional localization of learning, that is, localization of higher cognitive skills in a very general way. His theory was a compromise between the position held by his mentor, Karl Lashley, who proposed that it was the overall size, or total mass, of the entire cortex that mattered, and the earlier view of strict localizationists of the Wernicke school, who looked for specific brain "centers" for each putative higher mental function.

One of the many difficulties with testing Hebb's proposal is that it is hard to estimate the precise volume of association cortex in isolation from other cortical areas. This is not only because there are so many complex anatomical structures involved but also because the detailed functional neuroanatomy of "association" cortex suggests that it is not a homogeneous structure. When Jerison published his study of brain size in several hundred species, he did not try to quantify association areas separately; rather, he measured either total brain mass or total cortical mass, both of which can be determined fairly easily.

Passingham (1982) added considerably to the earlier analyses of Jerison by exploring the relative size of different subregions of the brain. His chain of reasoning was as follows: brain size has a reliable relationship to the relative size of various structures within the brain; for example, the ratio of total cortical volume to brain volume follows a predictable mathematical function; so does the ratio of the medulla to the rest of the brain, or the thalamus. Brain size also has a reliable relationship to the density of neurons in the brain; generally, density goes down as brain size increases, so that, for mammals, the number of neurons roughly doubles if the brain triples in volume. The number of "additional neurons" can therefore be predicted as a function of EQ, at least within the history of a given species. In the brain, as in a computer, the number of computational units supposedly determines the power of the device; and since the EQ predicts that number, it also predicts intellectual power.

This theory appears to agree with some of the facts: as we ascend the primate line from prosimians, through Old World monkeys, to New World monkeys and apes, the EQs and intelligence of species become greater. This increase has been shown both in the laboratory and in the field; the higher primates perform better at a variety of problem-solving tasks, live in larger groups, and demonstrate greater aptitude in event representation and self-representation. They also have a higher ratio of brain volume to medullar volume; monkeys have 50–75 percent more total brain than prosimians for the same medullar volume, and the great apes have 300 percent more (Passingham and Ettlinger, 1974). In other words, it is the more encephalized primates who have the widest behavioral repertoire, the highest levels of concept formation, the best ability to solve problems, and the general attributes of greater intelligence. The genus *Homo* is thus the latest in a long line of more encephalized, and therefore more intelligent, primates.

Passingham explored the many avenues open to the anatomist: the ratio of one structure to another within the brain; the ratio of each major structure to the whole brain; the relative size of each structure across various primate species, and so on. The uniform, and systematic outcome, of his studies is that the primate brain expanded accord-

ing to a certain logic when it increased in relative size. Certain structures expanded more than others; generally the brainstem (medulla)
did not increase much in size, the diencephalon (thalamus and hypo-
thalamus) showed a small but significant increase, the hippocampus a
greater increase, and the cerebellum and neocortex showed the greatest increase. This is shown in Figure 4.3, which is based on some of
Passingham's calculations (1982).

Moreover, Passingham found that within the cortex, the premotor
cortex expanded more than the sensory or motor cortex, and the association areas expanded the most, as Hebb might have predicted. In
fact, the pattern of human brain expansion is a direct extension of the
earlier primate expansion, from prosimians to apes: when the volume
of the neocortex is plotted as a function of the volume of the brain,
the human brain is seen as exactly what would have been predicted
by extrapolating the expansion pattern of the brain across the other
primate species. We possess, in other words, a typical primate brain,
proportioned the way a very large primate brain would be expected to

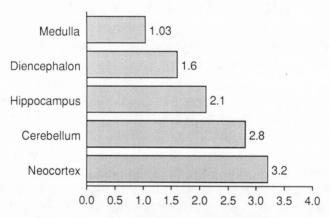

Figure 4.3 Evolutionary expansion of selected regions of the human brain. The
numbers, derived from Passingham (1982), express the differences in size between
areas of the modern human brain, compared with the values predicted for nonhuman primates of the same body weight. A value of 1.0 means that humans have the
same proportion of that brain area as primates; a higher value indicates an increase
in the relative size of that region in humans. The value of 3.2 given for neocortex,
for example, means that humans have more than three times as much neocortex as
a nonhuman primate of the same size.

be proportioned (Passingham, 1982). As Passingham himself admits, the fossil estimates are open to a large margin of error, since they are based on small samples; nevertheless, they point to our continuity with the ape brain and to our consistency with the statistical trend of brain expansion across the entire primate line.

However, this result does not rule out certain types of specialized adaptation. While it may be broadly true that relative brain size predicts the intelligence of a species, there is a danger in taking this approach too far, because it may ignore the many enormous and important differences that exist between species without an accompanying difference in relative brain size. For instance, dogs and prosimians have similar brain–body ratios but very different forms of intelligence. Dogs construct highly specialized olfactory maps of the environment, which serve a variety of purposes in defining territory, in mating behavior, in aggression, and in identifying friend and foe. Prosimians lack this ability but possess excellent binocular vision and manual skills completely lacking in dogs. Both skill systems obviously involve specific features of perception, memory, and action, none of which is evident in the overall EQ measures of canines and prosimians. It is still an open question whether these differences are reflected in more detailed anatomical ratios in the nervous system, but an overall encephalization index, such as the EQ, would sometimes fail to capture important differences in the forms of intelligence between species.

It may be countered that within the primate line such qualitative differences are less likely to muddy the picture. Passingham goes into considerable detail on the sensory, perceptual, mnemonic, and motoric similarities between apes and humans, to demonstrate precisely this point. Humans are primates, and our basic cognitive capabilities, as well as our bias toward visual–manual skills, are primate in derivation. Given our clear primate origins, and the predictably primate form of our brains, the only things that single us out are, physically, the enormous size of our brain and, behaviorally, our one special skill, language. The temptation to link language ability to brain size is understandable.

Passingham (1982) was fully aware of the perils involved in such speculation. His introduction contained a veritable panegyric on the virtue of scientific caution. Alas, on page 106 he revealed himself to be human, in stating his belief that brain size is the major factor in language development; and on page 234 gave in completely to an orgy of speculation on the origins of higher intelligence in brain expansion. Jerison (1973) fell prey to the same temptation, going so far as to place a number on the critical brain size needed for language—600 cc. Passingham avoided getting in quite that deep, but he and Jerison are both representatives of a school of thought that Hebb would have found compatible: they believe that some global learning or memory capacity, dependent on cortical size, underlies human cognition.

Habilines

The first major step in encephalization came with the advent of the genus *Homo*, and additional brain growth has accompanied virtually every step on the road to modern humanity. The most ancient examples of these encephalized primates were named *Homo habilis*, who appeared a little over 2 million years ago.

Habilines looked very much like australopithecines in both size and overall appearance; they weighed about 40 kilos, walked erect, and had a similar skull shape, except for a slightly less prognathous muzzle. However, they differed in two important respects: their relative brain size was slightly larger, and they seem to have manufactured simple stone tools, although it is still debatable whether they were the only users of the simple stone implements, about 2.5 million years old, found near their remains. (It is possible that the later australopithecines also used simple stone tools.) The EQ of habilines went up to 4.5, from the average of 3.4 for australopithecines.

Much has been made of the stone tools found by Leakey, and labeled Oldowan, after Olduvai Gorge, Tanzania, where they were first found. The tools have a characteristic shape and are very primitive, consisting of sharpened pebbles. Oldowan culture was the predominant tool culture for the period from about 2.5 million years to about 1.5 million, when more advanced tools appeared. The Oldowan toolkit survived until at least a million years before present. The main reason habilines are classified as *Homo*, rather than australopithecine, is the evidence for their possession of stone tools; otherwise they looked, and appear to have lived, much like australopithecines.

The habilines are a controversial species, because there are so few skeletal remains to use in their reconstruction. Brace (1979) considered them to be a later australopithecine subspecies, rather than *Homo*. Moreover, the link between them and the Oldowan tool culture is not nearly as strong as the links between later hominid species and later tool cultures. In fact, the link is simply the lack of a feasible alternative; the dates attributed to the earliest Oldowan tools correspond, roughly, to those of the earliest habiline remains, although the oldest finds are actually older than any known remains of *habilis*. The Oldowan tools are from the same geographic regions and strata as the habilines and later australopithecines. In Hadar, Ethiopia, Johanson and his coworkers discovered Oldowan tools at strata that were between 2 and 3 million years old, antedating the oldest *Homo* finds, though even Johanson was hesitant to suggest that the tools were used by australopithecines. However, there were australopithecine remains in the same time and place. Since they are a much older species, with slightly smaller brains, and did not possess manufactured tools earlier, Johanson considered them less likely to be the inventors of the tools than a new species, whose time of appearance coincides roughly with Oldowan tools.

How sure can we be that the Oldowan tools are in fact manufactured tools? This is not an easy judgment to make, since the artifacts identified as tools consist simply of small pieces of rock that are sharpened on one side. Could they not have been "found" tools, similar to those used by apes, but hoarded and accumulated in one place in a more systematic manner? Archeologists answer that the choppers and scrapers identified as Oldowan must have been manufactured because they are consistent in shape and size, appear to have been intentionally flaked to a sharp edge, and are fairly difficult to reproduce. They also tend to come in groups, as though the tools were left after having been used. But this would not necessarily indicate tool manufacture; found tools could have been collected, perhaps slightly refined by striking against a harder surface, and left behind in one place after use.

Regardless of whether or not they were manufactured, what were Oldowan tools used for? Electron-microscopic examination of habiline dentition (Walker, 1987) suggests that habilines were unlikely to have been mainly meateaters; their diet seems to have been largely fruit, like that of australopithecines. Therefore, the main use of stone tools may not have been for butchering meat, as has often been suggested. It seems they may have been used for sharpening wooden implements, splitting fruits and nuts, and only occasionally for butchering small game.

An additional complication is that the size of the habiline brain is not agreed upon. The few existing samples of habiline skeletons come from different times and places, and in some cases are so discrepant that there is no agreement as to whether they come from the same species. Estimates of increased habiline cranial size depend heavily on the famous skull 1470 discovered by Richard Leakey at Lake Turkana, Kenya, in 1972, which had a cranial capacity of 775 cc (Leakey, 1973). However, other skull fragments do not point to such a large average size for habiline brains. In fact, skull 1470 is the exception rather than the rule; most habiline crania look very much like those of gracile australopithecines and are closer to 600 cc in measured volume.

In conclusion, habilines appear to be a borderline hominid species, perhaps a transition type from australopithecine to *Homo erectus*. Their EQ is hard to establish but appears to be significantly greater than that of the australopithecines. Skull 1470, even taken in isolation, seems to confirm the existence of a large-brained hominid between 2 and 3 million years ago. The Oldowan tool culture is very primitive and, this, combined with their conservative survival strategy, does not point to a great cognitive difference from the australopithecines.

Homo Erectus

The situation, however, changes dramatically with the next species of hominid, *Homo erectus*. This is a very well-documented species, with many independently corroborated pieces of archeological evidence from a variety of locations. Virtually complete skeletons and crania, numerous tools and artifacts, and, for late specimens, even complete campsites have been found. Moreover, there is no possibility of confusing *Homo erectus* with a preceding species. *Homo erectus* is clearly different from australopithecines and habilines—much more human in appearance, brain size, stature, and culture. With this species, a major threshold had been crossed in human evolution.

Homo erectus appeared about 1.5 million years ago, and survived until several hundred thousand years ago. A number of major morphological changes marked the advent of this new species. The brain was enlarged, at first by about 20 percent, to 900 cc, but eventually, in late *Homo erectus*, to 1,100 cc, or about 80 percent of modern human cranial capacity. The face became flatter and the dome of the cranium higher. The height of the species increased considerably over the habilines, to slightly over five feet. In appearance, *Homo erectus* looked quite human but had a prominent brow ridge over the eyes, a forehead that was still sloped back when compared with modern humans, and a thickening of the skull at the inion, or back of the cranium. The face was still prognathous.

Endocasts have been made of several *Homo erectus* crania, and the chief exterior landmarks of brain anatomy are all there: the Rolandic and Sylvian fissures, the large temporal and frontal lobes, the expanded parietal lobe, and an enlarged cerebellum. The brain was expanding toward its modern human counterpart, without any drastic changes in overall shape. The EQ reached 5 and in late specimens is closer to 6 (Jerison, 1973).

The expansion of the brain in *Homo erectus* was accompanied by increased tool manufacture. The evidence in this case is extensive and solid, compared with the rather thin evidence on habiline tools. Some of the earliest *Homo erectus* finds are associated with Oldowan-type stone tools, but most are associated with a more sophisticated toolkit, which is identified with the Acheulian culture. The Acheulian toolkit remained in use for over a million years and has been found in various sites across Eurasia and Africa. The main tools are handaxes and choppers, but a number of other devices were also manufactured. The quality of the tools was much higher than that of the Oldowan kit and could be said to be qualitatively different in terms of the amount of planning and organization required to produce them. Without doubt these are manufactured tools, with considerable finish. Some authors have classified Acheulian tools into two or three categories, according to the degree of detail and specialization in their design. The most advanced category is found only in the later sites, suggesting a capacity for gradual cultural change.

Drawing appropriate conclusions about the cognitive capacities of *Homo erectus* is an important task, because advanced tools and greater encephalization appeared approximately at the same time and in proximity to one another on the archeological scene. *Homo erectus* is thought to have changed significantly in eating patterns and accordingly in overall survival strategy. The toothwear patterns of *erectus* indicate a marked change in diet, suggesting a meateating omnivore. The basicranium was more flexed than in the earlier species, indicating a slight descent of the larynx, but Lieberman (1984) did not conclude from this that *erectus* possessed any form of speech. Rather, he concluded that the slight descent of the larynx freed them from obligatory nose breathing, allowing the large intakes of oxygen through the mouth that large bipedal creatures required. This adaptation might have laid the groundwork for the later descent of the larynx and development of the supralaryngeal vocal tract in modern humans.

Stable seasonal home bases were found, even in some of the early *erectus* finds dating back over a million years before the present, suggesting that this species had a stable social structure from the start, probably building upon the survival strategies Lovejoy proposed for australopithecines and, by extension, habilines. Home bases served as the geographic focus of cooperative seasonal hunting of large game, which represented an improvement of the opportunistic scavenging thought to characterize habilines.

Later *erectus* sites in the caves of Choukoutien, China, have yielded evidence of continuous fire use in the same location 400,000 years ago. Earlier fire sites have been reported in China and in Kenya. The latter would push back the date of the first use of fire to well over a million years ago, but it is disputed whether those early fires were permanent firesites, deliberately set and maintained, or natural accidents used opportunistically.

Migrations were characteristic of this species, and over a period of several hundred thousand years *erectus* spread out of Africa to Europe and Asia, reaching sites as widely dispersed as China, Indonesia, and France. This migratory capacity testifies to a completely different creature, in terms of adaptability and cognitive resources, from any that had preceded. The main impression *erectus* leaves is one of a systematic, organized creature who was able to hunt and manufacture tools cooperatively, cook food, transmit skills across generations, and grow culturally, albeit at a snail's pace compared with modern humans.

Homo Sapiens

The last stage in the encephalization of hominids came with the arrival of *Homo sapiens*, about 200,000 to 100,000 years ago. The final increase in brain size entailed a further increment of about 20 percent in overall brain volume. At the same time, there was a continuous acceleration of the rate of cultural change. Tools did not change initially; the Acheulian toolkit survived until fairly recently. However, around 100,000 years ago, Mousterian culture emerged, and toolmaking became gradually more refined, until it was revolutionized in the Mesolithic and Neolithic cultures. The correlation with brain size breaks down during this period; cultural change, once it began to accelerate, proceeded without any further change in brain size or, as far as can be determined, brain structure. Whereas change had been agonizingly slow during the period of *Homo erectus*, it now became an increasingly visible characteristic of human society. Ritual, art, myth, and social organizations developed and flourished in rapid succession. A new cognitive factor had obviously been introduced into the equation. The human capacity for continuous innovation and cultural change became our most prominent characteristic.

The question is, what was changing? Toolmaking was only a part of it, albeit a particularly visible part, since stone tools survive so well in the archeological record. Extensive effort has gone into reconstructing the societies of the past using knowledge of climate and geography, diet, food sources, hunting techniques, and population movements. The one lasting impression a psychologist gets as he makes his way along this very well-traveled path is that the more sophisticated manufacture of tools, although critically important in allowing humans to adapt to their environment in new ways, was only the tip of the cognitive iceberg.

Speech: A Recent Innovation

Philip Lieberman, while acknowledging that EQ is important, has argued that human intellectual power resulted from a specialized adaptation, rather than from encephalization per se. He ceded the point that encephalization, bipedalism, and the various cultural changes accompanying them carried hominids to a level of primate culture never before achieved, but it was a primate culture nonetheless, and no more. Language changed that, and initiated the rapid, cumulative cultural evolution that has led to modern human society.

Lieberman's theory of language evolution (1973, 1975, 1984) tied the arrival of language firmly to a special linguistic adaptation, which was manifest both in gross anatomy and in the functional organization of the human brain. Lieberman holds that language came very recently, with the specialized vocal tract of modern humans, that is, with the advent of *Homo sapiens sapiens*. Through a complex chain of reasoning, he has taken a strong theoretical position, which he stated in a clear, testable hypothesis.

His basic proposal is that symbolic language originated with the recent arrival of a completely novel human trait: speech. Speech could not have evolved slowly by half-measures; it was too complete a break with our primate ancestry. Primates are remarkably poor at vocalizing, given the level of their visual–motor intelligence. This had been previously observed by the many comparative psychologists who tried to condition primate vocalizations, including B. F. Skinner (1957). The neural apparatus of human vocalization would have evolved in conjunction with the unique, and highly specialized, human vocal tract. The perceptual and motor devices that make speech possible evolved as a single, radical adaptation, according to Lieberman's hypothesis.

Lieberman attempted to construct a model of the vocal apparatus of early hominids, using skeletal remains and cranial endocasts (1973, 1975). His reconstructions suggested that even Neanderthalers lacked the elevated vocal tract that characterizes modern humans. The primary evidence for this, other than the absence of the large basicranial flexure mentioned above, is the relative height of the cervical vertebrae. The length and angle of the Neanderthal neck could not accommodate the human supralaryngeal vocal apparatus, which is elevated relative to the vocal cords. The supralaryngeal apparatus includes the soft palate and a highly flexible tongue, which permit changes in the place of articulation and the shape of the oral cavity. When a modern right-angled vocal tract was superimposed upon the outline of a Neanderthal head and neck, it simply did not fit. It wasn't even close; the modern larynx ended up in the Neanderthal chest. The lack of a modern vocal tract in a species as highly evolved as the Neanderthalers suggests that this anatomical change occurred during the past 100,000 to 200,000 years. Even if we are not directly descended from the Neanderthalers, we undoubtedly shared an ancestor with them within the past 200,000 years, and their immediate ancestor would necessarily have lacked the modern vocal tract as well.

The validity of Lieberman's anatomical reconstructions has been questioned. LeMay (1975), Burr (1976), and Falk (1976) have challenged some of the assumptions underlying his modeling process; and, as Passingham (1982) has pointed out, the elevated vocal tract did not necessarily have to accompany the first occurrence of spoken language. It is possible that hominids had some form of voluntary vocal communication system before the modern vocal tract evolved. Passingham favors some form of speech earlier in hominid development, as does Holloway (1974). Steklis and Harnad (1976) agree, believing that toolmaking would have required some form of symbolic representation, not just to coordinate manufacturing but also to transmit the skill across generations. Neanderthalers, and even earlier hominids, might have possessed partly evolved vocal tracts, that is, something other than the standard-plan primate vocal tracts assigned to them by Lieberman. The limited archeological evidence contains enough room for error to support this possibility, and Lieberman has acknowledged this. Lieberman did not deny the possibility of some form of Neanderthal speech; he only pointed out that in view of the differences in their anatomy, significant differences between Neanderthaler vocalizations and those of modern humans were likely.

However, attributing speech to archaic hominids would imply several successive major vocal-tract innovations during the course of evolution, and it would also leave unexplained why the sudden acceleration in the rate of human cultural innovation came so late. For these reasons, it appears more likely that speech per se is a very recent invention. Whether the vocal tract, and its corresponding central nervous system changes, started to change earlier and went through several gradual adaptations, we cannot know with certainty, at least not yet. But it is likely that the essence of Lieberman's message is correct, namely, that the modern vocal tract, and the more abstract aspects of language, came very recently with *Homo sapiens sapiens.*

Lieberman's theory holds that the new vocal tract necessitated a prior or concurrent change in the nervous system, in both sensory and motor areas of the brain. On the sensory side, Lieberman reviewed extensive evidence that human speech perception is specially "tuned" to discriminate certain features of sound that are central to speech. He goes so far as to postulate that there are specialized acoustic analyzers for each of the major speech sound parameters: voice

onset time; formant transitions; vowel sound parameters corresponding to particular places of articulation, and so on. The perceptual apparatus is thus constructed so as to map smoothly onto the motor
controls for phonological production. The new vocal tract involved
new muscles and a radically altered respiratory control system, both
of which needed appropriate central neural innovations. The overall
patterning and sequencing of high-speed speech sounds, which are
unlike anything that preceded them, required a systematic mapping
of the perceptual apparatus onto the new motor anatomy as well,
since reception and production are necessarily two sides of the same
functional system, especially during language learning.

What Lieberman proposed is a complete, self-contained vocalization system that was built according to the general principles of mammalian sensory and motor control but, like many other special adaptations, required its own special apparatus. Although he did concede
that some limited degree of voluntary control over vocalization might
have been achieved by *Homo erectus*, this control would not have
constituted anything like high-speed symbolic language.

The fact that human children learn speech so early, and so effortlessly, argues in favor of their possessing a special language apparatus,
not merely a more encephalized nervous system. The brain, especially
the association cortex, is nowhere near fully mature when speech is
acquired. Unlike general-purpose skills, which are acquired gradually
and late in both apes and humans, speech arrives early and rapidly. A
general-purpose capacity for language, such as might be implied by
Jerison's extra-neuron model or Passingham's encephalization model,
would not develop in this way; it would grow slowly during life, like
general knowledge and social skills.

Lieberman is not suggesting, as Chomsky did, that language is entirely different from what preceded it. Chomsky (1965, 1980) argued
that human children innately possess not merely a capacity to learn
language but an actual model of the linguistic universe against which
to test whatever information they pick up from the environment.
Lieberman, on the contrary, argues that language is built upon existing mammalian capabilities and falls into the category of a novel skill,
but one that still goes by the same rules that govern other skilled
motor behavior. Phonetic sequencing is fundamentally an extension
of our motor-sequencing skills, which were already rule-governed.

This argument bears a certain resemblance to Corballis's notion, reviewed in Chapter 3, that the left-hemisphere capacity for "generativity" originally developed for manual praxis and later generalized to speech, with the difference that Lieberman sees language as a more specialized generalization of vocomotor, rather than manual, skill. Even syntax, for Lieberman, is an extension of basic vocomotor skill. Corballis pointed out that skilled motor behavior in animals is usually not generative, that is, it does not involve categorical analysis and recombination of elementary parts, the way human praxis does. Thus, the line of descent was not directly from motor skill to language; it was from generative praxis to language. But both authors hold to the principle that there was some precedent in human motor behavior for the kinds of operation underlying language; language was not the qualitative break with all pre-existing skills that Chomsky had proposed. Thus, speech perception may involve special neural adaptations, but these are not qualitatively different from other perceptual adaptations for other purposes. Language is learned, just as other skills are learned; but it is learned with a special apparatus that evolved recently, rather than a general-purpose cerebral expansion that evolved over a period of millions of years.

What did Lieberman have to say about the preceding expansions of the human brain, as reflected in successive increases in the EQ of hominids? He held that these imply a continuing evolution of ape skills—an elaboration of the lines of cognitive development that would naturally develop, given the basic character of pongids. Thus, we might expect improvements in manual skills, tool use, spatial representation, memory, social cooperation, and any other skills that were already in place. These would occur with each successive new hominid species; but they would not inevitably lead to speech, because speech broke with the primate line in a more fundamental way.

Conclusion: The Question of Cognitive Stages

Both lines of reasoning about the origin of language have a basis in fact and logic. The existing evidence suggests both that the EQ increased in a series of successive changes and that the vocal apparatus did appear late in human evolution, coinciding roughly in time with a sudden, rapid acceleration in the rate of human cultural change. It is possible that encephalization alone could account for most of the physical basis of language, and that our special vocal apparatus was only the icing on the cake, serving to improve a capacity that had gradually emerged. It is also possible that the opposite was true: that hominid culture was essentially apelike until speech appeared on the scene, very recently, and that development of the special vocal apparatus that gave us the capacity for speech was the watershed event in our cognitive evolution.

This points to a dilemma confronting anyone who would try to construct a theory of language evolution without an underlying cognitive rationale. If the theorist relies entirely, or largely, on archeological and morphological evidence to reconstruct the stages of development, it is difficult to choose between the alternatives that emerge from the physical data. Lieberman, of course, relied on more than the physical data, and so did Passingham, but both approaches were largely biological rather than psychological. There was no cognitive theory driving either approach. Yet the cognitive realm is obviously the appropriate arena for the debate over the genesis of language. Language is a reflection of the intelligence of the speaker; it is limited by the memory, knowledge, and ability of the speaker. It receives its meaning from the broader cognitive realm of experience and context. Language is, in a sense, secondary to the evolution of fundamental cognitive skills. Thus, the debate over the emergence of language is also a debate about the emergence of *all uniquely human styles of representation*.

In this context, it is not reasonable to expect that the gulf between ape and human can be bridged by language alone. A great many cognitive differences between ape and human are not specifically linguistic in nature. In fact, relatively few of the many tasks that humans routinely master can be learned by apes. It isn't just that apes cannot master symbol-driven tasks like mathematics, musical performance, reading, and conversation. It is that they cannot master any number of nonverbal tasks as well. They cannot acquire our athletic or play skills, for example. Human children play rule-governed games by imitation, often without any formalized instruction. They invent and learn new games, often without using language. Apes, like other animals, cannot learn similar games; they are restricted to games that, by our standards, are very simple. Yet games, in human society, are at the bottom of the cognitive pecking order; they are identified with childhood. Most informal games have little or no verbal component, and deaf-mute children have no difficulty mastering them. Yet these games are already beyond the reach of apes. The problem of bridging from ape to human would thus appear to involve a great deal more than pinpointing the arrival time of vocal language, as though language could, in itself, explain our cognitive advantage over apes.

The definition of cognitive stages will thus require human culture to be reinterpreted in cognitive terms. The various known species of hominids and australopithecines have been defined primarily in terms of anatomy, but culture, and therefore cognition, has also been taken into account. This is especially true of *Homo habilis*, who would probably have been classified as australopithecine if it were not for the primitive stone tools of the Oldowan culture, which were associated with habiline remains. *Homo erectus* stands out as an anatomically identifiable species, but the culture of *Homo erectus* is also of central importance in defining that species. The culture of hominids showed signs of radical changes at certain points, and this implies, indeed insists upon, the existence of intermediate hominid cognitive abilities between that of the ape and our modern human culture. The foundations of these intervening levels of culture must have been successive underlying cognitive adaptations.

The successive cultures of hominid evolution have been reconstructed by archeologists and anthropologists on the best available evidence and constitute an important set of theory and data that has been largely ignored by psychology, presumably because it is not experimental in nature and is therefore seen as speculative. Yet the language of experimental psychology provides a useful set of concepts by which to interpret what little is known of our predecessors. If, for the moment, we suspend our dismissal of this evidence and try to understand it in our own terms, it may reward us with insights into the evolutionary history as well as the underlying structure of modern human cognition. To reiterate Darwin: Evolution produces vestiges. The history of the species lives on in the modern species, and one of the most important pieces of evidence for this lies in vestiges of previous adaptations that may prove redundant, or useless, or even maladaptive, today. Just as this is obviously true of human emotional expressions, it must also be true of cognition.

In conclusion, the chronological time-line of human evolution contains marker events, both anatomical and cultural, that suggest major changes at 4 million, 1.5 million, and about 200,000 years before present. Between these markers are long periods of slower change, during which less major variations in both the anatomy and culture of hominids occurred. The advent of australopithecine anatomy and culture 4 million years ago probably established some of the essential social structures that are still characteristic of human society, especially pair bonding to reduce intragroup aggression and social cooperation in obtaining food and caring for children. Without these changes in survival strategy, bipedalism would have exposed australopithecines to the risk of rapid extinction.

Starting about 2 million years ago, both a major expansion of the cerebrum and a radical cultural change occurred, whereby hominids commenced to accumulate skills that led to distinctive toolmaking industries, systematic group hunting, seasonal home bases, migration out of Africa, and use of fire. These cultural changes imply fundamental cognitive changes; thus, the first major cognitive marker should be placed around 1.5 million years, with the advent of *Homo erectus.*

The second major cognitive marker should be placed at the time of speciation of *Homo sapiens,* about 200,000 years ago, when both anatomy and culture again changed in a major way. The cerebrum expanded once again, to its present size, and the vocal apparatus changed. Cultural innovation became much more rapid, leading to Neolithic culture and eventually to the very high rate of cultural change we know today. There is general agreement that high-speed symbolic language is a recent innovation, and this implies that language-based cognitive abilities are of recent origin. There is considerable disagreement about what, precisely, the recent invention of symbolic language involved, in terms of basic cognitive and neural structures. This question is closely tied to the order of succession of cognitive changes.

How did language fit into the existing cognitive architecture of hominids? This major question depends on what cognitive capacities are attributed to *Homo erectus,* that is, on what the major intervening hominid culture was like. Conceding that australopithecines were probably still primatelike in their intellectual skills, and that archaic

Homo sapiens was biologically very close to us, the cognitive architecture of *Homo erectus*, and also of other intermediate species such as habilines, becomes a central issue. If there is an intervening layer of function—an archaic hominid brain, nearly but not quite human—what would its structure have been, and what did it enable *erectus* to do? What sort of culture would this prelinguistic human intelligence have developed? The starting point of such an inquiry must be the cognitive structure, and culture, of the great apes.

Primate Cognition: Episodic Culture

Ape Skill: An Overview

It is still possible, a century after Darwin's admonitions about our prejudices, to underestimate the intelligence of the great apes. This is perhaps because their culture resembles the cultures of other mammals more than it resembles our own. Apes produce no enduring cultural artifacts and, when transported into our culture, find it difficult to learn even the most elementary human tasks and habits. Nevertheless, a list of their highest cognitive achievements reveals that their intellectual skills are rich and diverse.

Munn (1971) reviewed the early literature on pongid higher learning, problem solving, and concept formation and tried to specify what was distinctive about primate cognition. He concluded that most laboratory procedures such as classical or operant conditioning, sensory discrimination, and maze running did not differentiate the higher from the lower mammals, although in some cases they could distinguish mammals from fishes and reptiles (Bitterman, 1965).

However, two procedures in particular were successful in demonstrating a significant superiority of apes over other mammals: the "insight" experiments of Wolfgang Kohler (1925) and the delayed reaction experiment, in many variations. The "insight" experiments required the animals to invent solutions to problems, usually of gaining access to food. In a famous example, chimps piled boxes on top of one another to reach a banana suspended from the ceiling. Lower mammals could not solve such a problem, given any amount of time. The solution of the problem depended upon the recognition of relational properties of the visual environment.

The delayed reaction test usually involved showing the animal where to find a reward, then removing the stimulus for a period of time before allowing the animal an opportunity to react. Apes are much more adept at delayed reactions than simpler animals, and this was taken as evidence that they possess an abstract central representation of the environment, or what Hebb (1949) called"autonomous processes." This was confirmed by their ability to learn the solution to a problem through observation and imitation. In other laboratory demonstrations, apes were found to be uniquely successful at solving visual–manual puzzles. They could combine objects in the laboratory to use them instrumentally. They were able to learn various tasks following a single demonstration. In all of these things they exceeded the capacities of lower mammals.

Their use of found tools is now well-documented. They can even combine tools, as shown in their use of natural hammer stones and anvils to shatter nuts (Boesch and Boesch, 1984). Their ability to employ name signs, like visual tokens or sign language for the deaf (ASL), has been demonstrated many times. To a limited degree, they can transmit some name signs to their offspring (Premack, 1987). In addition, Gallup (1970, 1982) has shown that apes possess the ability of self-identification in front of a mirror, an ability lacking even in simpler primates, like monkeys. They are very good at certain kinds of imitation; in fact, that ability is at the heart of their acquisition of sign language. Moreover, it is clear from the early studies of Kohler (1925) and Yerkes (1943) and from recent studies of ape representation that they possess what we would call an ability to perceive "events" and "situations." That is, they do not simply perceive discrete stimuli; they perceive objects in relation to context and discriminate easily between a wide variety of complex perceptual arrays involving the same subjects in different relationships—for example, different dominance relationships, or affective relationships. Premack (1987) devised a task requiring chimpanzees to make judgments of the similarity or difference of complex visual displays and showed that they can perceive abstract second-order relationships.

Moreover, to borrow a phrase from Schank and Abelson (1977), they behave as though they employ "scripts," or sets of rules and procedures guiding behavior in different contexts. Where humans may have scripts for typical behaviors like eating in restaurants, or attending a movie, apes have scripts for sharing food, or behaving in the trainer's office, or playing with offspring. An aspect of such behavior is the ability to perceive the "scenario" unfolding in a sequence of actions and to know exactly what to do as it develops. Note that many of these cognitive achievements are found not only in apes but also in other mammals, and all without the use of language. But apes seem to stand above most other species because of their capacity for signing and self-recognition and their increased tool use. And as Premack (1987) has pointed out, we should also concede to them some limited capacity for social attribution and delayed social imitation. Cooperative social behavior, including the pongid equivalent of a highly organized war against another tribe, is found among chimps in the wild (Goodall, 1971, 1986).

Given this impressive list of abilities, the question arises, why can't they represent their knowledge somehow? Why do they not develop languages of their own? Despite their cognitive achievements, apes in the wild do not possess even a rudimentary system of voluntary gestures or signs. The absence of language is their most obvious cognitive limitation and the most frequently cited difference from humans. But an even more striking limitation in ape behavior is that they have relatively few "skills." Their culture appears invariant and stereotyped compared with human culture, which seems infinitely variable in terms of play, custom, sport, craft, and expression. Goodall (1971, 1986) has shown that two different groups of chimpanzees may be significantly different in their collective behavior patterns, but for the most part the range of variation is small. Human groups can vary along many more dimensions of behavior.

Nevertheless, given their abilities, which are clearly superior to other mammals in many ways, one might expect apes to possess some form of voluntary communication system. Everything would appear to be in place: memory representations of the environment, highly developed visual–manual skills, problem solving, including tool use and elementary toolmaking ability, and a capacity for self-perception.

With this expectation, there have been many attempts to teach language to apes.

Training Apes to Use Signs

The earliest attempts to raise an ape in a controlled human environment, by Hayes and Hayes (1951) and Kellogg and Kellogg (1933), were largely unsuccessful in teaching language to apes. However, these pioneering studies did show several important things about the cognitive strengths and weaknesses of our nearest living relatives. They showed, for instance, that apes can imitate many daily activities of humans without special training (Kellogg, 1980). They also showed that apes can learn to understand many spoken commands and utterances. Unfortunately, apes are very poor at vocalizing; after years of special training, the most successful subject, the chimpanzee Viki, trained by the Hayes, only managed four spoken words, and with great difficulty.

The observations of these early experimenters on the gestural capabilities of apes were particularly interesting. They were not trying specifically to train gestural signs, as in later studies of chimpanzee communication. Therefore, they were observing what capacity the apes may have had for spontaneous gesturing in a supportive environment. They did not find anything that could be confirmed as an intentional signal, even though they observed a number of behaviors that served to communicate information. For instance, Viki would attempt to lead the experimenter by the hand (Hayes, 1951), and one of Yerkes' orangutans did the same. Kellogg (1980) reported a variety of pseudo-gestures produced by the chimpanzee Gua that were reliably paired with certain behaviors. These included climbing into the high chair (indicating hunger), removing her bib from her neck (finished eating), and holding the genitalia (need to urinate). As Kellogg pointed out, these behaviors might well be regarded more as anticipatory reactions regularly occurring during a common sequence of behavior rather than as communicative gestures. They superficially resemble many of the gestures used by human children, but they seem to lack the intentionality that is evident in the child's gesturing. This is emphasized by the absence of spontaneous pointing behavior, which is common in human children (Terrace and Bever, 1980; Bruner, 1986). This issue will resurface later; apes simply do not use intentional gestural signals in the wild.

The first ape to learn to use sign language was Washoe, a chimpanzee trained by the Gardners (1969, 1978). Washoe learned a modified version of American Sign Language for the Deaf, or ASL. Finger spelling, normally an integral part of ASL, was left out of training, as was the gestural component of ASL, which normally develops alongside the formal signs as an equivalent of prosody in spoken language. The prowess of Washoe has been well-documented; the main point here is that the use of manual signs allowed Washoe to progress much further than any previously trained ape. Evidently the limitations of Viki's and Gua's achievements were due in part to their difficulties with vocalization rather than to an inability to learn and use signs.

What was the nature of Washoe's achievement? Without doubt, Washoe learned a substantial number of gestural signals with specific meanings, or semantic content. Moreover, those signals were both comprehended and produced in appropriate contexts, and were generalized in meaning much the way young children initially generalize newly acquired words. For example, the ASL signal for "more" was initially learned as a request for more tickling; but it was quickly used to ask for more hair brushing, more games, extra helpings of food, and more entertainment (Gardner and Gardner, 1978). Washoe had a vocabulary of 34 ASL signs at the age of three years and often used them in combinations of two or three signs, much as young children do.

As Washoe grew older, there was a tendency for the strings of signs to get longer, so that combinations of three or four signs became more frequent, just as the mean lengths of utterances of young children become longer with increasing age (Brown, 1980). However, development stalled at that level; Washoe did not progress to the subsequent stages of human language development, where vocabulary expands rapidly into thousands of words and sentence structure becomes much more complex and subtle. Since the Gardners' experience with Washoe, many other apes have been trained in similar ways. The results have tended to confirm the Gardners' main findings: apes can learn a significant number of signs and use strings of two or three signs appropriately; but none of them has yet progressed beyond that limit.

Much debate has centered on whether Washoe and other recently trained apes have acquired "true" language. It is not difficult to find essential features of language that are lacking from their repertoire. Lenneberg (1980) set down a set of criteria for anointing an ape communication system as true language. He proposed a simple, minimal language, corresponding to the language of a young child, consisting of about 20 words, including object names (hat, shoe); actions (take, point); a few qualifiers (small, big); positions (up, under); and function words (no, yes, is?). To demonstrate mastery of this language, the ape would have to be tested objectively, using questions that required responses. The key requirements, in Lenneberg's view, are the use of sentences rather than single words as the elementary units of expression and the ability to answer essentially any question possible within this circumscribed elementary language.

Clearly, by Lenneberg's criteria apes cannot acquire language. Critical examinations of the utterances of ASL-trained apes fail to produce any convincing evidence for sentence construction or awareness of word order rules (Brown, 1980; Lenneberg, 1980). Strings of ASL signs may be selected appropriately to describe a situation, but there is no evidence of syntax. The words are strung out in no particular order, usually two or three at a time, such as "door open," "Washoe ball hit," and so on. Longer strings are rare, tend to contain many rote repetitions of previous words, and show no structure. The lists of linguistic games primates can't play is seemingly endless: Anderson (1983) singled out iteration and recursion, and Bronowski and Bellugi (1980) singled out the process of reconstitution of the environment.

There is no point in belaboring the point; human linguistic superiority appears safe for the moment, apes and their trainers have been put in their place, and splitting hairs will not change things. But the more important question is surely unresolved: these experiments have shown a signaling capacity that had not been previously demonstrated in apes. In a sense, apes trained to use ASL have been provided with a lexicon. Why is a similar lexicon, however limited, not invented or used in the wild? There are obviously no motoric or perceptual impediments to apes' using sign language, and they understand enough of the world to possess something to communicate about it. They are very close to a form of communication that, if not language, is surely proto-language. What is missing?

The answer may be contained in Terrace and Bever's (1980) critical analysis of Premack's (1976) book entitled *Intelligence in Apes and Man*. Premack was the first person to train apes to communicate with a system of sign tokens, or symbols, that were entirely exterior to the animal, much like various kinds of iconic symbols in human language. Plastic chips were used, and for the most part the form of the chips had no similarity to the things and ideas being represented. The chips represented words, rather than phonemes, much like the visual symbols we know as "emblems" (McNeill, 1985), except that they were much more constricted in meaning than the latter.

Just as Washoe had mastered hand signs, Premack's protegée Sarah mastered the use of plastic chips to communicate with humans. Unlike ASL signs, which had to be reproduced from memory, the plastic chips were always present, and this reduced the memory demands placed on the ape by the task. Thus, Premack reasoned, if the limitations of ape linguistic ability are partly or wholly due to memory limitations, this factor would be eliminated in his training regimen. Moreover, the rules for using the chips could be defined much more precisely than the rules for using ASL signs, since the chips were used in a fixed place, under virtual laboratory conditions, with controlled reinforcement schedules, and so on. On paper, Premack's plan had great appeal, as a means of testing out the communicative capacities of apes under controlled, almost unambiguous conditions.

Curiously, Premack's apes progressed much as Washoe had. They acquired a basic vocabulary of several dozen signs and then learned to use them in combinations of two, three, and four, to make requests, answer questions, and solve problems. However, in addition they appeared to acquire order rules, or syntax. That is, the apes apparently understood that the order in which the chips were placed on the table had significance. At first glance it appeared as though Premack had guessed right in using a controlled laboratory training situation with a low memory load. Sarah appeared to have some capacity for both understanding and producing syntax (Premack, 1976).

Terrace and Bever (1980) addressed this claim in their critical review. In particular, they focused on the training regimen used by Premack. Training had progressed through a series of stages, and the teaching of correct word order was gradually extended from two-word sequences through three- and four-word sequences. At each level the animal was only rewarded if the correct word order was produced. At the highest level, the four-word sequences, a sequence had to contain the name of the subject ("Sarah"), followed by a verb ("give"), followed by an object ("banana"), and finally by the name of the recipient ("Gussie"). If the order was wrong, for instance, the donor and recipient were interchanged, the reward went elsewhere or wasn't awarded at all. The sequence had to be correct. Since there were many different possible word orders, Premack built up from shorter sequences so that by the time the ape had to master four-word sequences, it had already been trained thoroughly at the three-word level, and thus effectively had only to learn where to place the fourth word.

Human children, of course, receive no such training. No one systematically reinforces correct sequences of words, and the child does not become locked into, or dependent upon, such a rigid, fixed schedule of rewards. One of the striking features of human language is its fluidity and the ease with which it is learned by children simply through observation. However, even if the apes were more limited and rigid than human children, it might be claimed that they were performing qualitatively similar syntactic tasks, provided that it could be shown that they truly understood the meaning of word order.

Terrace and Bever argued that there was no evidence to support the claim that apes understood word order. Premack's data presentation was incomplete, especially on the longer sequences; thus, it was not clear how well the apes did on the task, even within the constraints of his experimental design. Sarah's understanding of word order was never tested rigorously, since there was never more than one alternative response pattern available on any given trial. In fact, in teaching propositional signs for words like *on* and *in front of*, the signs were never contrasted with other prepositions. In terms of action, the ape had only one possible response; the only linguistic problem confronting the animal was to identify the correct object that had to be placed on top of, or in front of, the other object (Terrace and Bever, 1980). In other words, the task had been so streamlined and constrained that it did not appear to involve much more than a complex event discrimination.

More importantly, the referents of the iconic signs were usually present during a trial. That is, the names of the subject and object may have been the experimenter, the ape who was being tested (Sarah), or someone else who was in the room. Each of these individuals, including the apes, was wearing a tag containing the same iconic sign used on the problem board. Moreover, the motive in every case was the obtaining of reward, usually food reward. Terrace and Bever's objection, then, is that the ape could solve the problem of obtaining food simply by making a discrimination between two alternative sets of plastic symbols. Unless it could be shown that the apes could solve sentences that cut across many categories of possible responses, they could not be said to possess any syntactic competence. Rather, they had learned sequences of signs by rote.

The apes did at least appear to have the ability to use the plastic chips to represent things. But even their lexical mastery has been called into doubt. Savage-Rumbaugh (1980) stated that "there is no evidence other than anecdote to suggest that Washoe and other signing apes are producing anything more than short-circuited iconic images." It was not clear, she argued, whether signs could be used across randomly varied trials, for instance, naming a variety of fruits while eating ice cream. Moreover, the communications had always taken place in the presence of highly motivated and highly sophisticated human beings, who undoubtedly structured the situation and interpreted the signs of the ape.

Despite these objections, it is clear that Washoe and other signing apes used the ASL signs in appropriate ways, both in attributing properties to objects and in indicating actions. Premack's apes similarly have demonstrated considerable mastery of quasi-symbolic situations which, if not meeting all the requirements of linguists, were still sufficiently subtle and complex to suggest that apes were close to some form of signing.

Moreover, their uses of signs were not directed exclusively to humans; Savage-Rumbaugh and colleagues (1978, 1980) have demonstrated that chimpanzees can use signs to communicate meaningfully with one another. In a series of studies, they have trained chimpanzees to use symbols to request tools or to produce tools if requested. Initially, they were shown six simple tools: (1) a key, to unlock various locks on doors and boxes; (2) a sponge, to soak up liquids; (3) a wrench, to unscrew bolts in various locations; (4) a stick, to access food out of reach; (5) money, to operate a vending machine for food;

and (6) a straw, to drink liquids from containers out of reach. The apes learned to use the tools through observation and trial and had previously been trained to use Rumbaugh's symbolic communication system (Rumbaugh, 1974, 1977). After repeated trials, the apes were able to use the correct symbols to request the tools appropriate to the task at hand. They were also able to name the tools without immediate food reward. The naming of the tools, however, was difficult for them to learn, and in general their ability to remember labels for things in the environment was very limited and subject to forgetting, distraction, and confusion. The authors pointed out that their criterion of competence in naming was the presence of robust, virtually 100 percent correct performance, whereas the Gardners and Premack had employed a criterion of only one correct trial, making their chimps appear much more competent than they really were. Savage-Rumbaugh persisted in training until the chimps had virtually perfect performance on tool-naming tasks before proceeding to the next stage of the experiment.

Once the chimps could both make and reply to requests for specific tools with humans, their ability to communicate with one another was tested. Two chimps were placed in separate rooms, in visual contact with each other. One chimp had access to food but no tools; the other had the necessary tools to get the food. The first could ask the second for the appropriate tool, using Rumbaugh's apparatus. The chimps discovered this possibility rather quickly, once the human experimenter refused to play the role of tool provider. Within one day's training time, the chimps were communicating with one another using Rumbaugh's symbols.

This result is most interesting in the light of some earlier experiments by Meredith Crawford, as reported by Munn (1971). Crawford trained two apes to cooperate in obtaining food; they had to pull on ropes simultaneously to obtain access to a heavy platform containing food that was placed outside their cage. Although the apes learned that they had to pull together, they were apparently unable to communicate with one another in any specific way about their knowledge. When one ape wanted food and needed the cooperation of the other, Crawford reported that the first one would try to get the other's attention turned to the food platform by trying to turn its head, or pull its hand over to the bars. If this failed, the only other thing it would do would be to act excited, or signal its frustration in some nonspecific way.

Crawford's apes had not received any training in the use of either gestural or other forms of signing. Here was a clear case where the apes possessed the knowledge of what to do, the ability to cooperate, and the motivation to obtain food. But there was *no spontaneous signing*. Savage-Rumbaugh's apes were in a similar position, and with their previous training in the use of signs, they were able to see the uses of that skill in a relatively new situation. They were capable of using signs, but somehow the idea of *inventing* a signing system never occurred to them.

Invention was, of course, the key piece of the puzzle; the first user of specific gestural signs had to be able to invent them *de novo*. And invention is also a key aspect of human language capacity. Chomsky's (1965) emphasis on the generative nature of the human language acquisition device was unquestionably correct. Language would not have emerged in humans, and probably could not have been sustained, unless each succeeding generation was capable of reinventing it.

The Importance of Symbolic Invention

After twenty years of systematic experimental effort, it is evident that apes do not have anything like the human capacity for syntactically complex, high-speed communication. But we now know that apes have internal representations, perhaps even intentions, and some capacity to label their cognitions when a signaling system is tailored to their abilities. Although not quite language, these symbolic communications show how close apes are, and how far away. They also vindicate Chomsky's fundamental criticism of the behavioral theory of language acquisition (Skinner, 1957). Many of the sign-using apes of the last twenty years, and particularly those trained by Premack, have been perfect Skinnerian performers. But they cannot be programmed to produce any syntax other than simple two-word order rules (even Skinner acknowledged this limitation). Nor can they be led to fully represent what they evidently know about the environment. Most importantly, they cannot invent any sign or symbol spontaneously, even when the capacity for using signs in the Skinnerian manner is present, as it evidently is. And the Skinnerian approach cannot ever achieve a useful system of signs, let alone a fully developed language, in apes, despite the presence of the required discriminanda, operants, reinforcers, and raw "associative ability."

Of course, some attempts to train apes in symbol use have not used either Skinnerian techniques or the carefully controlled training situations used by the Gardners. For instance, Savage-Rumbaugh (1986) taught a bonobo (pygmy chimpanzee) named Kanzi in a quasi-natural environment on a 55-acre reserve. Kanzi learned mostly through imitation, and his human caretakers were free to use spoken language. Kanzi carried a laminated board containing over two hundred visual symbols; he or his caretakers could point to the board to communicate, even when moving about in the woods. This relatively unstructured situation worked very well; Kanzi was free to use other communication media than the visual symbols, and in fact occasionally did so, using both vocalizations and gestures. His vocalizations and gestures were somewhat reminiscent of those reported by the Kelloggs, except that fairly often Kanzi would combine a gesture (such as touching a person) with symbol pointing (usually to an action sign like "carry"). Such combinations were not specifically taught, although they were obviously reinforced: for instance, after producing a sign combination, someone would carry Kanzi around, a result that Kanzi found highly reinforcing.

Nevertheless, a significant percentage of Kanzi's outputs were spontaneous; and before he could receive reinforcement, he had to invent these combinations. Some of Terrace and Bever's criticisms of earlier studies, especially those of Premack, would not seem to apply here: Kanzi went beyond his specific training sessions to use symbols in self-initiated novel combinations, often with accompanying gestures. He appeared to be both spontaneous and intentional in his symbol use.

Greenfield and Savage-Rumbaugh (1990) have interpreted Kanzi's behavior as support for the argument for "true" ape language. Their position is that Kanzi's symbol use is similar to that of very young deaf children with hearing (that is, nonsigning) parents, as described by Goldin-Meadow (1979) and Goldin-Meadow and Mylander (1984). These children typically use two-sign combinations consisting of a pointing gesture (for instance, pointing at a door) combined with an iconic sign (as in a turning of the hand, for "open"). The parallel with the two-sign combinations of bonobos was clear. Moreover, the parallel extended to the level of what the authors define as elementary grammar: simple ordering rules. For instance, after four months of practice, the animals learned that action preceded object, a rule used by their human trainers but never specifically taught.

Kanzi was not as adept as human children, even children at the two-word stage, in several respects. First, he learned much more slowly. Second, he made very few observations, or declarative statements, producing mostly requests for action (97 percent). But most importantly, he remained basically locked in at the two-word combination level, failing to produce more than a few three-word combinations. The latter, moreover, were extremely simple in structure, as in "chase-bite-person," where "chase-bite" might be interpreted as a single lexical item describing a very common event sequence.

Without wanting to enter the treacherous waters of linguistic debates over what constitutes the minimum requirement for true "language," I would judge human language to be light-years removed from Kanzi's accomplishments. Kanzi was given a structured communicative device that he could use very effectively, but to call his simple ordering rules "grammar" is stretching the definition of the word. And although his symbol use was more spontaneous than that found in any previously trained ape, this may have been a testimony to the clever design of the signing device. Kanzi's sign use remained at the presentence level, and certainly at a prepropositional level. This is not to diminish his, or his trainers, remarkable accomplishments; but it is important to think clearly on this rather emotion-laden issue. As our definition of language is laid out in the next few chapters, it will become clear that Kanzi remains several crucial steps removed from human linguistic ability.

He is close, however, to a more elementary level of human culture. As discussed in the next chapter, pygmy chimpanzees, and possibly other apes, are close to the first distinctly human level of representation, but far removed from meeting the conditions for language. It is most important to remember that, although they could learn to use, and recombine, a symbol set given to them, *they did not invent the symbol set*. In contrast, somewhere in the course of human evolution, hominids did precisely that: they invented an infinitely expandable symbol set. To proceed from the limited representational capabilities of apes to the next level of symbol use required a qualitative cognitive change, a move towards symbolic invention, with the concomitant cultural change such a capacity would imply. In its initial form, such a capacity might still have fallen far short of human language. But it would have been a necessary first step up, and a necessary precursor to the much faster, complex, syntax-governed languages of humans.

Social Intelligence in Apes

Training apes to use signs and symbols may be the most direct attack on the problem of describing their linguistic skills, but language cannot easily be dissociated from social intelligence in the more general sense of the term. Since language is a social device first and foremost, it is logical to expect the growth of language to be tied to the evolution of social structure. As social groups increase in complexity and size, the control and stabilization of group behavior, as well as the sharing of knowledge, becomes important. Individuals who are part of a large social structure have to comprehend, and remember, a number of dyadic (two-way) relationships. In addition, rules and customs have to be understood. The usefulness of symbols in social regulation is self-evident.

Social intelligence has never received an adequate definition, perhaps because it is difficult for humans, whose thought processes are so dependent on symbolic expression, to conceptualize what nonsymbolic or presymbolic forms of social intelligence may be. But whatever forms preverbal social intelligence may take, it is clear that language was the final step, and that presymbolic forms of social intelligence must have been its foundation.

One index of social intelligence is the average size of the elementary social group of a species, that is, the weight of its social "atoms." Social complexity increases with group size in primates, because the social groupings are exquisitely structured, unlike, large herds of cattle or flocks of sheep, for instance, which possess simple structure. Dunbar (1988, 1990, in preparation) has proposed that relative brain size, measured by the EQ, in primates increases with the size of the social groups they form. As demonstrated in Chapter 4, the EQs of prosimians, simians, and apes range from 1, the mammalian average, to well over 2. Dunbar has shown that the EQ in this line of succession correlates highly with group size.* In fact, he argued that the primary selection factor driving up the EQ in primates was the intellectual ability required to service large groups.

The cognitive skill needed for servicing large groups is considerable. As group size increases, apes must understand and remember an increasing number of relationships. This is particularly true of groups with the leisure time to interact closely for long periods of time, for instance, while grooming; chimpanzees spend a great deal of time grooming one another. As Dunbar points out, large groups are unstable unless the individual dyadic relationships within the group are serviced regularly. This would require considerable memory capacity and the ability to comprehend social complexity.

Dunbar's data refute another popular theory, sometimes called ecological theory, which also tries to account for the evolutionary expansion of the primate brain. Ecological theories claim that primate intelligence was a response to the demands of the ecological niche in which primates live. Clutton-Brook and Harvey (1980) and Milton (1988) have attributed the greater cognitive skill of primates to the intellectual demands created by having a larger home range, which in turn is created by a dependence upon fruit, rather than more abundant vegetation, as a principle source of food. They cite evidence that frugivores have significantly greater EQs than folivores and that EQ correlates with relative home-range size.

The underlying cognitive hypothesis of ecological theories is that primates had to construct more elaborate cognitive maps of their environments in order to accommodate larger home ranges. In addition, they had to be able to observe and remember when and where certain types of trees came into fruit, since the fruit of any given tree is available, and edible, only for a short time. The cognitive hypothesis is closely tied to anatomical change; relative brain expansion is seen as the other side of the coin. The frugivore diet also enabled apes to

---

*There are exceptions to this trend, most notably in orangutans, who are believed to be virtually solitary. But Dunbar argues that so long as exceptions do not violate his cardinal rule that group size cannot *exceed* the limit set by EQ, they do not invalidate his theory. Moreover, he believes that the size of orangutan groups may have been underestimated by modern observers, and that orangutans may have degenerated recently in their social behavior, having once lived in larger groups. There have been reports of orangutans living in groups (Rodman, 1973; Mackinnon 1974).

support the larger brains that their changed diet required. Martin (1984) claims that a frugivore diet provides the surplus energy needed to support an energy-intensive organ like a large brain. This all appears rather circular: obtaining fruit over a wide home range demands considerable cognitive skill, which demands a large brain, which demands an energy-rich diet. But in evolution, selection factors are often interrelated and, if not circular, spiral upward. In this case the same selection pressures favoring larger-brained apes would also result in the cognitive and nutritional adaptations required to sustain the larger brain.

Dunbar (in preparation) believes that intense socialization was a more important selection factor than ecological niche. He therefore obtained data on both range size and social group size for a number of primate species. Since the data supposedly supporting the ecological hypothesis confounded range size with diet, he also entered an index of diet into the correlations. The diet index measured percentage of diet obtained from leaves, which is the converse of percentage from fruit. His EQ index was a modification of Jerison's (1973) index, using Harvey and colleagues' (1986) correction of the baseline exponent. His results show that group size is significantly correlated with EQ, while range size and leafeating are not. The social evolution hypothesis was strongly supported over the ecological theory.

Dunbar extrapolated his data to humans. This was difficult, since group size in agricultural and industrial societies is an artificial number determined by institutional bureaucracies and governments. To circumvent this problem, he estimated group size for the six surviving aboriginal societies that were still either nomadic or hunter–gatherer and were completely lacking in such superordinate structures. He then extrapolated the EQ–group-size function by entering human encephalization indices into the regression equation. Thus, the regression of EQ on group size was used to predict an average group size for human society. Finally, he compared his prediction to his obtained data on aboriginal group size.

He had initially predicted an ideal human group size of 223. The average estimated size of human hunter–gatherer groups turned out to be 155.8 and the size of the nomadic societies alone, 174.6. There were difficulties in determining human group size, of course, since several layers of social organization are found in human societies, even the simplest ones. Dunbar used as his index an intermediate level, equivalent to a clan or band. He dismissed temporary groups, like hunting camps, and large tribal structures where relationships between members were not intimate. He felt that intermediate human groups shared the properties of intimacy and permanence that characterize ape social groupings. His conclusion was that, prior to the development of agricultural society, human groups tended to be within the 95 percent confidence intervals extrapolated from primate data, as a function of their EQ. This supported his contention that it was social factors, rather than ecological ones, that drove up the EQs of primates.

There is always the possibility that both hypotheses are at least partially correct. The social-intellect hypothesis was supported by Dunbar's research. However, on the basis of his data, it is hard to dismiss completely the ecological variables that may have shaped the human brain. Factors like range size and frugivore diet may be complicated by any number of other relevant environmental factors, such as the complexity and density and variety of vegetation within the home range. To expect such a correlation to hold up across different species and ecologies might be expecting too much. In contrast, a measurement of group size is probably vulnerable to fewer such influences. This does not invalidate the strong correlations Dunbar found, but it does throw into question his interpretation of the weak ones. It remains entirely possible that both ecology and socialization were responsible for the continuous increase in primate encephalization.

In the final part of his argument, Dunbar proposed that human language evolved because of the cognitive demands of maintaining large group organizations. In effect, language allowed much more efficient servicing of social relationships. It enabled vicarious servicing; that is, instead of acting out relationships on a behavioral level, they could be explored, worked out, and controlled symbolically. The advantages of maintaining large groups, and exploiting the weaknesses of smaller groups, are self-evident: in competition for food and territory, and in self-defense, a well-serviced large group will dominate a smaller one. It is only the unstable larger group that can be overthrown or undermined. If the members of the group have the intellectual capabilities needed to solidify group organization, the group will remain stable. In this way, the adoption of larger group structures would lead to evolutionary pressures favoring large brains. Dunbar did not try to specify the precise nature of the intellectual capabilities needed to support larger groups.

Dunbar's theory raises important questions about what aspects of primate intelligence are crucial to socialization. How are social relationships encoded? What kind of knowledge is involved? What demands would large social groupings place on recognition memory? How is the overall complexity of a group likely to influence the need for communication? And finally, what specific aspects of this need would be addressed by language?

Self-Awareness and Consciousness

A very different approach to cognitive evolution in apes has been taken by a number of comparative psychologists and physiologists. Oakley (1983, 1985) modeled the hierarchical structure of intellect in primates in relation to higher cognitive processes in humans. His approach is representative of the physiological mainstream, and the focus is not on the sheer size of the brain but rather its structure. The brain evolved in a series of systematic structural changes, with the most "archaic" structures, like the brainstem, appearing in very primitive creatures like lampreys and sharks. The midbrain appeared later in evolution and remains the highest structure in most fish and amphibia. The hippocampus and cerebral cortex appeared even later on the scene and are the dominant structures in mammals. In the simplest possible terms, primitive nervous systems evolved first, possessed only the brainstem structures, and were not very intelligent. The higher mammals evolved last, possessed elaborate cerebral cortices, and were more intelligent. Because evolution always builds on what already exists, the latest nervous systems, like those of the primates, contain within them the archaic structures of their primitive ancestors. Thus, the brain is organized into layers, like an onion, that reflect the stages of its evolutionary history.

This linkage between anatomy, evolutionary chronology, and cognitive sophistication led McLean (1973) to coin the term "the triune brain," which referred to the three stages of evolution contained within the human brain: reptilian (brainstem and midbrain), mammalian (hippocampus and cortex), and human (the speech and association areas of the cortex). A similar proposal was made earlier by

Lindsley (1958), and this line of thinking dates back to the evolutionary proposals of Hughlings Jackson in the nineteenth century. Oakley has offered an updated, somewhat behavioristic version of this type of theory, which tries to take into account recent data on decorticate animals, human amnesics, and hypnotic dissociation.

Oakley's theory is illustrated in Figure 5.1, a simplified version of his own illustration. The drawing is quasi-anatomical, in that the two halves of the brain are reflected in the symmetry of the drawing, and the hierarchy from cortex to subcortex is also built into the scheme. Broadly speaking, the "triune" brain concept is reflected in this model: reflex systems, the basis of inframammalian cognition, are at the core of the model, in the brainstem. Homeostatic systems (instinctual drives and motives) and simple associative conditioning, typical of birds and lower mammalian species, evolved on this reflexive base and were located in the midbrain and limbic structures. Representational systems evolved most recently in higher mammals and are dependent upon the hippocampus and neocortex. These form the outer shell of the model and the highest level of control in the system. Even though the basic triune concept is continued here, Oakley's model is more specific than its ancestors, attributing uniquely human skills to a new representational system: self-awareness.

He distinguishes three levels of awareness in evolution. The first level consists of "simple awareness" and is a by-product of simple adaptation to the environment. At this level we have reflexes, homeostats, Pavlovian conditioned responses, habituation, and instrumental learning. The corresponding brain structures are subcortical, and the awareness of animals lacking a cerebral cortex is limited to this level. The second level is what he calls "consciousness" (note that his use of the term is unique), which implies an integrative cognitive strategy of handling information, rather than simple association. At this level, the environment is represented, that is, modeled, in the cerebral cortex. An ongoing biographical record of experience is constructed and stored within a spatiotemporal framework, the basis of our so-called episodic memory.

Consciousness evolved, in Oakley's view, with the cortex and hippocampus; therefore, its evolutionary origin can probably be best placed at the point of emergence of these structures, in the more differentiated reptiles. It developed further, if brain structures are taken as the norm, in birds and mammals. Since birds and mammals are

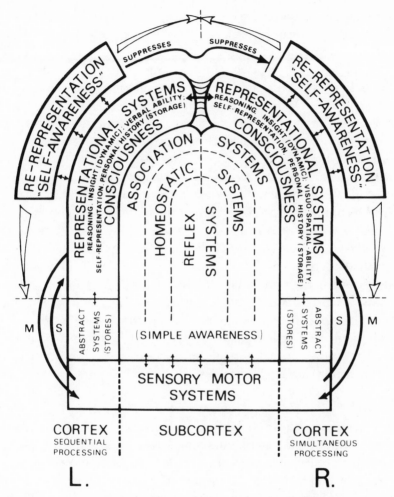

Figure 5.1   Oakley's (1985) anatomical model of systems of representation and awareness in the human brain. The nervous system is schematically represented as a bilaterally symmetrical layered structure, with the most archaic brain systems (reflexes and homeostatic functions) in the center, and the newest systems (self-representation) on the outside shell. The computational styles of left and right cerebral cortex are depicted as "sequential" and "simultaneous."

also more intelligent, by a number of measures, than fishes and reptiles, this proposition seems reasonable.

The third and highest level, in Oakley's view, is found only in certain of the great apes and in more developed form in humans: this level is called "self-awareness." Self-awareness involves the re-

representation of the contents of normal mammalian consciousness, in which attention can be focused not only on representations of the outside world but also on representations of the self. In other words, the animal becomes the object of its own representations. This activity is easiest to verify in humans through introspection and is co-extensive, in Oakley's terms, with subjective experience. Self-consciousness resides in a specific part of the brain, probably cortical but as yet unidentified, that is permanently held in reserve for the reworking of the contents of consciousness.

Self-awareness would have had high adaptive value for primates, since it allowed a much more flexible approach to problem solving and enabled planning and foresight. It freed the mind of its environmental boundedness. One of the key tasks of the third-order system is to form a model of the observer, a self-image. The same modeling processes that served to construct a model of the outside world would be used to construct a representation of the self—its relationships with the outside world, its physical and social properties, and so on. One interesting thing about Oakley's model is that language is not mentioned except as a reflection of, or perhaps a by-product of, these structural changes in basic cognition. Instead, self-representation is placed at center stage.

Self-Representation in Apes

Self-representation is defined differently by various disciplines. James (1890), attempting to describe the complexities of self-representation in humans, decided that there were many "selves" and that the self extended beyond body representation into personal property and the social realm. However, most of his definitions of self would be difficult to apply to animals, even to apes. Self-representation in animals is inherently limited by the absence of language, and therefore speculation about it is limited to evaluating the ability of animals to construct a perceptual–motor model of their own behavior. Toward this end there are two kinds of evidence, neurophysiological and behavioral.

Neurophysiology tends to view self-representation in terms of a body image, that is, a sensory–perceptual device that records the state of one's own body. Accordingly, it has mapped out the projections of the body senses on the brains of various species. Primates, like other mammals, have several detailed body maps in the region of the Rolandic cortex, one for light touch, another for deep touch, and separate maps for muscle, joint, and tendon sensation. Each map is topographical and proportional to the discriminatory sensitivity of areas of the body. But the body maps of primates, although more extensive and detailed for the hands, are not distinguished from those of other mammals by any qualitative markers. Thus, there is no hint in neurophysiology of whether, and how, primates might have any uniqueness in self-representation.

Behavioral tests prove more revealing. Most mammals possess at least a rudimentary body image, as testified by their ability to produce and learn complex coordinated actions. Such learning would not be possible without an integrated motor system, in which the various parts of the body involved in a movement are controlled by a superordinate motor field. Animals without centralized nervous systems, like insects with ganglionic nervous systems, are incapable of true coordinated action, because each appendage is under local control. The legs, for example, each work in parallel, and no part of the insect's nervous system has information on the state of all the legs. Therefore, an insect deprived of a leg may circle endlessly, where a more sophisticated creature, like a mammal, would drag itself with novel locomotor action patterns to its destination. To do this, the motor system must have an integrated cognitive map, including some sort of body map, at its disposal.

This, however, does not mean that the animal has that information consciously available. Many kinds of sensory information remain inaccessible to consciousness, even in humans: for instance, the eye muscles contain receptors that are used for guidance, but we are generally unable to "feel" passive eye movement. In fact, the output of the millions of muscle spindle receptors used in movement control are outside of perceptual awareness, as can be shown by passive movement of a muscle injured in an accident (Merton, 1964). The body maps employed in locomotion thus appear to have only limited value as a basis for conscious self-representation.

However, higher mammals behave in ways that intuitively seem to demand a different kind of self-representation, something approaching self-awareness. The problem is, how can this be tested objectively? The problem of assessing self-perception in animals was approached by Gallup (1970, 1982) in a simple but effective manner: the mirror-recognition test. In a series of experiments, Gallup placed apes and monkeys in front of mirrors for extended periods of time and observed their behavior. Some of them seemed to recognize themselves: they would use the mirror for grooming, for example. Others showed no awareness that the image in the mirror reflected themselves and treated the image as another individual, attacking it or ignoring it.

Gallup decided to test their self-awareness by secretly, under anesthetic, placing colored marks on their bodies outside of their visual field—for example, on the forehead—and putting them back in front of the mirror. Those with self-recognition immediately and repeatedly touched the marked areas upon seeing their reflections, testifying to their understanding that it was their own bodies being reflected. Among our closest relatives, the chimpanzees were fastest at learning to recognize themselves in the mirror, with orangutans somewhat slower, while gorillas failed the mirror test altogether. None of the monkey species tested by Gallup and his collaborators was capable of mirror self-recognition.

The fact that some apes, but not all apes, possess such a sophisticated self-representation testifies to the recency of this development. As pointed out by Oakley (1985), the mirror test is not entirely to be trusted. Congenitally blind humans who later acquire sight initially treat their mirror images as another person and, like chimps, take time to learn to recognize the images as their own (Von Senden, 1960). The congenitally blind evidently have a self-image, since they can do some of the things the rest of us do; but the mirror test requires specific visual skills in addition. It is possible that the gorilla, and by extension other animals who fail the mirror test, lack only those specific visual skills but possess some other sort of self-image (Oakley, 1985).

Fair enough. But it is surely no coincidence that mirror recognition occurs only in a few of our closest relatives. The question is, what is being learned in the mirror test? And what sorts of skills did the chimp acquire to become so adept at solving the test? One possibility might be considered: of those that failed the test, the gorilla is a knuckle-walker, and the monkeys are arboreal brachiators; both use their hands a great deal in locomotion. By comparison, the chimpanzee's hands are much freer for manual exploration as well as for manipulative acts. The chimp uses its hands for a wide variety of functions, and its remarkable tool-using abilities have been known since the time of Darwin. Although gorillas have some manual skills, theirs are not as developed as the chimpanzee's, particularly with regard to exploration. This suggests that chimps have a more sophisticated self-representational system, at least as it relates to manual exploration.

Note that hand control involves, for the first time in evolution, a coming together of visual, tactile, and proprioceptive feedback on the same action system. Hand control may be regarded as the crossing of a biological Rubicon, in that a dominant distal sense—vision—comes to control and modulate actions directly. By way of comparison, the type of visual locomotor control implied by the possession of a cognitive map is indirect and aimed not at the specifics of limb movement but rather at the animal's destination in *external* space (O'Keefe and Nadel, 1977). For the most part, the specific details of locomotor acts are mostly under the control of balance, proprioception, and touch, not vision. In these species, vision is a distal sense, reserved mostly for the external world.

However, visual hand control is present in some form in most monkeys as well as in apes. What might the chimp have that makes it successful at Gallup's mirror test, when monkeys are not? Presumably it is a computational strategy—a capacity to create a more complex perceptual model. When a chimp explores the mirror, it typically touches the mirror with its hands; this provides a link between an action system for which it has a visual map and the rest of its visual experience. Since, like all higher mammals, it has a capacity for comparing and relating two simultaneously present visual events, it has a starting point for making the connection between the visual image and *its own actions* (not merely between two external visual events). The point is, there is a step from the limited visually guided hand movements of most primates to the more general use of the hands in the exploration of both self and world that is found in chimps. Such exploration could have taken the next step, in hominids, to a completely new kind of self-representation, if extended to the whole body.

What relation does this have to the social intelligence theory of Dunbar? Is it possible that the cognitive adaptations that were needed to allow large groups to cohere were the same that enabled self-awareness? An evolutionary stance forces us to examine cognitive skills in relation to one another. At every point in evolution, the various capacities of a species must always combine to form a successful survival strategy and a coherent culture.

Episodic Culture: A Definition

It is difficult to account for the evolution of human cognition without considering the cultures of apes and hominids. As Lovejoy pointed out in his treatment of the bipedal locomotion of australopithecines, traditional one-dimensional explanations for the adoption of erect posture—that walking erect enables better hunting and self-defense in tall grasses, for instance—are inadequate; walking erect would have been a fatal adaptation without other major changes in the survival strategies of apes. These included changes in habitat, food sharing, rates of procreation, and child-rearing strategies. The result is a much richer, and more convincing, theory of how australopithecine *culture* evolved.

As pointed out in Chapter 1, cognitive skills are necessarily implemented in a culture, which we define as a collective system of knowledge and behavior. The culture typical of a given species reflects the cognitive capacities of the individuals making up the culture. It follows that, when considering the origins of a radical change in human cognitive skill, we must also look at the sequence of cultural changes. Cultural patterns constitute our most fundamental information on the cognition of animals, and ethological observation is a more accurate and more efficient method of mapping out basic cognitive features of a species than laboratory experimentation. Knowledge can be furthered by controlled experiment; but a rich ethological framework gives meaning and context to the experimental data. The same principle should apply to human cognition; there ought to be a cognitive ethology of human culture, a temporal framework for the emergence of mind.

There are no existing classifications of ape or human culture that are based primarily on cognitive considerations. Ape cultures are usually described in terms of feeding habits or territory, as in "frugivorous" or "arboreal" cultures. Human cultures are sometimes specified in terms of technology—for instance, Stone Age and Bronze Age cultures. Sometimes they are specified in terms of religion and myth:

Campbell (1959, 1988) refers to the cultures of "animal powers," referring to the common creation myths of certain aboriginal cultures. Sometimes they are named after their manner of obtaining food: hunter-gatherers, slash-and-burn agriculturalists, pastoralists, and so on. They have been labeled after their tools, trading practices, diet, level of social organization, and geographical setting. They have been labeled as repressive and post-repressive, warlike and peaceful, patrilineal and matrilineal. But they have not typically been classified in terms of their predominant cognitive features. In fact, the only such classification in wide use is along the dimension of literacy, which is not particularly useful in characterizing premodern cultures, since they were all preliterate.

Yet when we are trying to develop a strategy for bridging the gap between humans and the rest of the animal kingdom, cognition is the most important dimension along which cultures are distributed. A cognitive classification of culture could be built on any number of cognitive dimensions, but the most likely place to start would seem to be in the area of *representational strategy*. Modern humans have many apparently novel systems of representation in memory, and their development may be the central processes underlying our cultural evolution from the apes.

If apes are taken as the starting point, how might their overriding representational strategy be described? Despite their formidable skills, they lack language, and they also lack much of the nonverbal knowledge evident in humans who have been stripped of language. Their behavior, complex as it is, seems unreflective, concrete, and situation-bound. Even their uses of signing and their social behavior are immediate, short-term responses to the environment. In fact, the word that seems best to epitomize the cognitive culture of apes (and probably of many other mammals as well, although this is tangential to the argument) is the term *episodic*. Their lives are lived entirely in the present, as a series of concrete episodes, and the highest element in their system of memory representation seems to be at the level of event representation. Where humans have abstract symbolic memory representations, apes are bound to the concrete situation or episode; and their social behavior reflects this situational limitation. Their culture might be therefore classified as an episodic culture.

Although used here in a rather idiosyncratic sense, the term "epi-

sodic" is ultimately derived from the commonly used term for concrete or time-bound memory, which Tulving (1983) labeled as episodic
memory. Episodic memory is, as the name implies, memory for specific episodes in life, that is, events with a specific time–space locus.
Thus, we can remember the specifics of an experience: the place, the
weather, the colors and smells, the voices of the past. Typical examples
of episodic memories are found in the details of specific experiences:
a death in the family, first love, and so on. Such memories are rich in
specific perceptual content. By definition, episodes are bound in time
and space to specific dates and places. The important feature of this
type of memory is its concrete, perceptual nature and its retention of
specific episodic details.

The ancient foil to episodic memory is procedural memory. Procedural memory is quite different and structurally more archaic. For the
most part, procedural memories can be regarded as the mnemonic
component of learned action patterns. Simple organisms can learn
patterns of action without any detailed episodic recall; procedural
memory involves the storage of the algorithms, or schemas, that underlie action. Sherry and Schacter (1987) have observed that in terms
of its storage strategy, procedural memory is the opposite of episodic
memory. Whereas episodic memory preserves the specifics of events,
procedural memory preserves the generalities of action, across events.
Procedural memories must preserve general principles for action and
ignore the specifics of each situation. For example, in learning to catch
a ball one must learn the principle of tracking a moving object, no
matter what the speed of the object, the starting point, or one's initial
posture at the time it is thrown. It would be cumbersome to remember the exact speed, starting point and position of each successful
practice catch; a new throw is unlikely to match any specific counterpart in practice. Thus, learning a procedure, even on this level, involves setting parameters and forming general rules. Detailed episodic
recall would interfere with this process.

Episodic and procedural memory involve different neural mechanisms, as can be shown in birds, who will lose their songs (a procedural memory system) if lesioned in one nucleus, and their ability to
hide and relocate food (an episodic memory system) if lesioned in
another. The same distinction exists in humans, as seen in amnesics.
In the famous case of H.M., followed by Milner and her associates

(1966, 1975) for over twenty years, the patient developed catastrophic anterograde amnesia following neurosurgery. He retained a capacity for acquiring new procedural memories, that is, he could still learn new motor skills. But his capacity for new episodic memories was destroyed; he could not record any new events in his life. For instance, he had to be reintroduced over and over to the doctors who were treating him after his surgery. And although he acquired new motor skills, he could not recall ever having learned them. Thus, new procedural memories were acquired but the specific episodes during which they were acquired were not recalled.

Both episodic and procedural memory systems seem to be present in a variety of animals, including mammals and birds. Sherry and Schacter (1987) reasoned that episodic memory evolved separately from procedural memory for the very good reason that their storage strategies are mutually incompatible. Whereas procedural memories generalize across situations and life events, episodic memory stores the specific details of situations and life events. Thus, one memory system stores the generalities and discards the specifics; the other system, the episodic, stores the specifics but does not generalize. Obviously, the same neural mechanism would have difficulty doing both; therefore, two separate mechanisms evolved for the two types of storage, and the distinction has endured across many species.

Episodic memory is apparently more evolved in apes than it is in many other species, in the sense that apes are sensitive to subtleties of social and pragmatic situations that other animals cannot register. This reminds us that there are significant gradations in the computational power of the episodic memory system; the complexity of its contents may vary widely between species. This will become clearer in the following section, which deals with event perception. But even though it may vary tremendously in power, episodic memory differs fundamentally from procedural memory in that it involves a degree of conscious awareness; an animal lacking a capacity for episodic memory and restricted to the procedural level would be not much more than a stimulus–response organism, a high-level automaton of the sort favored by the early behaviorists.

Episodic memory also differs fundamentally from the dominant form of human memory, specifically semantic memory. The third category mentioned by Tulving in his original taxonomy was semantic memory, which is usually considered to be symbolic in nature and characteristic of humans. The kinds of facts usually tested on IQ tests or college entrance examinations involve semantic memory. The closest parallel to human semantic memory in animals might be found in the signing behavior of trained apes; but even apes who sign or use visual symbols do not appear to store up large numbers of facts and propositions about the world, the way humans do.

Since humans and nonhuman mammals, including apes, differ so fundamentally in the types of memories they can retain, it is possible to use this fact to characterize their two types of society. Most animals, including humans, possess procedural memories, and therefore the term is not particularly useful in characterizing the dominant cognitive feature of mammalian culture. Episodic memory is probably unique to birds and mammals, forming the basis for Oakley's definition of rudimentary consciousness. Humans possess both procedural and episodic memory systems, but these have been superseded in us by semantic memory, which is by far the dominant form of memory in human culture, at least in terms of the hierarchy of control. In contrast, episodic memory is dominant in most mammals, including apes. Animals do not seem to possess the systems of representation that would allow them to have elaborate semantic networks. Their experience, in this light, is entirely episodic. The pinnacle of episodic culture, the culture of the great apes, marked the starting point of the human journey.

The dependence of apes upon episodic memory throws light on their difficulty with sign language, even when trained by humans under extraordinarily favorable conditions. Signing has a procedural aspect, which is simply the motor "skill" of reproducing the movement that constitutes the sign. And it might be expected to have a semantic memory aspect, much like the human use of words. But the use of signing by apes is restricted to situations in which the eliciting stimulus, and the reward, are clearly specified and present, or at least very close to the ape at the time of signing (Terrace and Bever, 1980; Savage-Rumbaugh, 1980). Situational specificity is not typical of semantic memory. If apes possessed an abstract semantic representation to which the sign referred, this concrete, situational limitation would not apply.

The reason apes use signs in such a concrete manner is that they are using episodic memory to remember how to use the sign; the best they can manage is a virtual "flashback" of previous performances. Thus, their understanding of the sign is largely perceptual and situation-specific. To say their understanding is perceptual in nature could be misunderstood to mean that they only perceive simple features of the environment. In fact, event perception is the most evolved form of cognition and the basic component of episodic memory. The episode is the "atom" of ape experience, and event perception is the building-block of episodic culture.

Event Perception in Apes

Event perception is, broadly speaking, the ability to perceive complex, usually moving, clusters and patterns of stimuli as a unit. This property also characterizes object perception, but event perception resolves input into much more than a single object: motion and context are taken into account. A passing car constitutes a perceptual event, and so does a kick, or a threatening grimace, or the lifting of a spear, or a hand sign. Event perception is a subject of some importance in the fields of auditory perception, touch, and artificial intelligence research (McCabe and Balzano, 1986). Visual event perception, in particular, has been studied in recent years.

Poizner, Klima, and Bellugi (1987) have tried to model the signs of American Sign Language for the Deaf (ASL) as complex visual events. They have attempted to specify the physical parameters that might be applied to decoding ASL signs, and it is clear that such visual events are not simple. But they are indeed elementary when we consider what mammals can perceive in the environment; they can correctly perceive not only individual patterns of motion but also social situations in which many different agents and objects are involved over a significant period of time.

The perception of events is the ultimate objective of the perceptual process, at least in reasonably complex animals. Intelligence in animals might even be defined in terms of the complexity of events they can perceive. Animals that we call intelligent are those that respond to events of increasing complexity and abstraction. Apes can discriminate hand signs that are too complex or subtle for dogs; but dogs can read aspects of behavior that are missed completely by rats. Events can be arranged in a hierarchy of complexity; the simplest events are those that are closest to the level of object perception. A hand sign is an object in motion, and the perception of a hand sign or visual emblem as a unified event is well within the capacity of an ape. Yet, this level of event perception obviously does not suffice for language.

Complex events are made up of smaller perceptual segments. Higher mammals, including apes, have no difficulty in perceiving complex life events, even when the perceptual segments change. A dog can assess, quickly and effortlessly, a visual array that includes a female in heat, another dominant male, and the presence of various humans on the scene. His behavior will be affected by a variety of contingencies and variables, each of which can be understood only with reference to the other elements present in the situation. Substitute a different cast of characters, change a single element, including things as subtle as the weather conditions or the distances between the animals and people in the scene, and the dog's behavior will change. Perceiving and reacting to events of this complexity are the normal stock-in-trade of higher mammals.

Apes are particularly good at visual-event perception. In the "insight" experiments cited earlier, Kohler's monkeys and apes were asked to assess complex visual events in the laboratory; they proved more than capable of assessing the relationships between the various elements present in the room. Their capacity for event perception "solved" the problem of how to reach the banana; they were able to break down the perceptual components in the situation and imagine another arrangement of the same components. This type of complex event perception, which bears close resemblance to perceptual decomposition in Corballis's sense, appears to be the highest achievement of the episodic mind.

The difference between an ape and a human child apparently resides in how event perceptions are encoded and remembered. In the ape, where episodic memory is the dominant device, the event can only be remembered in a literal, situation-specific manner. Thus, the "meaning" of an ASL sign to an ape is simply the episodic representation of the events in which it has been rewarded. This is not qualitatively different from more conventional forms of operant conditioning; the only difference is in the degree of perceptual "intelligence" shown by the ape, that is, in the complexity of the events so encoded. The human child eventually breaks out of this episodic world and develops completely different semantic representations. Apes and other large-brained mammals such as whales, elephants, dogs, and cats may differ greatly in which perceptual events they can encode, but they share the same cognitive limitation: their dominant form of memory is episodic, and their cultures, whatever their individual features, are therefore episodic. In this respect, these cultures are globally different from human culture, in which the dominant forms of memory are semantic.

Although the mechanisms of event perception are not well understood, progress is being made in constructing models of simpler forms of perception, including elementary event perception. Perception, by its very nature, extracts generalities. Recent parallel distributed processing (PDP) models of early perceptual processing demonstrate this principle in a convincing manner. McClelland and Rumelhart's Interactive Processing model (1986) showed that, even on the most elementary level of processing, circuits tend to generalize their initially learned perceptual categories to new items on the basis of similarity. Perceptual learning, in other words, even in simulations, does indeed involve differentiation, as James Gibson wrote in 1950.

But it involves more than differentiation of the object. At the level of object movement the perceiver effortlessly extracts what appear to be perceptual templates of action. Johansson's (1968) classic experiments with the perception of dance steps illustrates this point. He placed a few luminous strips of tape on dancers, either on major limbs or on joints, and then had them go through various routines in the dark. Videotapes of these luminous patterns were immediately understood by observers; in fact, for those who were familiar with the dancers, it was even possible to recognize individuals from their style of walking and moving, with nothing but faint luminous lines and points as cues to their body shapes and positions. The only way one could explain this capability would be to posit a perceptual process that could immediately and effortlessly perceive the resemblance between the overall patterns of movement of the luminous lines and the movement patterns stored during normal perceptual learning. This implies a level of perceptual processing that integrates entire patterns of action. This is a prerequisite for elementary event perception.

Jane Goodall (1971) has described the complexities of chimpanzee social life, which depends on the ability of the individual chimp to comprehend such relationships and events. Two examples, drawn from many possible ones, illustrate the subtlety of chimpanzee event perception. One of Goodall's most memorable descriptions was the story of Mike, an imaginative male who started at the bottom of the dominance hierarchy and worked his way to the top by developing an improved method of display. Male chimps normally display their aggression by swinging tree branches back and forth, occasionally hitting their opponent, and making hooting sounds at the same time. Males will do this to one another until a dominance hierarchy is established; the submissive "loser" of a display contest will then approach the winner in a submissive posture and offer to groom him. Mike was the usual loser of such contests.

But one day Mike made a discovery, one that was no trivial intellectual achievement for a chimp. He discovered that two kerosene cans made a much more effective display than branches. One day he screwed up his courage, grabbed two kerosene cans, and charged the other males in his group. The display, and especially the racket, was so intimidating that the others ran away. One by one they returned to indicate submission, and Mike ended up with three placated rival males grooming him. He maintained his dominance even after the cans were removed from the scene.

A number of cognitive factors had to enter into Mike's use of the cans. He was not the first chimp to pick up a can, with the ensuing racket. Many of the others had done the same. But to use the can, not by accident but by calculation (and, Goodall added, without immediate emotion) in a series of displays, he had to generalize from a number of event perceptions, and then interrelate them. First, obviously, he had to have a perceptual template for display behavior, like any male chimp. Then he had to recognize the correct conditions for such displays, again like any healthy male chimp (displays in inappropriate situations could be fatal). He also had to recognize that the sound of clanging kerosene cans can be intimidating and that this might make them useful in displays. The latter could not have been realized in so many words, but prelinguistic children also involve themselves in rivalries and don't have difficulty making similar judgments. Therefore the chimp's ability to achieve this type of insight in the absence of language should not surprise us.

Another of Goodall's observations shows more clearly than any laboratory experiment could how social event perception lies at the core of chimpanzee social intelligence. Young chimpanzees play together in an endless series of games. Typical games involve running around trees, climbing branches, chasing, swinging, wrestling, and tickling one another. Most of this activity serves to develop procedural skills that become useful later in locomotion and fighting. But as Goodall points out, in the process a great deal of social learning goes on. The young chimp quickly learns which of its peers is stronger or weaker, faster or slower, dominant or submissive. It even learns whose mother will retaliate, who will be intimidated by a display, and who will call a bluff. All of this depends on its ability to perceive social events accurately and to remember them.

The young chimp thus maps out a complex set of relationships in memory—a network of knowledge that will serve as the basis of later social prowess. This is the type of knowledge Dunbar had in mind when he wrote of the importance of memory in servicing social relationships. Complex societies demand a tremendous memory capacity, and the type of memory that is important in social relationships is, above all, episodic memory. Episodic memory is little else than a storage system for event perceptions, and thus there is a close tie between episodic memory and the capacity for social event perception.

Culture and Evolution

Cognition and culture both depend on the learning that takes place during the lifetime of the organism. Episodic culture is heavily dependent upon such learning. But how can cognitive and cultural factors come to influence the course of biological evolution? Would they favor general-purpose adaptations, or would they favor highly specific patterns of change? This question, which is at the center of cognitive biology, has been addressed by Plotkin and his colleagues (1982, 1987, 1988), who have examined the basic assumptions of evolutionary theory as applied to intelligence. Evolution for them involves the acquisition of "knowledge," and what is conventionally known as individual intelligence is only a subset of the total knowledge of an evolving species.

Plotkin and colleagues have opted for a four-level process of evolution, in which each level is sensitive to events of different frequency. The base level (genetic) is sensitive only to very slow-moving events. However, three levels of subsidiary evolutionary factors (developmental, learned, and sociocultural) speed up the selection process, by allowing information that is obtained from more rapidly changing areas of influence to affect survival and reproductive fitness. They propose that the entire hierarchical evolutionary network acquires "knowledge," and that much of cognition and learning can be encompassed within a neo-Darwinian framework.

Intelligence, in this context, becomes much more than the capacity and skill of one individual mind. Rather, it includes the entire knowledge-structure of the species, as it is stored across the various levels available to the multilevel evolutionary process. Variation is generated at all four levels; the variants are tested, and the successful variants are selected and then regenerated.

The generality of this type of model is such that behavioristic learning theory can be absorbed as a subset of the evolutionary growth of intelligence. The capacity to learn is basically an adaptation that enabled organisms to respond to environmental variations in a time frame not accessible to level one (genes) and level two (development). As such, learning would always be *tailored to the needs of a particular species.* One advantage of this model is that it accounts for species-specific constraints on learning more effectively than other approaches (Plotkin, 1988). And since variation and selection at each level is an essential part of the process, the inventive and generative aspect of learning and cognition is better accommodated than in a traditional framework.

Plotkin discusses the case of manual dexterity in chimpanzees. The unique manual skills of the chimpanzee were themselves a product of evolution at levels one and two. But manual skills allow them to change the environment at a high rate, and these changes cannot be tracked by levels one and two. In Plotkin's (1988) words, "The chimpanzee has therefore evolved modules of knowledge gain that extend to the third level and hence can operate at the required frequency— but the changes . . . are specific to hand use." The intellectual skills of chimps are thus not a unitary nonspecific learning capacity but very closely tied to manual skills. Volume conservation and analogical reasoning in chimps might be direct, specialized products of their visual–manual skills (Pasnak, 1979; Gillan, Premack, and Woodruff, 1981).

The same logic holds for the specific nature of human intelligence, which depends to a great extent on the use of language. Humans can track, and elaborate upon, the very rapid rate of information exchange made possible by language. This is a clear case of a generative, inventive mechanism whose outputs are continually put to the test and regenerated in precisely the manner specified by Plotkin's model. Language would also have led to innovations at level four, culture, since it would have increased the pace and complexity of sociocultural development.

Plotkin's message is especially relevant to the construction of an evolutionary scenario for human language and intelligence, because human cognitive innovations appear to have been focused mostly at levels three and four. Plotkin takes a position that is in many ways the opposite of the mainstream; there is no room for all-purpose, nonspecific skills in this picture. Thus, concepts like a "critical mass" of extra neurons, proposed by Jerison, would appear too vague and unlikely if a multilevel evolutionary model is accepted. A multilevel evolutionary model, however, provides a mechanism and a rationale for the evolution of highly specific adaptations such as advanced forms of social event perception in primates. Cultural changes are thus closely tied to basic cognitive abilities in Plotkin's model of evolution. And the extraordinary event perceptions of higher mammals make them increasingly sensitive to fast-moving variables at levels three and four, which in turn could exert selection pressure for further developments in this direction.

Conclusion: The Episodic Mind

In conclusion, apes represent the pinnacle of the episodic mind: procedurally skilled, as in the case of most animals possessing central nervous systems, and extremely adept in the realms of event perception and episodic storage. Where mammals in general may excel in both these areas, apes have an increased capacity for self-representation and better problem-solving skills than other mammals, and even than other primates. In our closest relatives, chimpanzees, social structures depend upon the servicing of numerous relationships, and this requires both complex event perception and episodic memory. Unlike complex insect societies, for instance, which operate on pheromones and automatic, stereotyped, reflexive principles, chimpanzees remember large numbers of distinctly individual learned dyadic relationships.

From a human viewpoint, the limitations of episodic culture are in the realm of representation. Animals excel at situational analysis and recall but cannot re-present a situation to reflect on it, either individually or collectively. This is a serious memory limitation; there is no equivalent of semantic structure in animal memory, despite the presence of a great deal of situational knowledge. Semantic memory depends on the existence of abstract, distinctively human representational systems. The cognitive evolution of human culture is, on one level, largely the story of the development of various semantic representational systems.

As a result of the success of attempts to train chimpanzees and gorillas to use sign language and other forms of communication, it is generally acknowledged that they are able to use symbols, in the critical sense that they can use them as substitutes for their referents. There is no room for debate; they are obviously able to learn to use Saussurian-style signs and signifiers. But they are incapable of symbolic invention; and therefore they have no natural language of their own.

It is difficult to say whether those apes that have learned to use signs have true semantic reference systems; in a sense they have. After all, it cannot be denied that they possess a lexicon, in the strictest definition of the term; they have a set of "words," albeit a limited one, which refer to a set of "meanings." Yet there is very little evidence that they can use their hard-won symbol sets to construct any kind of useful semantic memory system. Human semantic memory systems may contain thousands of easily accessible labels of things and events, but they contain a great deal more than that. The labels are, in a sense, the least important element in the human semantic memory system, as will become clear. Protolanguage would involve more than labels.

Premack (1987) wrote of protolanguage and its social context. He argued that language should not necessarily be seen as *the* human adaptation. Consciousness, pedagogy, social attribution, and aesthetics are also uniquely human; in his view, our obsession with language obscures this fact. He was willing to concede to apes some limited capacity for delayed social imitation but none for systematic pedagogic training of the young. And he also conceded to apes a limited degree of social attribution, about at the level of a three-year-old human, where a child can understand that another shares a similar belief. But Premack did not concede significant linguistic skill to apes. In fact, he pointed, with some puzzlement, to the fact that animal call systems have not increased greatly in complexity from those of bees, which possess about 20 signs, to those of apes in the wild, with about 35. But obviously in their individual cognitive capacities apes are incredibly more evolved than bees. In other words, despite the absence of linguistic or symbolic evolution, social and cognitive evolution continued.

In apes, language was still at a standstill, despite great leaps forward in the realms of procedural and episodic memory and self-recognition. The distance from episodic culture to our own modern human culture is so vast that it is difficult to conceive of a single adaptation that would result in so great and so rapid a change. But after considering both the archeological and neuropsychological evidence, it appears unlikely that human cognitive culture was the product of a single innovation. As Premack wrote, language was part of the change but not the whole of it. At the very least, there appears to have been a major break in the culture of *Homo erectus*. Whatever that culture was, it must have been logically intermediate between the episodic mind and the symbolic representational systems that characterize modern humanity.

First Transition: From Episodic to Mimetic Culture

Mimetic Culture: The Missing Link

In this chapter I propose a category of archaic but distinctly human culture that mediated the transition from ape to human. This mediating, or intervening, layer of hominid culture is called mimetic, on the basis of its dominant or governing mode of representation. Although evidence for it is indirect, I feel it is persuasive; in fact, an intervening layer of cognitive culture is a logical necessity in constructing a credible evolutionary scenario for human emergence. The specific notion of a self-contained mimetic culture is consistent with the idea of cognitive vestiges; I will argue that the use patterns of mimetic representations in modern human society have remained distinct from the uses of our later cognitive acquisitions. In effect, there is still a vestigial mimetic culture embedded within our modern culture, and a mimetic mind embedded within the overall architecture of the modern human mind.

An archaic human cognitive adaptation serves several important functions in the chain of evolutionary reasoning. On the one hand, it can be systematically "built up" from the episodic culture of apes. On the other hand, it can be made to dovetail with our knowledge of the structure of modern human cognition and culture. Perhaps equally important, it should provide a solid basis for later cognitive innovations in the human line. Accordingly, there should be as smooth a blend from archaic to modern human culture as can be achieved. Given these constraints, three broad sources of evidence must be considered: the starting point (episodic culture), the finishing point (biologically modern humans), and archeological evidence that bears directly on the cognitive capabilities of human ancestors.

The first hominid species to appear in the evolutionary record was *Australopithecus afarensis*, which was present in Africa about 4 million years ago and was believed to be ancestral to all later australopithecines. None of the current reconstructions of *afarensis* habitat and culture suggests a major cognitive change, at least not of the magnitude that would be needed to break the mold of the episodic mind. *Afarensis* stayed in a small region of Africa, fixed in as narrow an ecological niche as any ape. There are no signs of manufactured artifacts, long-distance migrations, or base camps. In the absence of any evidence for major cognitive change, the culture of *afarensis*, on present evidence, must be regarded as episodic.

Later australopithecines, and so-called *Homo habilis*, are a bit more problematic. Loosely associated with their remains are crude artifacts, the basis of Oldowan culture, a very primitive toolmaking culture. But they were still narrowly fixed in geographical location and, by implication, in their ecological niches. If it were not for the existence of the famous (and anomalous) habiline skull 1470, with a cranial capacity approaching 800 cc, it would be easy to dismiss the idea that *Homo* had appeared so early in the archeological record. Without trying to specify too much on the basis of this enigmatic fossil period, one must concede that something was beginning to change. Primitive stone tools, larger brains—these were at least signs of a transition in progress. But the evidence is insufficient to draw clear conclusions about this uncertain era.

With *Homo erectus*, the uncertainty disappears. *Erectus* developed a variety of sophisticated manufactured tools and spread over the entire Eurasian landmass, adapting to a wide variety of climates and living in a society where cooperation and social coordination of action were central to the species' survival strategy. Such achievements were well beyond those of apes or australopithecines and, on current evidence, those of habilines.

Could an elaborate ape culture, that is, an extended and improved episodic culture, have achieved what *erectus* achieved? Those who propose a linear, unitary process of encephalization and intellectual evolution for this period might run into difficulty here. Some features of the culture of *erectus* suggest qualitative changes in cognition, rather than more of the same. Their systematic tool technology alone would place demands on the intellect that go beyond the concrete, literal, time-bound episodic mentality. Widespread tool manufacture required both an elaborate mechanism for inventing and remembering complex sets of procedures and the social skills to teach and coordinate these procedures. *Erectus* also cooperated in seasonal hunting, migrated over long distances, used fire, cooked food, and evolved a brain that eventually reached 80 percent of the volume of the modern human brain. What sort of mind, what sort of culture, could account for this?

A review of conclusions drawn from previous chapters might help here. One conclusion was that, on anatomical grounds, high-speed vocal language was a relatively recent invention, unique to *Homo sapiens*. Since rapid cultural change is so closely tied to symbolic language, it seems unlikely that slow-changing *erectus*, essentially stable for a million years, would have possessed any form of symbolic language. Nevertheless, *erectus* seems to have broken free of some of the constraints of episodic culture and moved a major cognitive step forward.

Another conclusion was that, prior to the evolution of a system as revolutionary as human language, the cognitive stage had to be set. There had to be an immediate rationale for the emergence of language; the mechanisms of evolution do not possess foresight. The immediate adaptive pressure for this new trait had to be closely tied to structures already in place in the preceding culture. Therefore, the culture of early hominids must not only have been built upon the episodic cultures of apes and australopithecines; it must also have had properties that served as the basis for the later arrival of human language.

To create a picture of this adaptation—the archaic, original form of hominid culture—it is not enough to single out some feature of the human mind or brain (for example, encephalization, cerebral laterality, serial motor control, phonological skill, consciousness, generativity, syntax, or intentionality) and try to build a model around that feature. We should be looking for a complete, self-sufficient pattern of adaptation intermediate between the modern human mind/culture and the episodic mind/culture that preceded it.

Another conclusion, derived mostly from neuropsychological evidence, was that language is an isolable skill, even in modern humans. Brother John was perhaps the most dramatic clinical case in which the human mind, stripped of a capacity for speaking, reading, writing, and internal speech, could be observed. But there have been other instances as well where partial language loss revealed something of the nonlinguistic mind. And there are the intentional skills of the prelinguistic child, and the illiterate deaf-mutes of past societies, where glimpses can be caught of the human mind without symbolic language. These glimpses all point to the same conclusion: without language, the human mind is still far superior to that of the ape. It has properties that clearly break the mold of episodic culture.

There is, of course, a danger in assuming that the modern mind without language is equivalent to the archaic hominid mind to which language was added. The acquisition of language may have been accompanied by other late cognitive changes, and precursors of language may have developed in earlier hominid cultures. But one of the basic principles of evolution is the conservation of previous gains in adaptation. The human sensory and motor apparatus has remained essentially similar to those of primates presumably because the primate sensory apparatus continued to serve its purpose perfectly well. Changes in our brain, by contrast, were driven by a different level of selection pressure. A qualitatively new cognitive culture like the one that must have characterized *erectus* would have left vestiges in its descendants; and a cognitive culture that was successful in inventing, transmitting, and maintaining complex social and technological skills would continue to be useful, even after language had been adopted. Thus, we should be able to find cognitive vestiges of *erectus* in modern human society.

The idea of vestigial structures suggests one of the important sources of evidence that must be considered in the next few chapters: the architecture of modern cognition. Are there characteristics of the modern human, isolable from language, that are logically intermediate between episodic and symbolic culture? Could those features have formed the basis of a well-integrated adaptation, and could they have accounted for the achievements of early hominids, and especially of *Homo erectus?*

Human Cognition without Language

The prelinguistic child, the illiterate deaf-mutes of history, and Brother John during one of his long paroxysms share one interesting characteristic: they are not incomplete. They are quite unlike most neurological and neurosurgical patients who lose a specific function and appear as though they are missing a cognitive "part." In all three cases, they are able to cope with the demands of living, provided the use of symbolic language is not involved. Episodic memory continues to function, skills are retained, general knowledge of the environment remains in effect, and the individual is able to cope with complex social situations. Even while isolated from symbolic use, they appear to be, quite literally, employing the cognitive skills of a different cognitive culture, one that, in its own way, is self-sufficient.

In contrast, the neurological patient with global aphasia is usually disordered in many other ways; thought, learning capacity, memory, knowledge, affect, and movement are also likely to be affected. Neurosurgical patients with a major resection of cortex—for instance, the split-brain cases studied by Sperry and Gazzaniga—or the many patients with temporal or frontal lobectomies usually present a very peculiar cognitive profile. There may be a discoordination of different areas of mind, or clear dysfunctions of basic cognitive skills such as memory or consciousness. The isolated right hemisphere in the split-brain patient, so widely cited as a source of wisdom about laterality, is perhaps the worst case in this regard. Judging from the heated exchange between the principals in the field published in the *American Psychologist* (Gazzaniga, 1983; Zaidel, 1983), it seems impossible to draw conclusions about the prelinguistic brain from these cases. The reasons for this are complex and are related both to the crudity of the physical lesions and to the disruptive effects of such lesions on the functioning of the whole brain. It is very difficult, in such patients, to dissociate language from thought, or the control of action from some form of symbolic control. The same may be said of the vast majority of aphasic patients.

In the three cases mentioned above, however, the individuals can perform a variety of difficult cognitive functions without language, without even the possibility of internal speech. The range of their cognitive competence is impressive: it includes intentional communication, mimetic and gestural representation, categorical perception, various generative patterns of action, and above all the comprehension of social relationships, which implies a capacity for social attribution and considerable communicative ability.

The cognitive style of people stripped of symbolic language is familiar to us, because a significant part of normal human culture functions without much involvement of symbolic language. Examples are found in trades and crafts, games, athletics, in a significant percentage of art forms, various aspects of theatre, including pantomime, and most social ritual. Illiterate deaf-mute adults, studied intensively in the nineteenth century (Lane, 1984), proved to be fully aware in every sense of the term and could comprehend and remember events with great accuracy. Thus, their episodic memories were intact, and consciousness was not impaired. Their tendency to invent gestures and mime on the spot, in highly creative ways, to communicate their thoughts demonstrates that they possessed excellent intentional communicative skill. Their ability to operate machines and invent solutions to practical problems is evidence that they possessed generative praxic skills. They could recognize the functions of objects, emotional responses, and social relationships, and their event perceptions seemed intact. They were able to teach and communicate skills, manage socially complex scenarios, such as holding a job (usually a menial one, unfortunately), and even produce works of art. They were fully able to cooperate with and recognize the intentions of others.

Rudolph Arnheim wrote a chapter in his book *Visual Thinking* (1969) entitled "Words in Their Place," in which he argued that language was not indispensable to artistic thought; on the contrary, in the case of the visual arts it was largely irrelevant. Coming as the book did at the peak of the popularity of transformational linguistics and artificial intelligence, Arnheim's claim was largely ignored. But Arnheim's view of visual metaphor and the nonsymbolic forms of representation has endured and won out: visual thinking is now seen as largely autonomous from language. The same may be said of nonliterate, or naive, musical invention like much of early jazz, of most sports, and of the most ancient human crafts like pottery and weaving, which are learned, to this day, largely without language.

When humans lack language, provided they do not suffer some brutal brain lesion that robs them of other, more fundamental, cognitive skills, they can continue to participate in all those forms of human culture that do not require language. Surely these abilities in themselves, even in abbreviated form, might constitute the basis for a culture.

It is evident that some form of archaic hominid culture must have existed that set out the fundamental cognitive architecture of the modern mind. For over a million years that culture constituted the main line of evolution toward modern symbolic cultures. If symbolic language came late, as an isolable skill, and was preceded by a culture of some intellectual power, it stands to reason that the forms of archaic hominid cognition should bear some fundamental resemblance to the extralinguistic features of the modern mind. *Erectus* may not have had all the cognitive skills of modern humans without language, but this would not have been necessary, if all the achievements of their culture could be accounted for with less.

Mimetic Skill

Mimetic skill or mimesis rests on the ability to produce conscious, self-initiated, representational acts that are intentional but not linguistic. These mimetic acts are defined primarily in terms of their representational function. Therefore, reflexive, instinctual, and routine locomotor acts are excluded from this definition, as are simple imitative acts and conditioned responses, whether classical or operant in origin. Also excluded, in the context of modern human society, are gestures that occur in a linguistic context or serve a linguistic function; these may sometimes have a mimetic origin, but their method of representation is too closely interleaved with language to be labeled as "purely" mimetic.

A distinction can be made between mimicry, imitation, and mimesis. Mimicry is literal, an attempt to render as exact a duplicate as possible. Thus, exact reproduction of a facial expression, or exact duplication of the sound of another bird by a parrot, would constitute mimicry. Many animals possess some capacity for mimicry, usually of conspecifics, although birds can mimic other species. Imitation is not so literal as mimicry; the offspring copying its parent's behavior imitates, but does not mimic, the parent's ways of doing things. Imitation is found especially in monkeys and apes. Mimesis adds a representational dimension to imitation. It usually incorporates both mimicry and imitation to a higher end, that of re-enacting and re-presenting an event or relationship. Some gestures are mimetic, at least in origin—for example, holding the heart or covering the face to indicate grief. The gesture may have originated in someone's imitating the actual grief reaction of someone else, but upon its being used for representation, we would have to call the same act mimetic. It may have been culturally diffused by simple imitation; but its original invention was a mimetic act.

Thus, mimesis is fundamentally different from imitation and mimicry in that it involves the *invention* of intentional representations. When there is an audience to interpret the action, mimesis also serves the purpose of social communication. However, mimesis may simply represent the event to oneself, for the purpose of rehearsing and refining a skill: the act itself may be analyzed, re-enacted and reanalyzed, that is, represented to oneself. Mimesis is not absolutely tied to external communication.

Mimesis can incorporate a wide variety of actions and modalities to its purpose. Tones of voice, facial expressions, eye movements, manual signs and gestures, postural attitudes, patterned whole-body movements of various sorts, and long sequences of these elements can express many aspects of the perceived world. In an ancient game like charades, which is still played by humans for fun despite their possessing speech, the innate mimetic capabilities of people, even young children, become evident. Most of the actions in such games are untrained and unrehearsed and are unlikely to have a simple imitative derivation. They are creative, novel, expressive acts.

Mimetic representation is still a central factor in human society. It is at the very center of the arts. In some cases, art form is purely mimetic: for example, pantomime or ritual dance. Certain forms of drama that have little or no dialogue and are based on visual tableaux, such as some medieval European plays, are largely mimetic. Early Chinese and Indian dance, Greek and Roman mime, and many other forms of mimetic representation in human civilization trace their origins back to prehistory. Archeological artifacts have verified that the civilizations of Australian and Tasmanian aborigines have remained unchanged for tens of thousands of years; and these groups have maintained autochthonous dance rituals that are still essentially mimetic: each dancer identifies with, and acts out the role of, a totemic animal. (These traditional dancers also understand parody: when Australian "corroborees" were first filmed by modern anthropologists, the performers added an unexpected bonus: a merciless satirical mime of the film crew.)

Most modern art forms, even those that depend heavily on oral or written language, are cognitive hybrids. Opera and theatre are good examples, where the prosodic aspects of acting and singing, the facial expressions and gestures, and the interrelationships between principals are mimetic, whereas the lyrics and script are linguistic in content. Cinema, which started out in imitation of theatre, has become overwhelmingly mimetic in style; very little of what a good film communicates is capturable in words, although viewers commonly spend hours afterward trying to represent it in a form compatible with language.

The great literary theorist Eric Auerbach, in his book entitled *Mimesis* (1953), used the word in the context of theatre, without making the distinction I am making between the *purely* mimetic and the verbal, since it may be argued, correctly, that words are also used mimetically, in the broader sense of formulating the shape of a myth or story and its attitude. In such cases, it is as though the mythic theme was, at a deeper level, driven not strictly by verbal rules and ideas but rather by an underlying mimetic form in which language is embedded.

The point I would extract from the work of Auerbach, and from that of a number of analytic thinkers in the field of literary theory, especially Northrop Frye (1957), is that literary forms are driven by what they call the underlying literary "imagination," a capacity to reconfigure the world in certain stylistic ways. The term used by Frye is "mythic modes," the forms into which literary productions may be cast; these modes include the ironic, the heroic, the satiric, and the high mimetic. They are literary *stances* that influence many aspects of a play or poem; they may determine the style, the kind of outcome, the ways in which characters and plots unfold, and so on. The purely linguistic act of instantiating a given literary idea under the guidance of a particular mythic mode may be more a technical, or "front-end," matter, but the driving forces behind the products of the literary imagination go much deeper, to the author's perceptions of large-scale literary themes, whose cognitive roots are ultimately mimetic. These issues will be expanded in later chapters, as part of a discussion of the later integration of mimetic into oral culture. For the moment I will deal with the prelinguistic origins of mimetic representation, looking only at those aspects of mimesis that do not involve language use.

Although it is logically prior to language, mimetic representation has characteristics that are considered essential to language and would thus have set the stage for the later emergence of speech. The important properties of individual mimetic acts include intentionality, generativity, communicativity, reference, autocueing, and the ability to model an unlimited number of objects.

Intentionality

Mime is intentional; its objective is the representation of an event. Intentional communication in modern humans is not restricted to language; it precedes language in ontogenesis. The first evidence of intentionality in children comes with pointing behavior, which was first studied by Vygotsky (1934, translated 1962), and more recently by others (Weiskrantz, 1988). Intentional pointing first emerges at about fourteen months, following a period during which children have learned to direct their gaze toward a point in space where their mother's gaze is fixed. To achieve this, the infant must have been able to calculate the vector, and find the target, of the mother's gaze. Children at this stage can also estimate the projected point of disappearance of an object. This is more than a spatial skill; the most important variable is the child's ability to attribute intention to the mother's gaze. Chimpanzees lack this central component of intentional gazing and pointing: the ability to realize the intentions of others. Understanding the intentions of others requires a break with cognitive egocentricity; similarly, the desire to communicate one's own intentions is a break with the egocentricity of the episodic mind. Archaic hominid mimetic skill must have started on this level, beginning with elementary intentional attribution, long before graduating to more exotic communicative media.

Generativity

Mimetic representation involves the ability to "parse" one's own motor actions into components and then recombine these components in various ways, to reproduce the essential features of an event. The perceptual component of this skill is present in apes but cannot be transformed into representational acts. In human children's play, however, the practice, rehearsal, and refinement of action takes on a generative property; the same elementary actions (lifting, smiling, hitting, falling) may be combined and recombined into sequences that represent events. The most common form this takes is the re-enactment of a situation or event. A child is spanked, then goes to its room and spanks its doll, or tries to spank the dog. In Piagetian terms (see, for instance, Piaget, 1980), the child is "assimilating" the event; but from our viewpoint, the remarkable thing is the child's ability to reproduce the essential mimetic features of the event. Human children routinely re-enact the events of the day and imitate the actions of their parents and siblings. They do this very often without any apparent reason other than to reflect on their representation of the event. This element is largely absent from the behavior of apes.

Communicativity

Although mimesis may not have originated as a means of communication, and might have originated in a different use of reproductive memory, such as toolmaking, mimetic acts by their nature are usually public and inherently possess the potential to communicate. A mimetic act can be interpreted by others who possess a sufficient capacity for event perception. Given the pre-established primate capacity for event perception, the presence of mimetic skill would inevitably lead to some form of social communication.

Reference

The act must be distinguished from its referent. A dog might perceive the "meaning" of a fight that was realistically play-acted by humans, but it could not reconstruct the message or distinguish the representation from its referent (a real fight). In contrast, in mimesis the representational act must be distinguished from its referent. Trained apes are able to make this distinction; young children make this distinction early—hence, their effortless distinction between play-acting an event and the event itself.

Unlimited Objects

Mimetic representations can model an unlimited number of individual perceptual events. There are serious limitations on the devices that mimetic culture can employ in representation, and therefore on the kind and complexity of the relationships it can express: mimesis is inherently a concrete, episode-bound medium of representation. But there are no apparent limits to the number of individual perceptual events that can be represented, or the physical modality of their representation, once the ability is in place.

Autocueing

Mimetic acts are reproducible on the basis of internal, self-generated cues. This permits voluntary recall of mimetic representations, without the aid of external cues—probably the earliest form of representational "thinking."

Social Consequences of Mimetic Representation

The presence of mimetic skill in the members of a group would immediately alter the array of available action patterns and collective cognitive skills available to them. Reciprocal mimetic interactions would ensue, leading to collectively invented and maintained customs, games, skills, and representations. Mimetic skill, added to a pre-existing episodic culture, would necessarily lead to cultural innovation and new forms of social control. In effect, the mimetic "customs" of a group would serve as the collective definition of the society. The forms of shared mimetic representations may be summarized as follows:

Modeling of Social Structure

Mimetic skill results in the sharing of knowledge, without every member of a group having to reinvent that knowledge. Although a precursor to this can be found in the accumulation of customs and skills evident in gorillas and chimpanzees, these species do not represent what they know. Mimetic skill, extended to the social realm, results in a collective conceptual "model" of society, expressed in communal ritual and play, as well as in social structure. Social roles, in a complex society, can only be defined with reference to an implicit model of the larger society. Mimetic representations would thus be tremendously important in building a stable social structure.

All higher mammals possess social knowledge; young chimps learn about dominance hierarchies in their play, for instance. But chimps only learn how to react to each individual in the larger group; human children model the *group* structure. A significant part of childhood is spent rehearsing and modeling society, and children can act out not only their own roles but those of other players. Human children can "model" an interchange between parents, for example, taking either role, or play-act a game with friends, taking various sides. This is clear evidence that they are implicitly modeling the larger social structure. Once again, this demands a break with an egocentric episodic view of the world.

Reciprocal Mimetic Games

Reciprocal mimesis is one of the sequelae of acquiring mimetic skill. If the members of a group of mimetically skilled individuals can re-produce or re-enact their creative behaviors, by definition they can reproduce similar acts generated by others. This leads to reciprocal games, where one player acts and another replies with the same, or another, act. Simple games with sticks and balls, common to all human societies, provide a good example. Someone invents a move; the next player imitates it and perhaps adds something new. And so on.

Ritual dance is perhaps the best example of a reciprocal mimetic game. People in every culture, especially very young children, frequently indulge in this type of play. Mimetic games are universal in human youth culture, often help to define roles, especially gender roles, and can be played in the absence of language. Nonsigning deaf children at play illustrate this point; they play essentially the same games as hearing children. With mimetic games, it is possible to "model" the behavior of adults, including social relationships, roles, and activities, without recourse to symbolic language.

Conformity and Coordination

Mimetic games, even in children, soon lead to conformity, and eventually can lead to regular, repetitive patterns of group behavior that resemble ritual. Children's games are highly ritualistic, involving rigidly enforced rules and an authoritarian structure: there are usually leaders and followers, "sides" and teams. Undoubtedly language has led humans to invent much more complex games than would be possible without it, but the spontaneous games of prelinguistic and deaf children nevertheless have the same properties. Language is not necessary for the development of complex social roles and rules, but mimesis is essential.

The final phases of the Acheulian culture, which persisted until the advent of *Homo sapiens,* had the earmarks of organized group activity. The campsites left in the later periods of *erectus'* tenure on Earth provide evidence of a society in which there must have been some specialization of function, especially between the sexes, and group coordination. The principal method of hunting larger animals, for example, consisted of driving the game, say mammoths or deer, into bogs or over cliffs, where they could be butchered—a hunting strategy that requires considerable coordination of activity, with various specialized roles. Hunting is widely believed to have been a male domain; some of the men—perhaps the majority, since they had to disperse—must have driven the animals from behind with noisemaking devices; others would have constructed blinds and hid behind them to channel the stampeding herds; and yet others would have been responsible for dispatching and butchering the animals so caught. Remnants of cutting and chopping tools, as well as the bones bearing the marks of chopping, have been found around such sites, dating back hundreds of thousands of years. Such an operation is not conceivable without leadership and implies a coordinated hierarchy dependent on the dominance, experience, and knowledge of one or a few individuals in the group.

It is likely that mimesis was the basis, if not the only formative element, behind this new cooperative, specialized social organization. Even if mimesis had initially served solely as the basis for toolmaking (an unlikely proposition, since mimetic skill would have been driven simultaneously by many different selection pressures), it would have enabled purposive signaling. Once hominids could self-generate a variety of representations, they possessed the essential cognitive support system for a larger, more complex society.

Group Mimetic Acts

When mimesis takes the form of a collective, or group, action, one common outcome is ritual. Ritual, and its derivatives in theatre, differs from most other forms of mimetic representation in that it is a collective act in which individuals play different roles. A well-documented, widespread example from human Paleolithic cultures is the acting out of conquest, often without the use of any words. A mimetic representation of the enemy is accompanied by chanting, drumming, and communal mimetic dance. The essence of the mimetic act in this case is not in the action of a single individual but in the orchestration of several actors. Such representations are coordinated social efforts, dependent upon the actors', and the audience's, sharing a global cognitive model of the society. This is generally true of ritual. Even in modern society, the power of mimetic forms in the control of large crowds is much greater than that of language. The carefully orchestrated action of a large crowd becomes a mimetic act in itself, representing, for example, the consensus, fury, or power of the group. Thus, the roots of public ritual are mimetic.

Innovation

A purely mimetic culture can evolve; the implicit model of the social world projected by the customs and rituals of a society may change. But it would tend to change slowly, precisely because mimetic group behavior, even in modern humans, tends toward rigid conformity and conservatism. In the complete absence of language, the complexity of the social ritual and structure that could be achieved would undoubtedly be quite limited. It is difficult to estimate the outer limits of mimetic culture, but in the absence of words and symbols, thought is pretty well limited to the act of event modeling and representation; no further analysis is possible. This point will be elaborated upon in the next chapter.

Pedagogy

In a mimetic culture, the acculturation of the young would be complex and would require time. Practical skills would be the basic pedagogical target: the use and manufacture of domestic tools, methods of hunting, construction of simple shelters, fire, weapons, and fighting. But rituals and games, and folkways and mores, would also require a systematic transmission of knowledge by means of mimesis.

Much of the education of children in simple societies is still mimetic in nature. The basic vehicles of such training are reciprocal mimetic games and the imitation and rehearsal of skills. Children mime adults in every respect, including mannerisms, posture, and gesture, they learn the customs and scenarios associated with each principal arena of action, and they acquire the manufacturing and survival skills essential to the tribal way of life. In addition, children learn a series of subtle limitations on impulsive behavior in a variety of contexts; this very basic type of learning is difficult to achieve in primates. Lovejoy hypothesized that the period of childrearing was already being extended with the australopithecines, but it surely would have become even more extended with *erectus*. A capacity for pedagogy in adults would be crucial in guaranteeing the child's acculturation into a mimetic society.

One necessary aspect of pedagogy is the ability of the teacher to understand the limits of the pupil's knowledge. This idea was central to Vygotsky's (1962) notion of the Zone of Proximal Development, or ZPD. Instruction is aimed at the ZPD; if it is aimed too far away from what the pupil already knows, learning will not take place. Premack (1987) listed the capacity for pedagogy as one of several distinctly human properties, since he was not able to observe any such ability in apes. Pedagogy requires not only some form of mimetic skill but the ability of the adult to sense what the child can, and cannot, learn—in other words, to judge the ZPD. This skill is also mimetic, inasmuch as it involves developing a "model" of the child's current knowledge. Evidently the same nonlinguistic skills that lead to the construction of a model of society, the attribution of social roles, and the development of ritual would allow the modeling of the pupil's ZPD in a mimetic society.

Modalities of Mimetic Expression

The primary form of mimetic expression was, and continues to be, visuomotor. The mimetic skills basic to childrearing, toolmaking, cooperative gathering and hunting, the sharing of food and other resources, finding, constructing, and sharing shelter, and expressing social hierarchies and custom would have involved visuomotor behavior. This includes virtually all forms of hand and limb movements, postural attitudes, and locomotor movements. Invention, rehearsal, and transmission of these mimetic skills would have been primarily under visual control, although visual control is correlated in complex ways with skin sensation, joint and muscle sensation, and balance, or vestibular sensation. Facial expression and other aspects of emotional expression, such as a variety of calls and cries, and strictly prosodic aspects of voice modulation would also have fitted well into a purely mimetic culture; this special case will be discussed separately.

Primate behavior in general is characterized by the integrated use of hand and limb movements, along with a combination of facial expression, sound, and gesticulation. Goodall's (1971) descriptions, which are the most detailed available accounts of primate behavior in a natural setting, suggest the degree to which this integration is typical. When Mike wanted to ascend the dominance hierarchy, he did not restrict his behavior to one modality or to the stereotyped displays usually reported. He employed manual gestures, whole-body loco-motor movements, threatening postures, sound (both his own hooting and panting and the sound made by his "instruments," the kerosene cans), and facial expression.

It is important to note that his motor acts were not neatly compartmentalized into separate systems for separate social purposes. That is, he did not restrict his gesticulating to the hands and arms, or his emotive expressions to the face and voice. His whole body was thrown into the performance, much the way a human actor will employ any available modality to achieve a given purpose. This kind of flexibility in motor behavior indicates that the programming of Mike's displays occurred at a supramodal level—that is, at a level in the nervous system that sits above all of the sensory and motor modalities and therefore can exert control over all lower-level inputs. In Fodor's (1983) terms, this is a property of an unencapsulated "central" system, rather than a property of specific modular input systems. Thus, the first mimetic expressions were not necessarily manual or facial; they could have involved posture, locomotion, or any of the available motor repertoire. We might conclude that mimetic skill could not have evolved solely or specifically for fine hand control, any more than it might have evolved solely for vocal control. It evolved for the purpose of re-enacting events and representing their structure. It must have been built upon the highest control centers of action, integrated at a supramodal level.

Nevertheless, it is still possible that mimetic skill was supported by additional special, modular adaptations at a lower level, for particularly important aspects of mimetic behavior. Refined eye–hand coordination, which started with primates and advanced with austra-lopithecines, would have been central to hominid culture. Given a supramodal mimetic adaptation, essential even to the simple tool-making skills attributed to *Homo habilis*, advanced manual dexterity would have become even more important. Selection pressure would thus have developed for a further specialized improvement in the neural apparatus controlling hand skills.

The example of toolmaking shows how these general and special skills might have combined to enable an important new mimetic skill. Improved toolmaking stands out as one of the primary early adaptive advantages of mimetic skill for *erectus*. Toolmaking is primarily a visual–manual skill, but it also involves obtaining the necessary materials, fashioning the appropriate tools at the right time, apportioning responsibility, and so on. Modeling this larger-scale scenario would require a more general, supramodular mimetic capacity. Innovative tool use could have occurred countless thousands of times without resulting in an established toolmaking industry, unless the individual who "invented" the tool could remember and re-enact or reproduce the operations involved and then communicate them to others.

The stone tools of *erectus* required expert fashioning; archeologists require months of training and practice to become good at creating Acheulian tools. They have to learn, and remember exactly how to strike a sharp edge and not break off the finished part with the next blow. The appropriate materials have to be remembered; to flake or chip a stone, two stones of relatively different hardness must be employed, the harder one as the shaping tool. The blows have to be modulated as a function of the type of stone; certain stones break in such a way that their edge is sharp and elongated, others flake in a different way. Such skill would not have been restricted to the use of stone: since there were so many other easier, but perishable and breakable, materials available, tools were almost certainly made first from materials like bones, teeth, shells, skins, and wood.

Toolmaking was probably the first instance of behavior that depended entirely upon the existence of self-cued mimetic skill. The reproduction could not be dependent on immediate environmental reinforcers or contingencies. Tool manufacturing is usually done at a time and place remote from those where the tool is finally used. By contrast, apes use as tools objects they find in the immediate temporal and spatial vicinity of the task. Tool manufacture, in other words, demands an ability to self-cue and reproduce or re-enact the scenario leading to the tool's manufacture in the absence of immediately present materials or even an immediate need for the tool.

Facial and Vocal Mimesis: Special Cases

Aside from advanced manual skill, there is another aspect of human mimetic behavior that appears to need a special, modular neural apparatus in addition to a general mimetic capability: facial and vocal mimesis, which combine in the expression of emotion. The combination of vocal and facial emotional expression might have played a paramount role in mimetic culture, as it still does in modern society. As Darwin suggested, it makes good sense to place the evolution of emotional expression early in human evolution, as part of an overall pattern of mimetic social adaptation. In comparison, it does not make a great deal of sense to place emotional expression in a later adaptation, along with symbolic language, which enters into a very different cognitive realm. The controlled use of emotional expression is an integral part of mimetic behavior, even in today's society. The face in particular is still one of the most widely used mimetic devices in the human cognitive repertoire.

Izard (1971) observed an increase in the importance of facial, as opposed to postural, emotional expression in the most highly evolved mammals. Facial expressions are absent in the lower vertebrates, for the very good reason that they lack the complex superficial facial muscles needed to express emotion. Monkeys, on the other hand, have a significant range of facial expressions. However, as Hinde and Rowell (1962) showed, monkeys usually combine a facial expression with a postural signal as well. In anthropoid apes facial expression is more independent from postural display than it is in monkeys (Van Hooff, 1967). In humans, facial expression emerges as a modality on its own, with a far greater range of modulation, particularly voluntary modulation, than in apes (Izard, 1971).

Facial expression, often combined with vocalization, is a medium of emotional communication and also of subtle, intimate expressions of feeling. There are many surviving vestiges of the output of this expressive subsystem, all uniquely human: laughter is one of the paramount examples, but humans also cry, growl in anger, gasp in aston-ishment, talk in "motherese" with infants, and so on. Young lovers giggle and make cooing sounds, groups express hatred and derision with public laughter and shouting, and individuals express disgust with exclamations and a typical screwing-up of the face, in all human cultures (Ekman et al., 1969).

These new vehicles for emotional expression would have been a useful visual device for social communication in the intimate family and tribal groups thought to characterize early hominid culture. The trend might have started with the australopithecines' move toward social stability and larger group structures, as Lovejoy proposed. The expansion of the range of facial expression, in itself, would not have represented any fundamental break with previous primate evolution, but the introduction of a voluntary component would have been a bigger change. Emotional expression is generally thought of as invol-untary, at least in infrahuman species (Izard, 1971).

Voluntary expressive use of the human face and voice seems to have advanced in two separate phases, the first coinciding with mimetic culture and the second arriving later, with language. Some of the ma-jor muscles underlying the flexibility, and voluntary control, of hu-man facial expression are clearly tied to the modern vocal apparatus, which appears to have evolved much later than *erectus*. As a result, the lower half of the modern human face is extraordinarily mobile, and our facial expressions are consequently much more variable, and under more conscious control, than those of apes. Thus, it is possible that some of the voluntary component of facial control was a by-product of acquiring the speech musculature and was delayed until the arrival of sapients.

Yet there is indirect anatomical evidence for an early increase in the range of facial–vocal expression over that of australopithecines, an increase that coincided with the advent of mimetic culture and *erec-tus*. Lieberman (1984), points out that the facial musculature of *er-ectus* appears to have been somewhat different from that of the aus-tralopithecines. The basicranial line, described in Chapter 4, was more flexed than in apes or australopithecines and would have had different insertion points for the muscles of the lower face. This difference would substantially alter the distribution and size of the musculature in this area, probably increasing the range of hominid facial expression and possibly indicating a corresponding vocal adaptation as well, since the two tend to go together. It is possible that some of the expressiveness of the lower face, and changes in the vocal tract, occurred as part of the overall move to mimetic culture.

Unlike the hands, limbs, and trunk, one's own face cannot be seen directly. Facial mimetic control is thus not directly dependent upon an associative linkage of facial motor control with visual feedback. Human children engage in reciprocal mimetic exchanges of smiling with their mothers, even though they can have no visual image of their own faces. Voluntary control of facial expression is thus more likely to depend on tactile and proprioceptive sensation (body sense) than on vision; and this justifiably raises some doubt about the extent to which emotional expressions are voluntary. Modern human actors have to depend upon extensive mirror practice to achieve good voluntary facial control. However, actors develop a very atypical degree of control; in most people, facial and vocal expressions appear to be more emotional and less voluntary than other mimetic acts. The degree of voluntary facial control evident in modern humans is still quite limited and usually confined to modifications of stereotyped response patterns.

Whereas voluntary modulation of facial expression would have followed an evolutionary trend already visible in primates, improved voluntary vocal expressive skill, by contrast, would have been a considerable break with the primate line. As we have seen, primates are remarkably lacking in vocal skill, considering their other attributes. However, just as there appear to have been two adaptations involved in facial expression, there is good reason to account for human vocal skill with two adaptations rather than one. Voluntary control over vocalization seems to involve at least two functionally independent areas of the human cortex. *Prosodic* control may be lost after damage to the right temporal region (Ross, 1981), while *phonetic* control may be lost after damage to the left. The first deficiency can occur independently of the second, and vice versa.

Prosodic control of the voice—that is, regulation of volume, pitch, tone of voice, and emphasis—is logically more fundamental than, and prior to, phonetic control; it is much closer to the capabilities of apes than phonology. It is close to what Darwin thought might have been the origin of the speech adaptation, a kind of rudimentary song. The limited vocalizations of apes are testimony to the great distance that primates had to travel to reach a human level of vocal competence. Primates produce a variety of vocalizations—a few dozen categories at most, according to Passingham (1982)—and these are all in emotional or manipulative contexts. Some chimpanzees, like Goodall's Mike, might well possess a rudimentary kind of voluntary vocal control, in their calculated, unemotional use of sham emotional signals.

According to stimulation-mapping studies of the macaque and squirrel monkey cortex, control of vocalization in primates seems to reside mostly in the limbic system and cingulate gyrus; hence the expression "limbic speech" (Ploog, 1981; Sutton and Jurgens, 1988). Humans still possess limbic speech; it has an emotional quality and is restricted to emotional utterances. It is often spared in nonfluent aphasia, testifying to its independence from the phonetic control of vocalization. (There is some question as to whether this is not really right-hemisphere speech, but this question does not affect the present discussion, since the important issue is the distinctiveness of emotive vocalization, not its precise localization.) The logical first step up from limbic speech to some form of voluntary voice modulation would have been improved prosodic modulation. The early emergence of prosodic modulation would have helped set the stage for high-speed phonetic control when language evolved. It is difficult to imagine the entire human vocal apparatus evolving in a single step.

Voluntary control of vocalization may thus have evolved as part of a general adaptation for bringing emotional responses under the control of the voluntary movement systems; for this specific purpose, prosodic modulation would have sufficed. Major modifications to the supralaryngeal vocal tract, so important in speech, would not have been necessary at this stage.

Following the same definition used for visual–motor mimesis, vocal mimesis would have allowed the self-cued regeneration of remembered utterances and their subsequent modification. It would necessarily have included a specialized vocal reproductive memory system, capable of rehearsal. Thus, response patterns that were originally involuntary would be stored for later voluntary reproduction; laughing, crying, gasping, and so on, with their appropriate facial expressions, would be integrated into the mimetic repertoire. Such a system could, in theory, be used in the same manner as visual–motor mimesis to model not the specifically visual but the emotional aspects of the human world.

In the society of *erectus*, the most obvious initial use for vocal mimesis would have been to supplement existing vocal emotional expressions, which presumably at that time still resembled those of primates. One obvious application would be to elaborate upon existing human cries and calls, using the prosodic modulation of features already in place: for example, a whining or begging sound could be sustained or made more urgent. Another use, which would have involved a more fundamental break with the past, could have been the *unemotional* use of cries and calls for instrumental reasons, mainly to manipulate the behavior of others. A primate who could act excited or submissive or dominant might enjoy some advantage from this. As pointed out above, there are reports of limited manipulative uses of emotional displays by modern apes.

Once mimetic culture was in place, vocal mimetic skill could have found important new uses, and a good case for continued pressure toward improved vocal modulation can be made. A more refined vocal imitative skill that could go beyond modifying existing calls would have required a radical redesign of the primate vocal apparatus, and this did not happen in *erectus*, if the anatomical evidence is to be believed. But even rudimentary imitations of animal and environmental sounds would have very good uses, particularly in a hunting–gathering culture that lacked speech: simulated bird calls and animal cries are still used in hunting. Some of these might have been possible without our elaborate vocal apparatus. Additionally, a flexible distal call system, however simple, would have been a valuable asset for achieving group coordination.

Vocalizations could also have supplemented mimetic re-enactment. The mimetic reproduction of the emotional "tone" of an event is still largely dependent upon the conscious use of face and vocal mimesis. Greek tragedy and European opera, as well as modern cinema, have used these dimensions to great effect. In addition, reproducing certain relevant environmental sounds would have helped to punctuate and elaborate the mimetic representation of events and situations. And vocal mimesis, once established, would have been useful in extending the range of intimate communication within tribal groups and families. Considering all of these potentially useful functions—supplementing mimetic representations, using sound instrumentally in social control, and increasing the power and range of emotional expression—the mimetic utility of nonlinguistic vocal modulation is evident.

In conclusion, expanded hominid control over emotional expression probably evolved in *erectus* (and possibly started even earlier than this, in australopithecines and habilines, although there is no strong anatomical indication in their case suggesting altered facial musculature), as part of the overall adaptation that established mimetic culture. This would make mimetic culture a very complex adaptation, from a neuroanatomical point of view. But adaptations driven at this high level of selection pressure are inevitably complex and may well involve a variety of specific subcomponents. For example, Isaac's and Lovejoy's proposals with regard to australopithecine culture involved a number of complex anatomical and social changes, all driven by the same high-level, iterative selection pressure.

Besides, neuroanatomical considerations should not dissuade us from facing up to the complexity of human cognitive evolution. Brain anatomy must have evolved as part of a much larger pattern of adaptation, and in accepting the theory of evolution in principle we have no choice but to accept that some extremely radical and complex changes took place during the emergence of hominids. A review by Armstrong (1981) on evolutionary influences on the hominid thalamus concluded that thalamic nuclei evolved in a "mosaic" pattern, that is, with different nuclei progressing at different rates. This would suggest they did not evolve in unison as part of a diffuse increase in brain size but differentially as part of a highly specific pattern of change. Since mosaic evolution sometimes proceeds at an accelerated pace, especially at times of speciation, this suggests that many different brain structures may change simultaneously, or at least concurrently, as the evolving population encounters complex selection pressures.

Plotkin's notion of evolutionary epistemology, or some similar notion, must apply here: given selection pressure at the highest-rate-of-change, or cultural, level (a likely level of pressure with the emergence of mimetic culture), many separate and specific anatomical subcomponents are likely to be recruited into an adaptation. Australopithecines changed radically from their predecessors' primate skeletal anatomy, but simultaneously they changed in terms of the female reproductive apparatus, diet, use of hands and feet, brain size, and emotional behavior; their adaptation was extremely complex. The cognitive adaptation leading to the emergence of *erectus* was no more complex, although its neural aspects, taken in isolation, were possibly more so.

Integration across Mimetic Modalities

Mimetic capacity, in modern humans, is an integral skill, coupling a variety of expressive subsystems. A single mimetic performance might include manual signals, postural attitudes, facial expressions, nonverbal vocalizations, and various forms of gesture, including metaphoric gesture. Mimetic ability undoubtedly evolved quickly into a system of standardized gestural signals, the implications of which will be discussed at length in Chapter 7. Suffice it to say, at this point, that any theoretical proposals about the neural systems controling mimetic expression must take this integrative capability into account.

One good illustration of the integrative capability of mimesis is found in the human capacity for tracking, and creating, rhythm. Rhythm is an integrative mimetic skill, related to both vocal and visuomotor mimesis. Rhythm is a uniquely human attribute; no other creature spontaneously tracks and imitates rhythms in the way humans do, without training. Rhythmic ability is supramodal; that is, once a rhythm is established, it may be played out with any motor modality, including the hands, feet, head, mouth, or the whole body. It is apparently self-reinforcing, in the way that perceptual exploration or motor play are self-reinforcing. Rhythm is, in a sense, the quintessential mimetic skill, requiring the coordination of disparate aspects and modalities of movement.

Rhythm is therefore evidence of a central mimetic controller that can track various movement modalities *simultaneously* and in parallel. It can interrelate the activities of these modalities in an incredibly subtle and rapid manner—for instance, by maintaining different rhythmic components for the hands, feet, and voice in the same performance, or by maintaining different rhythms for the left and right hands, or even by moving the whole body to one rhythm while beating out another in another modality. Rhythmic games are widespread among human children, and there are few, if any, human cultures that have not employed rhythm as an expressive device.

Rhythm, however, is different from other mimetic devices; it does not directly reproduce, represent, or signify any aspect of the natural world. It has to be regarded as a mimetic game, usually a reciprocal game, which is played for its own sake. Given the ubiquity of mimetic games among humans, from the facial–vocal games of mothers and children to the confrontations of whole societies in contest, it is not surprising that a supremely integrative perceptual–motor game like rhythm would emerge in hominid society, given a supramodal mimetic capacity.

Without a linguistic environment, it is likely that the uses of rhythm and vocalization would be largely supplementary to the major governing modes of mimetic representation, which are visually dominated. Vocalization does not usually correlate well in time with visual–motor experience. Often the auditory aspect of an event is inessential and secondary to its mimetic representation. This observation holds true for most of the self-generated actions that lie at the heart of visual mimetic skill; they have virtually no important or essential auditory aspect. They have a tactile, or body-sense aspect, and a proprioceptive aspect, but sound often has very little representational role to play, other than communicating emotion.

Despite a secondary role in the governing forms of mimesis, vocal mimetic skill in a social setting would provide an inevitable, and ultimately very important, offshoot: vocomotor games. Just as rhythm would have been a by-product of general mimetic skill, vocal games would be by-products of prosodic mimetic skill. Given a capacity to generate self-cueable vocal imitations, there would be a tendency to play with them, inventing and experimenting with new sounds. Vocomotor games would probably have led to new forms of emotional expression. It is possible that human laughter, especially group laughter, represents a vestige of this early vocomotor mimetic skill. It can be used in cruel and entirely voluntary ways to express group aggression. Combined with pointing, slapping, and other mimetic expressions of dislike, it can still be a very effective device in social control.

Possessing a range of mimetic modalities, *erectus* would have been capable of representing many aspects of the world, primarily toward the concrete, pragmatic ends of a fragile, survival-oriented society.

But the collective system of representations, or meanings, built up gradually in a purely mimetic culture would have been confined, by definition, to modeling event perceptions. It would have been a means of representing the outputs of the archaic episodic mind, embedded within the wider embrace of mimetic culture.

Mimetic representations are still prominent in human society. No better evidence of this been found than in the observations of Eibl-Eibesfeldt (1989). He has documented the commonalities of human nonverbal expression and has shown that many patterns of communication—tactile, visual, and vocal—recur in all human societies, from hunter–gatherer to modern. The mimetic level of representation underlies all modern cultures and forms the most basic medium of human communication.

Eibl-Eibesfeldt and his colleagues, using high-speed cinematography to reveal subtleties of expression and gesture that are impossible to observe casually, found that many patterns recur across cultures. For instance, in many cultures mothers discipline their infants by showing the whites of their eyes (without raising the eyebrows), and there are many other patterns of eye use that are universal. Many tactile expressions, found in reciprocal stroking, embracing, cuddling, and hand touching, are universal. Tongue displays are also universal: flicking the tongue can be friendly, aggressive, or flirtatious. Eyebrows are also a major human communication device; and as Eckman (1969) has shown, their pattern of use has some culturally universal components. Other examples include "opening" the face as affection, and "closing" the face as rejection; the mocking expression of distorting the features and curling the lips; pointing with the index finger as a threat; "display strutting" accompanied by chest pounding, in young males; and expressive patterns such as teasing, pouting, and head shaking.

Most human expressions have multiple uses and occur in complex social scenarios: for instance, two children might interact in play over a few minutes with a whole complex of nonverbal communicative exchanges, including aggression, rejection, reconciliation, and intimacy. The resemblances of such expressive scenarios, filmed in vastly different cultures, are uncanny. The resemblances extend from the expressions themselves to the actual sequences of their use.

Are all such expressive uses intentional and mimetic? Not necessarily; apes have somewhat similar expressive exchanges, yet remain limited to episodic culture. However, *human expressive exchanges can be made to serve a representational purpose.* Even young children can use their expressive repertoires in a characteristically human, intentional manner. They can manipulate, re-enact, mime, and play-act various scenarios at will. And as already pointed out, they are able to break out of an egocentric orientation and assume different roles in such scenarios. The elaborate games of human children grow out of such reciprocal mimetic interactions. Social roles are played out; rules are made; hierarchies are created.

Of course, in modern human culture mimetic exchanges usually occur within a larger semiotic framework that includes linguistic expression, but *words don't substantially change the nonverbal elements of the exchange.* Language seems to serve a different communicative purpose and carries on in parallel, without disturbing the fabric of spontaneous mimetic expression. Eibl-Eibesfeldt's evidence demonstrates how the mimetic layer of representation survives under the surface, in forms that remain universal, not necessarily because they are genetically programmed but because mimesis forms the core of an ancient root-culture that is distinctly human. No matter how evolved our oral–linguistic culture, and no matter how sophisticated the rich varieties of symbolic material surrounding us, mimetic scenarios still form the expressive heart of human social interchange.

Properties of the Mimetic Controller

The cognitive basis of mimetic action was the extended representation of self and the consequent improvement of conscious motor control. Sophisticated event perception was also an essential component, but it was already highly developed in apes, to the level of perceptual metaphor, as Premack has shown (1987). The major break with primate capabilities would have been in the way *the individual's own body,* and its movement in space, was represented in the brain. The essence of mimetic skill is thus to combine the power of primate event perception with an extended conscious map of the body and its patterns of action, in an objective event space; and that event space must be superordinate to the representation of both the self and the external world. Thus, there is a need for a new structure, which may be called the mimetic controller. The controller stands at the peak of a hierarchy. This is illustrated in Figure 6.1.

Self-representations can be evaluated and classified with reference

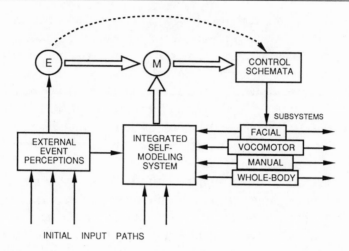

Figure 6.1 The mimetic controller (M) juxtaposed with its principal input paths from episodic memory (E), and self-representational systems. M controls and unifies the outputs of various motor subsystems involved in mimetic action; it also serves as a comparator-modeler for external events and actions of self.

to other self-representations, or to external events, in the mimetic controller. For example the event /begging/, which is widely understood by mammals in a variety of forms, could be metaphorically related to one's own self-representation of begging; this would enable "mimesis" of the event /begging/ and variations thereof. The first product of evolving mimetic skill would have been increased variability, novelty, and reproducibility of self-movement patterns. For this, *conscious* modeling of the self in action, and continuous rehearsal and refinement of movement, was a necessity.

Mime, play, games, toolmaking, skilled rehearsal, and reproductive memory are thus manifestations of the same superordinate mimetic controller. The range of acts that can be produced by such a system is limited only by the discriminatory power of event perception and the accuracy of the self-model constructed by the mimetic actor. Basically, anything that the event-recording episodic system can perceive and store, the mimetic controller can model, within its inherent limits.

The major direct inputs to the mimetic controller are (1) self-representations and (2) the outputs of episodic memory. The latter, containing a record of external event perceptions, was the governing memory system in primates. The mimetic system therefore superceded, and encompassed, the episodic system in the cognitive hierarchy. Episodic mind has no access to mimetic representations; but the mimetic controller has access to episodic representations. Where the outputs of episodic memory were perceptual and literal, the mimetic controller could incorporate those outputs in action metaphors. Event perceptions thus became event reproductions and re-enactments.

As hypothesized in this theory, mimetic representation was built upon, and superseded, its episodic foundation. It follows that, whereas episodic memory had been at the cognitive pinnacle in primates, it was later encapsulated by mimetic cognition. The mimetic controller thus became the unencapsulated central system, a position it must have held in the hominid mind until the evolution of language. And in the absence of language, the mimetic controller remains the dominant representational device.

Parsing and Modeling Event Sets

In principle, perceptual subcomponents can vary independently for any complex event. Thus, the commonly perceived mammalian event /begging/ encompasses many potential instances in which the subcomponents of the event will vary. Subsets of /begging/ might include: (1) different agents; (2) different actions: begging on the knees, with the eyes, crying and whining, and so on; (3) different contexts: begging in public, in groups; (4) different attitudes: defiantly, desperately, with threats. Begging serves to elicit a constant response pattern (giving or yielding) in the perceiver. But how can such varied patterns of behavior elicit common response tendencies in the first place?

The essential elements in recognizing /begging/ include not only aspects of posture, expression, and sound but also perceived dominance relationships and, most importantly, the perceived *intention* of the agent. Begging is a very powerful stimulus: it is found only in higher mammals and is easily understood despite its abstract nature. This testifies to the complexity of prehuman mammalian event perception, even before the evolution of mimetic skill.

Since event sets are processed at a high level of abstraction, there is no reason to postulate a simple perceptual prototype or "template" of an event set such as /begging/. There may be templates of various specific perceptual items that contribute to the perception of /begging/: a specific posture, a specific sound. However, these items are usually ambiguous in themselves. A whining sound, for instance, might indicate pain as well as begging and thus might also be part of an event set called /experiencing pain/. In the example made famous by Hebb (1949), a female monkey drove a hungry male mad with repeated begging for his food, until at last he threw the food against the cage wall. The male had to know the purpose of the female's actions—leading to his understanding that it was his food that she wanted—since identical behavior in a different context might have had other purposes. Thus, the perception of /begging/ involves a complex and abstract perceptual judgment and a degree of social attribution.

Event sets do not appear to be hierarchical, or pyramidal, in structure. There is no reason to assume a tree structure such as the one commonly proposed for symbolically encoded concepts. A better analogy than a pyramid might be a statistical "cloud" or cluster of data in multidimensional space. The recognition of events by animals is fuzzy, fluid, abstract, and entirely nondenotative; accordingly, even simple imitative behavior has this characteristic (Moerck, 1989). Mimesis, which goes far beyond simple imitation, nevertheless works on a metaphorical principle, a principle of perceptual resemblance: the differentiation of event classes is generally based on repeated exposures to the class of event, and episodic event prototypes incorporate more and more instances. The mimetic representations of others may be evaluated as events in their own right.

Control of the Mimetic System

The differentiation of *self*-representations follows the same course as the differentiation of external events, depending upon the resources available for creating a self-image. Parts of self are differentiated; classes of actions are differentiated; and the entire body is differentiated from the environment. This computation is unified; that is, the individual can envisage any part of the body in its surround and, at least in imagination if not in practice, alter and move the body image into a virtually unlimited variety of positions. The unique feature of self-representations in humans is that they are much more extensive than the self-representations of apes, and they are reproducible because they are incorporated into voluntary action schemas. The ultimate example of this skill might be in the human high-diver, whose extraordinary body control ultimately resides in an executable self-image.

The representation of event sets in terms of actions of self requires a parallelism between models of self and models of event sets and a common destination whereby they can be compared and evaluated. When a mimetic mind "interprets" an event in terms of action, this comparator process is at the heart of the interpretation. Self-conscious action is thus the basis of mimesis. Human motor control systems, which allow people to innovate and play with the modeling of episodic experience, do not have to be special in themselves, as motor control devices. They are distinguished by the direct awareness of the human actors of their output and by the supramodal modeling capability of the mimetic controller. In effect, the mimetic controller is the high-powered successor to the self-awareness system Oakley suggested in his models of the primate brain (Chapter 5). It is an *additional* level of representation, capable of integrating models of self and the external world and expressing these relations through the movement systems.

To reiterate: mime, play, games, skilled rehearsal, nonlinguistic gesticulation, toolmaking, other creative instrumental skills, many nonsymbolic expressive devices used in social control, and reproductive memory in general are all by-products of the mimetic system, as it continuously models the episodic world. In effect, this means that the mimetic mind models, in action, the outputs of the episodic mind. The mimetic system is thus a seminal hominid cognitive innovation, a mode of cognition that remains dissociable from language even in modern humans, and is the logical basis of the first truly human culture.

Neuroanatomical Considerations

When regarded as a complex adaptation, mimetic culture clearly could not be attributed to any single anatomical change in early hominids, any more than australopithecine culture could reasonably be attributed to any single anatomical change. The larger brain of *erectus* would have been consistent with a great expansion of reproductive memory and the invention of a new level of representation, with a concomitant increase in the general demand for memory storage. But under the surface of that larger brain, the pattern of expansion would have been far from diffuse and general: on the contrary, mimetic skill is made up of a complex of highly specific adaptations that would demand equally specialized brain adaptations.

These questions have not been explored as explicitly as they might, even in modern humans. Nevertheless, from existing published case material it can be said that in modern humans at least, the subsystems underlying mimetic skill seem to be distributed through several brain regions. For example, as reviewed in Chapter 3, emotive uses of vocal and facial mimesis, including prosody, appear to depend heavily upon right-hemisphere regions, while certain imitative (and possibly mimetic) visual–manual skills depend more on the left cerebral hemisphere. Rhythm might also be controlled more from the right than left; but whole-body mimesis, including postural and gestural signaling, might well be bilaterally controlled, since this skill appears to be highly robust after brain injury, even after hemiplegia. Mimetic uses of the face are similarly robust after brain injury, even in cases of facial hemiparalysis; bilateral, and possibly subcortical sources of control must be considered here as well.

Thus, some mimetic skill systems seem to be distributed primarily in the left hemisphere, some in the right, and some bilaterally and subcortically. The cerebellum, which expanded dramatically in size, was also undoubtedly an important structure in the evolution of human fine motor control, including mimetic action. In other words, if *erectus* is assumed to be the epitome of mimetic culture, the brain of *erectus* should have shown fairly diffuse expansion, when viewed in terms of gross anatomy, while differentiating tremendously in the detail of its motor organization. In principle, the subsystems of mimesis—emotional expression, manual coordination, self-representation, and so on—might be localized anywhere, so long as the outputs of these subsystems could gain access to the central mimetic controller, and so long as the outputs of episodic event-perception systems could have access as well.

The location of the mimetic controller is problematic. To my knowledge, there have been no reports of generalized "dysmimetic" syndromes in the neurological literature. Moreover, if such syndromes do exist, they might easily be labeled as generalized dementia, since according to our precepts, a purely dysmimetic person would lack the most elementary human representation system and even in the best of circumstances would be reduced to an essentially primate level of behavior. When there are failures of specific manual, vocal, or postural mimetic skills, as there are in some apraxias, whatever intact motor skills the patient is left with seem to suffice for a degree of mimetic expression. Presumably this is because, in the latter cases, the mimetic controller is intact. But if the mimetic controller were destroyed, there would be no superordinate event-modeling system to regulate the use of these subsystems, and the patient would revert to rigid, stereotyped behavior.

It is important to remember that the mimetic controller is conceived of as a supraordinate control center and is not directly motoric: it is a very abstract programmer of action. Its most logical location is the prefrontal cortex, or immediately adjacent regions of the dorsolateral frontal regions, whose functions have never been well understood. Its largest input must be from the event-representational systems of the brain, which are very likely to reside in specific parietal and temporal association areas. There are rich anatomical pathways from the latter to dorsolateral frontal and prefrontal regions; thus, a large bilateral frontal region might serve as the basis of the objective mimetic event space, with its largest input path coming from the parietal cortex. This circuit, especially on the right side, is sometimes invoked in theories of spatial attention in humans. But, as Hebb (1949) observed long ago, attention is nothing but the directional aspect of any organized behavior; there is no need for theories of a reified capacity called "attention." More probably, the frontal–parietal circuit is central to the objective representation of the active self in three-dimensional space. And since most formal psychological tasks activate this circuit in one way or another, it could be confused with an "attention center." But these questions must remain open until more direct tests of the present hypothesis can be made.

Some theories of hippocampal function have suggested that the hippocampus serves as a "cognitive map," that is, an objective spatial coordinate system onto which objects in the environment may be mapped (O'Keefe and Nadel, 1977) and which serves as a working memory field (Olton, 1977). These models were constructed on rodent and cat brains, and it is not clear how comparable simple spatial locomotor tasks are to human mimetic skill. But in humans the hippocampus is apparently part of a circuit related to episodic memory storage (Milner, 1966); in effect, event perceptions have to be processed through the hippocampus to be registered permanently. If the hippocampus plays a role in mimetic behavior, it is probably indirect, inasmuch as it plays a role in event perception and the episodic storage system.

Poizner, Klima, and Bellugi's (1987) theory that the self-image, or intrapersonal space, is represented on the left while the external world, or extrapersonal space, is represented on the right appears attractive on initial examination. But the mimetic skills of left-damaged people, other than their hand-signing skills, are far from destroyed, and this fact does not agree with a strong version of their hypothesis. The complex human self-image must occupy a major part of the brain and must, by its very nature, have a number of component subsystems; therefore it could be widely distributed and resilient after brain injury. Accordingly, the mimetic brain, prior to language, should have shown an expansion, relative to primates, of both right and left, frontal and posterior cortex. And there is no reason to think mimetic skill was strongly lateralized to the left or right, except possibly in the subsystem for hand control.

One final point: localization is not crucial to any argument of this type. It is not clear that the mimetic controller must be localized in any single anatomical structure, although it must have *functional* unity. There are conceptualizations of functional unity that could eschew easy neurological localization; for instance, in a hierarchical chain of command, mimetic control might devolve from one structure to another, depending on circumstance. Or it might be somewhat differently distributed in each individual; this possibility will be discussed in Chapter 9. Finally, mimetic representation in modern humans may be distributed differently from the way it was upon its initial evolution, whether in *erectus* or in an earlier species. In conclusion, available evidence suggests a wide, complex anatomical base for an adaptation like mimesis; and the concept of mosaic evolution would appear appropriate here.

Conclusion: Mimetic Culture as a Survival Strategy

On the principle of the conservation of previous gains, episodic culture would have been surrounded by, and largely preserved, within the larger context of mimetic culture. *Erectus,* and modern humans, retained the basic primate cognitive features that characterize episodic culture; there is no reason to believe that any higher cognitive skills were lost in the transition. For the most part, considering that the new adaptations were entirely within the constraints established in the primate line, the transition to mimetic culture involved adding to the cognitive architecture already in place. The elements of the case for an archaic hominid mimetic adaptation are summed up in Table 6.1.

Could mimetic skill not have arrived simultaneously with speech, as part of the same adaptation? For that matter, could it not be more recent than speech? The strongest reason for putting mimesis first is that it is inherently a more primitive form of representation. Mimesis is closely tied to the concrete, episodic memory of an event and does not involve the invention of an arbitrary set of symbols. It is much closer to the cognitive culture of apes than it is to spoken or written language. Even in modern humans, mimesis is usually an elaboration of, or a summary of, episodic experience. The mimetic representation of an emotional event such as triumph or anger revolves around postures, facial expressions, and gestures that epitomize the event. The representation of skills, whether in crafts or athletics, involves an episodic re-enactment. In modeling social roles, events are assembled in sequences that convey relationships. They resemble the events as they occur in the real world; in fact, they could be seen as an idealized template of those events.

Mimesis is thus a much more limited form of representation than symbolic language; it is slow-moving, ambiguous, and very restricted in its subject matter. But it blends in well with the level of event-perception capabilities of the primate. It is logically the next step up, one of the few available paths up the evolutionary scale for primates. Episodic event registration continues to serve as the raw material of higher cognition in mimetic culture, but rather than serving as the peak of the cognitive hierarchy, it performs a subsidiary role. The

Table 6.1   Elements of an archaic hominid mimetic adaptation, in which
language is completely absent. In a group context, the presence
of mimetic skills in individuals would combine and interact to
produce mimetic culture.

EPISODIC CULTURE (Primates)
+
MIMETIC SKILL

- intentional representations
- generative, recursive capacity for mime
- voluntary, public communicative system
- differentiation of reference
- unlimited modeling of episodic events
- voluntary, autocued rehearsal

+

SOCIAL CONSEQUENCES
- shared modeling of social customs and hierarchies
- reciprocal mimetic games
- enhanced conformity and coordination
- group mimetic acts
- slow-paced innovative capacity
- simple pedagogy and social attribution

=

MIMETIC CULTURE
- toolmaking, eventual fire use
- coordinated seasonal hunting
- rapid adaptation to climate, ecology
- intricate social structure
- primitive ritual (group mimetic acts)

highest level of processing in the mimetically skilled brain is no
longer the analysis and breakdown of perceptual events; it is the modeling of these events in self-initiated motor acts. The consequence, on
a larger scale, was a culture that could model its episodic predecessors.

Mimesis can be cleanly dissociated from the symbolic and semiotic
devices upon which modern culture depends. It serves different functions and is still far more efficient than language in diffusing certain
kinds of knowledge; for instance, it is still supreme in the realm of
modeling social roles, communicating emotions, and transmitting rudimentary skills. It is also dissociable in terms of its brain representation. Mimetic skill usually survives the disruption of symbolic lan-

guage. Shallice (1988) reports a case in which a patient with a devastating semantic disorder of language, unable to use words at all, was nevertheless usually able to mime his intention. In the clinic it is fairly common to use mimetic forms of expression to communicate with the brain-injured patient who has lost speech and writing. Brother John was able to employ mime to communicate with others during his paroxysms. In general, the destruction of general mimetic skill is extremely rare except in demented patients.

The latter fact serves to emphasize how fundamental mimetic representation remains in the operation of the human brain. When mimetic representation is destroyed, the patient is classified as demented, out of touch with human reality. The proper use of language is always a casualty of lost mimetic representation. But when language alone is lost, even completely lost, there is often considerable residual representational capacity left.

The logic of evolutionary biology gives us another compelling reason to place mimesis ahead of language in the evolutionary succession. Any change or variation needs an appropriate setting for its adaptive advantage to become evident, or the individuals exhibiting that variation will not be successful. In its first form a variation may not serve the same function it ultimately serves; feathers first achieved thermoregulation, not flight. Language is a very exotic adaptation indeed; there is no precedent for it in other species. And language is a social skill, dependent on the availability of a linguistic milieu for its development in the individual. Bruner (1986) has written convincingly of the need for an external support system, a linguistic milieu, for the acquisition of language. Such a milieu could not have existed when human speech began to evolve and probably would not have developed for many generations after the first individuals with a capacity for speech came into existence. Speech provided humans with a rapid, efficient means of constructing and transmitting verbal symbols; but what good would such an ability have done if there was not even the most rudimentary form of representation already in place? There had to be some sort of semantic foundation for a speech adaptation to have proven useful, and mimetic culture would have provided it.

In conclusion, mimetic skill represented a new level of cultural development, because it led to a variety of important new social structures, including a collectively held model of the society itself. It provided a new vehicle for social control and coordination, as well as the cognitive underpinnings of pedagogical skill and cultural innovation. In the brain of the individual, mimesis was partly the product of a new system of self-representation and mostly the product of a supramodular mimetic controller in which self-actions may be employed to "model" perceptual event representations. Many of the cognitive features usually identified exclusively with language were already present in mimesis: for instance, intentional communication, recursion, and differentiation of reference.

Mimetic culture had its pragmatic successes in toolmaking and socially coordinated activities like hunting, maintaining a seasonal home base, and using fire. But its greatest importance would have been in the collective modeling, and hence the structuring, of hominid society itself. Mimetic culture was a successful and stable adaptation, a survival strategy for hominids that endured for over a million years. It provided the basic social and semantic structures on which symbolic language was later to be added. The brain structures supporting mimetic action, unique to the human line, constituted the archaic human brain, the brain that would be further modified to incorporate linguistic skill into its armamentarium of systems and modules.

S econd Transition: From Mimetic to Mythic Culture

Language and the Rise of Human Culture

Human culture, in its most basic manifestations, is an integrated pattern of adaptation, a complete survival strategy. It forms the larger framework into which various cognitive components of that culture, including language, must be fitted. This idea is difficult to keep in perspective, given the dominant role language usually plays in models of human cognition. Language is usually placed at the top of the cognitive pyramid; but language evolved in, and continues to be employed in, a wider cultural context. To understand what adaptive value language had in its initial development, we must consider the overall pattern of aboriginal human culture. The problem is very much like the problem of understanding the ramifications of bipedalism or mimetic skill. Neither would make sense in isolation; they were part of a larger pattern of adaptation. The same is undoubtedly true of language.

In human culture, language dominates: yet language is not used equally in all areas of activity, nor is it the only means of communication and thought. It is possible that language is a "dedicated" system, that is, a specialized system devised for special applications, rather than a general-purpose device. The same applies for thought skills associated with language.

Mimetic skill has many of the properties of language. If we place mimetic skill in an intermediate position between episodic cognition and language, what, precisely, would language have added? How would the overall pattern of adaptation have changed? Like earlier cognitive adaptations, language would have been grafted onto an existing culture and an existing cognitive architecture. Presumably it did not replace what was already functioning well; it developed in response to new selection pressures. It was part of a complete cultural adaptation and part of a larger cognitive architecture. From the viewpoint of cognitive science, the essence of language lies in the ways it changed the fundamental structure of mind.

What were the first uses of language, and how did linguistically dependent cultures differ from earlier ones in terms of their overall survival strategy? A direct, experimentally based answer is not available, since we cannot get more than a reconstructive glimpse of archaic *Homo*. But existing archeological and anthropological evidence cannot be ignored simply because it is incomplete; some knowledge is better than none. If the problem of emergence is not addressed knowingly, it will be addressed inadvertently, since, as we have already argued, an evolutionary scenario is implicit in any model of human cognition.

The strategy adopted here must be, in part, reconstructive; early human cultures can be evaluated not only from archeological data but also from the extensive literature on surviving human hunter–gatherer societies. The salient features of a distinctively human cognitive adaptation can be seen most clearly in these cases, when there is minimal "noise" introduced by the accumulated effects of later cultural and technological change.

Transition Period

Mimetic skill is autonomous, isolable from language, and sufficient for the cultural achievements of *erectus*. As such, mimetic culture would have been the precursor to all modern human cultures. The reign of *erectus* lasted until about 300,000 years ago; in fact, virtually all hominid remains from most of that time appear to be from this species, suggesting that earlier hominids simply could not compete in the same niche. Between the reign of *erectus* and the appearance of modern humanity about 50,000 years ago, there was a long period of transition that is not well-enough documented, at least in archeology, to yield a clear line, or lines, of descent. However, the exact line of descent is not important here; what matters is whether there were two successive major cognitive adaptations in modern human speciation or only one.

The possibility of two cognitive adaptations is often raised because European archeology indicates two phases of late hominid evolution: a Neanderthal phase, from 150,000 to 35,000 years ago, and a modern phase, from about 45,000 years ago to the present. Neanderthals underwent a drastic, rapid extinction between 45,000 and 35,000 years ago, and concurrently were replaced by modern Cro-Magnons. This seems to imply two successive adaptations: first an archaic human adaptation about 250,000 to 150,000 years ago, followed by one about 50,000 years ago. But recent genetic studies suggest a common human ancestor for all modern humans, about 200,000 years ago. This suggests that truly modern humans might date back further than the archeological record indicates, and it opens up the possibility that a single fundamental human cognitive adaptation stretched out over several hundred thousand years. This would simplify the evolutionary scenario somewhat and remove the need to explain how modern humans could have emerged, with radically new cognitive abilities, so quickly.

The problem is, perhaps, our dependence on evidence from Europe. Modern-looking humans appeared very suddenly in Europe; skulls preceding them in Europe seem to belong either to Neanderthal or to predecessors of Neanderthal. And the late Neanderthal era is so well documented that it skews our perception of the entire transition period. But it now seems likely, even probable, that modern humans evolved before Neanderthalers dominated Europe. Neanderthalers may have been a conservative subspecies of sapient humans, evolving alongside us. It would be more parsimonious to propose that there was only one fundamental cognitive breakthrough in human speciation, and that language, including speech, had started to evolve much earlier than 50,000 years ago. In fact, there is evidence in favor of this view.

Dating from about 400,000 years before present, new varieties of hominid are found in Africa, Asia, and Europe that appear to be transitional types. They have some modern features, such as a larger, rounder cranium, and some archaic features, such as the brow ridges and prognathous facial profile found in *erectus*. One of the oldest such skulls is from Petralona, Greece, and dates to a period of 300–400,000 years ago. Although those dates are controversial, it is clearly from the early transitional period, and it contains a mosaic of features of both *erectus* and modern sapient humans. A number of European skulls from that transitional era appear, judging by their features, to have belonged to the predecessors of the Neanderthalers; brain size had already increased into the modern range, and Neanderthal brains were virtually equal in size to those of modern humans. The shape of their crania, however, and their facial features retained many characteristics of *erectus*, including an occipital thickening of the skull, a ridge along the top of the cranium, heavy brow ridges, and a prognathous profile. They were taller than *erectus*; and judging from their skeletal remains, their physiques were extremely robust and heavily muscled. As mentioned in Chapter 4, the base of the Neanderthal cranium lacked the angled flexure of modern humans.

Lieberman's measures of cranial endocasts, which are a clue to the shape of the archaic larynx, suggest that some of the oldest African and European skulls, notably those from Broken Hill, Zambia, and Steinheim, Germany, both about 125,000 years old, had modern basicranial flexures. These skulls, at least as old as the oldest Neanderthal skull, look modern in other respects as well. This implies there may have been two separate but parallel lines of human descent, one terminating in Neanderthalers, the other leading to modern humans.

Studies of mitochondrial DNA point to a common ancestor for all modern humans about 150–200,000 years ago, possibly in Africa. Thus, the emergence of modern humans may have been under way as long as 150–200,000 years ago. But the accelerated period of speciation preceding the proliferation of modern humans, like the one preceding *erectus*, evidently produced considerable anatomical variety in hominids; and the archeological record, with the exception of the Neanderthalers, is rather hard to read.

During the Middle Paleolithic period, 150,000 to 40,000 years ago, Mousterian tool culture appeared in the archeological record, in a pattern consistent from Central Asia and the Near East to France and Spain. Mousterian tool culture involved several new techniques and included smaller, finer, pointed tools and burins. The remains of Neanderthalers are coextensive with Mousterian tool sites. Neanderthalers also had clothing, body decoration, burial sites, and religious ritual. They appear culturally somewhat similar to the earliest modern humans of the Upper Paleolithic (Hadingham, 1979). They lived in caves, hunted large game, and constructed warm shelters from animal skins.

Exploration of Neanderthal caves at Shanidar, in Iraq, yielded a burial site in which the grave was decorated with several different kinds of spring flowers, as though the flowers held an emotional significance; this is not clear, however, since similar flowers have been found elsewhere in the cave. At Le Moustier, France, a boy was buried in a sleeping position with a beautifully crafted handaxe beside him, as well as the bones of wild cattle, as though meat had been included to sustain him in a journey. Evidence at this site suggests their ritual and symbolic importance. Other Neanderthal burial sites have revealed a different side to their grave rituals: at one site in France, a child's decapitated body was buried deep, with the head placed above the body, under a limestone slab. Some burials show signs of cannibalism on the remains; for instance, the skull had been opened as though to remove the brain, or a femur was broken as to remove the marrow.

A comparison of Neanderthalers with their predecessors leaves no doubt that Neanderthalers were advanced in cognitive skills. However, they were apparently less adaptable than modern humans, who co-existed alongside them for 5,000 to 7,000 years in some locations, competing for the same resources. Contemporaneous Neanderthal and Cro-Magnon campsites in Europe show that while Cro-Magnon culture was evolving at a steady rate, Neanderthalers were not changing. Within a relatively short time after modern humans appeared, Neanderthalers were extinct.

Remains of modern humans, classified as *Homo sapiens sapiens*, have been reported in Africa from about 125,000 years ago and are evident in European, Asian, and Australian archeological records from about 45,000 years ago. The earliest North American remains appear to be about 25,000 years old, but there are some claims of older finds; South America was apparently occupied around the same time.

Considering how slowly the face and cranial shape of humans had changed from early *erectus*, 1.5 million years earlier, to the Neanderthalers, modern humans come as a bit of a shock. Their skull shape is very different from other hominids, including Neanderthalers: the dome of the cranium is high, rising straight up from the orbits of the eyes, and the occiput is flat, producing a square-shaped, thin-skulled head. At the same time, the jaw is shortened, with a pointy chin, producing a flatter facial profile. The basicranial line is highly flexed, reflecting the changed shape of the jaw, nasopharynx, and upper vocal cavity. All existing humans share these characteristics, from Australian aboriginals and pygmies, to modern Europeans, Asians, and Africans.

All modern humans possess also speech and highly developed semiotic skill. The latter may be defined generally as the ability to invent and use signs to communicate thought. Speech may be thought of as a specialized subsystem of this ability, which allows more rapid, portable, and extensive semiotic use. It is not clear how rapidly speech and other semiotic devices might have come into use during the period from 200,000 to 50,000 years ago; but their continued employment throughout the modern human era, that is, during the past 50,000 years, seems uncontestable.

Even the most isolated modern human groups, some of whom were still living in the Stone Age when first "discovered" by Europeans, possessed highly elaborated systems of religion, myth, and kinship relations at the time of their discovery. For instance, the original Tasmanians, who were driven into extinction in 1876 and whose cultural artifacts had not changed for 35,000 years, had produced at least five distinct linguistic dialects, a rigid tribal structure with clearly defined territories, religious ritual, and various decorative arts, including elaborate body decoration, and rock engravings. The Tasaday of the Philippines, the pygmies of the African rain forest, and the Bushmen of southern Africa when first contacted by Europeans still had the same type of tool culture associated with the very earliest modern human remains. But they all possessed elaborate spoken languages and highly developed tribal structures, ritual, myth, and religion.

The cultural period associated with the arrival of modern humans in Europe is called the Upper Paleolithic. This period extended through several periods of differentiation and refinement until 10,000 years ago, the start of the Mesolithic era, when agricultural societies started to develop. The Upper Paleolithic cultures of humans were not initially very different from that of Neanderthalers, and their uniqueness was first evident in their rate of change. Whereas the Mousterian culture had not changed in tens of thousands of years, Upper Paleolithic humans kept innovating, so that every few thousand years human culture, judged by its artifacts, had moved up another notch. The rate of change varied in different geographical locations; in some isolated territories humans were still maintaining a late Upper Paleolithic lifestyle until recently.

It is difficult to know why Neanderthalers were replaced so rapidly by modern humans, but within 7,000 years of the appearance of modern humans, Neanderthalers disappeared from the archeological record. Some authors, notably Lieberman, have concluded that speech constituted the principal advantage of modern humans over Neanderthalers. This is possible; but, as he acknowledges, it is difficult to rule out completely the presence of some linguistic skill in Neanderthal society. It is more likely that the long period of transition during which the speciation of sapient humans occurred produced variants of *Homo sapiens*, including Neanderthalers, who shared a fundamental sapient cognitive adaptation that led to language, with varying degrees of proficiency.

The size of the Neanderthal brain is impressive, and the high state of development of Neanderthal culture places it close to the early culture of *Homo sapiens sapiens*. If the latter possessed a special cognitive advantage, high-speed speech is the most likely candidate. But it is possible that Neanderthals shared an earlier vocal adaptation, a protolanguage, that culminated in the high-speed speech of modern humans. Thus, Neanderthalers may not have been as proficient at speech as modern humans, but there is no reason to deny speech to them altogether.

The entire transition period, from 400,000 to 50,000 years before present, remains somewhat mysterious, but its end-result, the aboriginal human culture of the Upper Paleolithic, is not inaccessible or mysterious; there are numerous anthropological studies to draw upon. Humans by this time had taken on their modern biological appearance, and human culture had all of its characteristic earmarks and was just starting its rapid growth toward larger cognitive and social structures. The biological transition from *erectus* was complete;

and the complications introduced into the human cognitive profile by literacy and modern information technology had not yet muddied the waters.

Although the precise evolutionary events leading up to the modern human during the transition period are not yet known, the outcome is well-known. The Upper Paleolithic, and many still-extant Stone Age cultures, have been well-documented in a vast and detailed literature. For the present purpose, what we need to determine is the essential cognitive change, and biological adaptation, that might have been fundamental to preliterate human culture, and that would have differentiated it from the mimetic culture it superseded. A major part of that adaptation was undoubtedly speech, but speech did not appear in isolation. The first objective is thus to conceptualize the shift from the mimetic culture to the first well-documented modern human cultures. A closely related issue is the question of how to go about developing a cognitive classification of human culture.

Selection Pressures

The first appearance of linguistic skill, or its precursor, must have produced a tangible advantage in the struggle for survival and reproductive success. A common hypothesis about the selection pressure that produced *Homo sapiens sapiens* rests on the stresses introduced by the fourth glaciation. The onset of the last Ice Age (75–10,000 years ago) placed great stress on hominid populations and might have demanded a new adaptation. The last great glaciation covered huge areas of the northern and southern hemispheres but also opened up new migration routes by soaking up large amounts of sea water and exposing land. In Northern regions the water level dropped several hundred feet, exposing a vast land bridge of frozen tundra a thousand miles wide—and populated with various grazing animals—from Asia to the Americas. In the Southern hemisphere, Australia and Malaysia were virtually connected to Asia through a series of land bridges. Humans are thought to have migrated all over the globe during this period, presumably following the herds of grazing animals in search of food sources. Conditions of life were extremely difficult, in both north and south. Survival in this climate, under the arduous conditions of a nomadic existence that was dependent on the hunting of large game, would require better social coordination and planning. Enter language.

The difficulty with this idea is that some of the earlier migrations of *erectus* also took place under arduous conditions and evidently did not result in selection pressure for language. Stress can drive a new adaptation; but it can also extinguish a species. In this case there is no good reason to expect that hominids who became extinct, like Neanderthalers, were any less well-adapted to the cold climate per se than those who survived. In fact, Neanderthalers were much stronger and more robust than modern humans and seemed to thrive in cold European climates until their rapid disappearance about 35,000 years ago. Besides, modern humans are usually thought to have come out of Africa, where the fourth glaciation had a positive effect on the climate, at least in some regions. Finally, if modern humans emerged much earlier, as the genetic evidence suggests, the fourth glaciation could not have been the initial source of selection pressure that led to cognitive advance.

No convincing geographical or climactic conditions could have produced enough selection pressure to account for the emergence of modern humans. Hominid culture was already able to cope with a variety of climates. Although climate may have played some role, other forces must have been at work. The driving forces behind the new adaptation probably had more to do with *relative* success—that is, with competitive advantages over rival hominids who were battling for roughly the same, huge ecological niche. It is surely no coincidence that only one subspecies of the entire hominid line has survived; most other species of mammals have at least several co-existing subspecies, each occupying a special niche. But not humans. Apparently only one hominid can occupy the human niche for any substantial length of time.

Within such a species, able to adapt to virtually any climate ranging over the whole globe, major advances would probably be driven by this competition between subspecies. Moreover, competition would no longer be between the survival strategies of individuals so much as between the survival strategies of groups, and would test particularly their ability to act as a cohesive society. Thus, the evolution of humanity is likely to have been driven at the level of cultural change, and evolutionary pressure might have emerged when a cognitive innovation granted one group of hominids, as a group, a significant cultural advantage over another.

The competitive advantages of human culture over that of *erectus* are easy to illustrate. A direct conflict over food sources between mimetic and semiotically skilled cultures would not last very long. Humans are better and faster at everything: social coordination, tool manufacture, systematic war, finding and building shelter, gathering and hunting food. Archaic sapiens must have outstripped *erectus* rather quickly, because this enormously successful species disappeared long before traces of modern humans appeared in the archeological record, after having ruled the hominid world for over a million years. What supported this competitive advantage? Mimetic communication had already sufficed for a great deal of coordinated social effort: *erectus* had mastered the coordinated hunt several hundred thousand years earlier. Sapiens must have possessed a more fundamental advantage.

This returns the argument to its main focus: the underlying nature of the initial human cultural adaptation. Abandoning the idea that humans developed cognitive skills specifically in response to the fourth glaciation leaves the question of dates wide open. If intraspecies competition was the main source of pressure for cognitive–cultural change, the potential advantage of more powerful representations could have come into play at any time.

Mimetic versus Upper Paleolithic Culture

The shift from mimetic culture to the first truly human culture probably occupied most of the transition period from *erectus* to modern humans. The fundamental question is how the constraints on mimetic culture and its representations—which were imposed by the episodic referents of mimetic acts—came to be broken. Representations can be verified only with reference to their objects, and mimetic culture was highly concrete, tied as it was to episodic experience. Mimetic culture introduced a number of novel features, but because of its fixation on an episodic data base, it had a very slow rate of change when compared with our own culture. The present evidence suggests that *erectus* took half a million years or more to domesticate fire, and three-quarters of a million years to adapt to colder climates.

The contrasts between mimetic culture and the earliest human cultures are considerable. The general characteristics of Upper Paleolithic culture are fairly well known and appear equivalent to the cultures of the early Stone Age. They fashioned clothing, the extent and complexity of which depended upon the climate; they wove fabric and sewed garments and other items made of hide. They developed methods of transporting heavy objects, constructed various forms of shelter, and made a wide range of implements and weapons. They understood a great deal about the growth, selection, and preparation of food; they were remarkable navigators; and they probably used limited graphic devices for self-decoration. These same early human cultures had a rich social and religious life, marked by the use of dance, chants, masks, and costumes for various ritual performances. And above all they had capacious verbal memories, capable of long, highly formalized verbal exchanges. They had a political structure, although not as developed as their later Mesolithic cousins. They used various semiotic devices to indicate clan, status, and totemic identification.

All the basic properties of the human mind were present in early human cultures: thought, group decision making, and verbal problem solving; a shared lexicon or vocabulary; linguistic conventions, including phonological rules and rules of grammar; an elaborate verbal semantic memory; spatial and constructive skills. What sort of adaptation could possibly explain the explosion of tools, artifacts, and inventions of all sorts for all sorts of applications, and the eventual creation and maintenance of tribal political and social structures, which regulated everything from marriage to ownership, from justice to personal obligation? What change could have broken the constraints on mimetic culture with such a vengeance, leading to the fast-moving exchanges of information found in early human culture?

Speech and language are the obvious candidates to single out for these roles. They could solve the problem of explaining virtually all the new cognitive features of oral-semiotic culture, since these new features all rely on language to various degrees. Words allow the sharing of highly specific information, the rapid collection of new knowledge, and the regulation of all aspects of behavior. Indeed, one could argue that with language added to a pre-existing mimetic culture, many aspects of the previous culture could be elaborated and improved, from social complexity to manufacturing, while totally new cognitive features, like vast amounts of shared, rapidly changing knowledge, would become possible.

But what is language? What is speech? What kinds of structures would have to be added to a mimetic mind/culture to allow languages to develop and flourish? Words and grammars are mysterious achievements without precedent in evolution. We do not know what they are. Are they primarily memory-access devices? Do they map directly on to mimetic representations? Or are they new representational systems in themselves? It is an empty statement to say simply that language is the crucial adaptation that led to the first semiotic cultures.

The cognitive adaptations underlying language did not necessarily develop for the specific purpose of enabling what we call speech. There would have been no existing linguistic environment and, in Bruner's (1986) terms, no language acquisition support system at the time of its arrival to help individuals develop or learn language. It is important, therefore, to consider carefully the kind of advantage that speech, or any other linguistic skill, in its earliest form, might have conferred upon the first articulate hominids.

We have to assume that the descent of the larynx and the refinement of the supralaryngeal vocal apparatus in biologically modern humans signaled a very major cognitive change. It is difficult not to make this assumption, because our vocal communication system is such a radical break in the primate line. But the main importance of the vocal apparatus does not reside in how it might have enabled speech; in itself, it couldn't have. Like bipedalism in the australopithecines, it is probably only the tip of the iceberg, anatomical testimony to a more complex pattern of adaptation. The accompanying conceptual changes are more basic, and even these formed only part of a larger cultural change.

Early human culture was characterized by more than the possession of speech and other novel vocal skills. Simultaneously with the appearance of speech there appeared a whole constellation of thought skills that are associated with language and are, broadly speaking, linear, analytic, rule-governed, and segmented. Semiotic cultures also triggered completely new forms of information processing and storage: semantic memory, propositional memory, discourse comprehension, analytic thought, induction, and verification, among others. Determining whether these skills were present in early human culture, or were a product of later transitions, is important to our understanding of the emergence of modern mind.

Integrative Modeling and Myth

One way of examining how language evolved is to look at the earliest uses to which it was put. Language has become so central to the functioning of recent industrial societies that we too easily assume its equally wide use from the earliest times. Yet the evidence belies this idea; in surviving Stone Age cultures language has been used in some areas of endeavor much more than in others.

Stone Age cultures demonstrate how far language development initially outstripped technology. Technology in these societies is primitive, while language in social contexts soars to great heights. Language is the ultimate social arbitrator. It is used for watching the activities of others, keeping records of interpersonal relationships, regulating interactions, coordinating people, sharing practical knowledge of things like food sources and neighboring human tribes, and making collective plans and decisions. The use of language in tool technology, by contrast, is limited; most trades and skills are transmitted by apprenticeship, that is, by mimetic modeling. Many customs and mannerisms (including accents and linguistic gesticulation) are passed on mimetically. Skill with weaponry, cooking, and so on has a minimal verbal component.

The most elevated use of language in tribal societies is in the area of mythic invention—in the construction of conceptual "models" of the human universe. Even in the most primitive human societies, where technology has remained essentially unchanged for tens of thousands of years, there are always myths of creation and death and stories that serve to encapsulate tribally held ideas of origin and world structure. Stories about seminal events in history—attempts to construct a coherent image of the tribe and its relationship with the world—abound. These uses were not late developments, after language had proven itself in concrete practical applications; they were among the first.

Lee and De Vore (1976) have documented the culture of the !Kung Bushmen, whose technology, if uncovered by an archeologist and taken in isolation, would place them in the late Stone Age. The Bushmen apparently descended from one of the most ancient African populations, forced into the interior of the continent by pressure from both Caucasoid and Congoid competition. They lived, and continue to live, by hunting and gathering, and they use tools made mostly of wood, hide, gourds, stone, and bone. They also employ very effective, and deadly, poisoned arrows in hunting.

Their social structure, language, ritual, myth, and religious life are highly evolved. Their society lacks very little, from the viewpoint of a cultivated social life. Myth and religion permeate every activity, and these modes of thinking are not dissociated from daily life or compartmentalized, as they have become in our society. This may be seen in their attitude toward hunting the eland, classed in their mythology as the Master Animal, the first creation of their principal divinity. The eland must only be killed with a special poisoned spear or arrow. As the animal slowly dies from the poison over a period of hours, the hunter identifies himself with the victim and observes behavioral taboos that are thought to speed the action of the poison by virtue of mystic identification. In a ceremonial masquerade of identification with the eland, a dancer wears a headdress of eland's horns, to harness mystical energy, for which the mimed animal is seen as the vehicle.

The religious beliefs of the !Kung people are taken so seriously that, as Campbell (1988) pointed out, they are extremely reticent to talk of such matters, and if they do, it is only in hushed voices. Marshall (1962) reported that they were afraid even to utter their gods' names. As in most early religions, their god myths are closely tied to their idea of causality: gods cause pain and death, create life and the heavens, cause rain and thunder. The eland, identified with the moon through a creation myth, was also important in the ceremonial celebration of a young girl's first menstruation, since they knew of the correlation between the duration of the menstrual cycle and the lunar cycle.

Their mythical thought, in our terms, might be regarded as a unified, collectively held system of explanatory and regulatory metaphors. The mind has expanded its reach beyond the episodic perception of events, beyond the mimetic reconstruction of episodes, to a comprehensive modeling of the entire human universe. Causal explanation, prediction, control—myth constitutes an attempt at all three, and every aspect of life is permeated by myth.

Every hunter–gatherer society appears to have an elaborate mythological system that is similar in principle. Myth permeates and regulates daily life, channels perceptions, determines the significance of every object and event in life. Clothing, food, shelter, family—all receive their "meaning" from myth. As a result, myths are taken with deadly seriousness: a person who violates a tribal taboo may die of fear or stress within days, or be ostracized, or put to death.

Thus, although language was first and foremost a social device, its initial utility was not so much in enabling a new level of collective technology or social organization, which it eventually did, or in transmitting skill, or in achieving larger political organizations, which it eventually did. Initially, it was used to construct conceptual models of the human universe. Its function was evidently tied to the development of integrative thought—to the grand unifying synthesis of formerly disconnected, time-bound snippets of information. Where mimetic representation was limited to concrete episodes, metaphorical thought could compare across episodes, deriving general principles and extracting thematic content.

The myth is the prototypal, fundamental, integrative mind tool. It tries to integrate a variety of events in a temporal and causal framework. It is inherently a modeling device, whose *primary* level of representation is thematic. The pre-eminence of myth in early human society is testimony that humans were using language for a totally new kind of integrative thought. Therefore, the possibility must be entertained that the primary human adaptation was not language *qua* language but rather integrative, initially mythical, thought. Modern humans developed language in response to pressure to improve their conceptual apparatus, not vice versa.

When the Gestalt movement attacked structuralism in the early twentieth century, its most compelling point was that the psychologists of perception were looking at the wrong level. Structuralism believed that perceptions were constructed out of mental atoms, in a *pointilliste* strategy, like a Seurat painting. For the structuralists, a picture would have been assembled from points of light; thus the percept of a picture was constructed first by analyzing the individual details of the array and then synthesizing the overall pattern contained therein. The Gestaltists objected, pointing out that the primary perceptual impression was always holistic; for instance, upon entering a room, the *first* impression is its general shape, its light, and the major objects in it. Details are secondary and are perceived later. On a very brief exposure, the details, the points of light, are rarely perceived at all. In other words, the nervous system is designed to extract the Gestalt first; it gets around to the details later. The structuralists had it reversed.

What the Gestaltists said about perception might also be true of language. The rule structure of language, the unique phonetic properties of speech, and the apparently impossible complexity of linguistic constructs at the level of word and sentence might well be secondary phenomena. The primary objects of language and speech are thematic; their most salient achievements are discourse and symbolic thought. Words and sentences, lexicons and grammars, would have become necessary evils, tools that had to be invented to achieve this higher representational goal. In this view, language would have represented not an end in itself but an adaptation that met specific cognitive and cultural needs, that is, ultimately for the formalization and unification of thought and knowledge. It was not so much a communication system as an integral by-product of a new, much more powerful method of thinking.

Above all, language was a public, collective invention. Thus, the emergence of a new peripheral adaptation such as the modern vocal apparatus must have been contingent upon a corresponding change on the level of thought skills, a fundamental change that enabled, and then accelerated, linguistic invention.

Symbolic Invention

Every symbolic device humans collectively possess, whether it be a word, an ideogram, or a grammar, was originally an invention. The invention of words is sometimes regarded as the basic requirement, the *sine qua non* of language. We have hypothesized that archaic hominids did not invent true language, despite their capacity for mimetic representation. What modifications to mimetic skill might have allowed the invention of words? The idea of an archaic mimetic culture forces us to reconsider what language is. Mimesis is intentional, representational, communicative, generative, and it creates a semantic reference system. It assumes a degree of social attribution, some skill at pedagogy, and both social coordination and collective knowledge. Many of these attributes have, in the past, been attributed uniquely to language, and yet we have argued that mimetic culture, possessing all of them, could not have led to the invention of language. How can this be? If intentionality and semantic reference do not suffice to create the cognitive underpinnings of language, what is missing?

What the Language Adaptation Was Not

Apes have been taught to use arbitrary symbols; they use "words" as substitutes for their referents and possess a "vocabulary," in some cases, of 50 or 100 or entries. The specific forms of ASL signs taught to apes by the Gardners, or the visual symbols preferred by Premack and various other psychologists, prove that apes can learn a basic lexicon that possesses the arbitrary sign–signifier relation characteristic of true symbols.

However, apes have never shown signs of *inventing* symbols, and the depth of their understanding of symbols, other than pragmatically or instrumentally, is questionable. For example, Premack used a familiar plastic symbol for eliciting the judgment "same" in his apes. It was one of the most important symbols they learned, and served as the basis of further training. The symbol Premack chose to use for "same" judgments was the traditional equal sign, similar to that used in mathematics. Humans did not invent that particular sign until at least 40,000 years after they started to speak, and yet an ape could be taught to use it; how can this be? Why wasn't it, or its equivalent, invented earlier?

The confusion results from confounding the external form of the symbols themselves, and the behaviors associated with their use, with the type of intelligence informing their use. It may be true that an ape can be taught to use an equal sign in a conventional way, or at least in line with the inventor's intention. The equal sign was invented by Robert Recorde in 1557, one of the more useful conventions to emerge from British mathematics. The ape uses the sign to indicate sameness; so did the inventor. Recorde wrote that he invented the sign "to avoid the tediouse repetition of these words: is equalle to." The ape uses the symbol to indicate whether two things fall into the same or a different category: alive, red, edible, and so on. This seems very much in line with Recorde's intention.

But we are intuitively suspicious of the ape's competence in using this sign (for one thing, apes don't do math). A hint of the gap between the ape's use and the inventor's is given by reading a little further into the inventor's thoughts about his invention. He chose as his symbol a pair of parallel lines of equal length, "because no two thynges, can be moare equalle." So the symbol, in its invention, was not originally arbitrary; the two lines refer obliquely to a very abstract conceptualization of the symmetry of equivalence relationships. Moreover, the symbol, although useful, was an invention that took some thought and care; its simplicity concealed a great deal of knowledge.

For the ape (and for most of us) the equal sign's linkage with the concept of equivalence is arbitrary, since we did not know the inventor's intention. But in this case the arbitrariness of the symbol signifies only the inferiority of our grasp of the symbol's meaning; and it allows us no advantage in the process of representation. The point is, the invention of new symbols is not a trivial matter; even in literate Elizabethans, a new idea, or at least its encapsulation in symbols, was not to be taken lightly.

What is missing from the ape's use of the equal sign? Nothing, on one level. In the strict, curiously behavioristic, Saussurian definition of a symbol, apes can obviously be taught to use symbols to signify things. But this ability did not get them a language; it did not even get them a mimetic culture. The ability simply to link an arbitrary sign with its signifier evidently does not lead to language of any sort. In fact, the linking of signifiers to signs is not evidence of much more than the presence of conditioned discriminations, or of elementary event perceptions. The linkage of sign to signifier is a job for episodic cognition, and in the realm of episodic cognition, apes are masters.

The invention of symbols seems to have required a great deal more than a capacity for linking signs and signifiers. The equal sign might seem to be an esoteric case; but even the first word inventions of young children, when they are in the earliest stages of acquiring language, have the quality of an intellectual adventure. Capturing a chunk of episodic experience, or a concept, with a word requires experimentation. The use of the word reflects a process of sorting out the world into categories, of differentiating the things that may be named. The term "definition" is a particularly elegant invention in this regard: symbols "define" the world (rather than vice versa). Previously fuzzy properties become sharper after symbolization.

The invention of a symbol requires a capacity for thought; a symbol is an example of what Gregory (1981) has called a "mind-tool." Symbols are invented to facilitate a cognitive operation or purpose; and the purpose, and its solution in symbols, must somehow occur to the inventor. The ape lacks this competence and cannot invent a mind-tool. The equal sign does not represent "equivalence" to an ape; it is a pragmatic device, a given of the environment. Seven-year-old human children learning mathematics seem able to use the same symbol in a somewhat wiser way, and many adult humans can comprehend the symbol perfectly. But prior to its invention, few humans could have arrived at a level of understanding of equivalence relations that would allow them to perceive the value of inventing the equal sign, as Recorde did. In other words, the possession of symbols alone, defined in the traditional way, would change nothing. It is the representational intelligence underlying the symbol that defines its power and leads to its invention. It is thus the nascent mental model that cries out for the perfect symbol, the appropriate device, to express its as-yet-uncaptured concept.

Symbolic invention is a creative act, the elemental component of human model building. Symbols require rules for their use, and these rules are also inventions. The rules for calculus, and new computational algorithms, are obvious recent rule inventions. Many terms and conventions in modern English are of recent origin; the English language itself is a very recent innovation. By extension, the very first languages were also inventions, in all aspects. This suggests it would be wrong to attribute the power of the initial human speech adaptation to the acquisition of arbitrary symbols, the possession of a lexicon per se, or the evolution of an associative device for linking symbol to referent, such as the "second-signaling system" of the neo-Pavlovians (Vygotsky, 1962). This is putting the cart before the horse. Symbols could not have come first and triggered language and thought by their invention. The invention of symbols, including words, must have *followed* an advance in thought skills, and was an integral part of the evolution of model building.

Similarly, it is even more unlikely that the emergence of a high-speech vocalization apparatus in the brain and vocal tract of an ape, or even of *erectus*, in itself would have led to the invention of symbols or to speech. It would have led only to imitation and further pragmatic use of imitative vocalization. This is not to devalue anatomical evidence; major changes in anatomy probably, in fact inevitably, reflect functional change. But vocal skill would not have driven cognitive change; it would have followed, or at least paralleled, a fundamental change in the modeling intellect that made vocal skill useful. The most important source of selection pressure for an improved vocal apparatus would have been a mind that needed the features of vocal language for its modeling agenda.

Archaic Semiotic Invention: Gesture

How then did hominids evolve from mimesis to language? The first step could not have been simply to acquire phonological skill and construct an oral–verbal lexicon, grafted onto a mimetic brain. Nor could it have resided in the sudden acquisition of intentionality, reference, or any of the other features already present in mimetic culture. The most likely initial source of arbitrary symbols in mimetic culture would have been in the standardization of mimetic performance— that is, in gesture. Some degree of semiotic invention was inevitable in a mimetic culture, in the form of gesture.

As shown in the last chapter, some use of gesture was well within the capabilities of mimetic culture. Would the elaboration and standardization of gesture have served as the precursor of the more advanced semiotic inventions underlying speech? Or is the modern human use of gesture part of the same biological adaptation as speech? The best evidence on this question comes from recent experimental and observational data obtained from modern humans.

David McNeill (1985) has argued that gesture is closely tied to speech in modern humans; in fact, certain kinds of gesture are found only during speaking. McNeill argues that gesture and speech share a computational stage. But this does not imply they evolved together. "Sharing a computational stage" does not even necessarily mean that gesture is generated on the same cognitive level as speech, or that it could not exist entirely on its own. This is a critical issue; it is important to establish the identity, or independence, of these two levels of symbolic expression.

McNeill realized the independence of certain gestural elements from speech; these are what he calls "emblems," which are stereo-typed signs that can be interpreted outside of language. He gave the example of gestures for "okay" (thumbs up) or "psychologically pe-culiar" (index finger circling the temple), which can be used entirely on their own. There are many such emblematic gestures, and some are still in use. Some cultures use emblematic gesticulation more than others, but all cultures have some system of expression involving fa-cial and postural attitudes to express approval and disapproval, anger and sadness, puzzlement, derision, and so on. Many of these gestures are stereotyped and expanded variations of what were once sponta-neous emotional reactions and are well within the capacity of mimetic culture.

Emblematic gestures can be used for a variety of purposes and may be combined with simple semiotic devices in other sensory modalities. Andeans use hand-to-mouth whistling to signal across from one mountain to another; the Plains Indians of North America used a hand-signaling system for silent communication while hunting; other systems include smoke signals and drums for long-distance communication. Such forms of expression are probably elaborations of the most ancient human uses of symbols. Gestures may have started out as variants on emotional expressions: these would appear to be the most rudimentary symbols, that is, the symbols closest to a mimetic level of representation, one step removed from the mimetic scenarios documented by Eibl-Eibesfeldt (1989). Gestures of this sort include shrugging the shoulders, pursing the lips, and shaking the head to indicate puzzlement; waving and smiling to greet someone; saluting or salaaming to indicate submission; shaking hands, or re-fusing to shake hands, to indicate attitude. Putting the fingers to the lips and producing a "shh" sound indicates silence both in modern Western culture and in the Kalahari Bushmen. Pointing, combined with various facial and postural attitudes, can indicate accusation, in-terest, acceptance, or a command to depart.

Emblematic gestures very in detail from one culture to the next, and this demonstrates their arbitrariness; the same representational functions are served by a variety of gestures, even at this rudimentary level. Such gestures are not only useful in simple exchanges but as supplements to larger mimetic representations. In group ritual, em-blematic gesture is often combined with re-enactive mime, group chanting, and dance; but it does not represent much of a break from mimetic cognition. This just reiterates what was said earlier: the invention of symbols, per se, would not necessarily revolutionize thought.

McNeill pointed to a different class of gesture, closely tied to human speech and supposedly not intelligible without a linguistic context: linguistic gesticulation. The use of this type of gesture is sometimes lost in aphasia, whereas the use of mime and emblematic gesture are seldom lost in aphasia. Linguistic gesticulation is complex, and includes three subtypes. The first type is called iconic gesturing, of which exemplars are labeled "iconix." Iconix are images, traced in air, which relate to the meaning of an utterance. For instance, while saying "he picked up the box," the speaker might hold the hands apart as if holding a box. McNeill and Levy (1982) studied this type of gesturing in six narrators, producing a complex of gestural types that had 44 movement features and 38 meaning features. The correlation of meaning with movement pattern revealed that there were distinctive gesture profiles for each meaning. Thus, the gestures were not completely arbitrary and had a predictable relation to the meaning of the utterance.

A second kind of gesture is metaphoric gesturing, which McNeill called metaphorix. These are also manually traced images but have a more abstract relation to what is said. Mathematical metaphorix are common in technical discussions; for instance, McNeill gave the example of how the concept of limits was expressed by two of his subjects. The utterance "this gives a direct limit" was accompanied by the following metaphorical gesture: left index finger slides along right, and comes to stop just beyond the tip. "Conduit" metaphorix are even more abstract; the speaker creates the image of a container, such as a cup, while discussing an abstract concept. Montaigne's account of Zeno's division of the faculties of the soul, 2,300 years ago, pictured the following conduit gesture: "The hand, spread and open, was appearance; the hand half shut and the fingers a little hooked, consent; the closed fist, comprehension . . . , his fist still tighter, knowledge."

Metaphoric gestures seem to abound in aboriginal cultures. McNeill examined more than 100 gestures recorded in an anthropological film about the Turkana and found numerous metaphorical gestures. For example, the speaker (in what appears to be a liberal translation) says, "These Europeans want to extract all our knowledge—pft!" while producing a short sequence of accompanying gestures: hand moves to brow and appears to remove something, fingers create small object rising upward, object seems to fly away on its own.

A third category of linguistic gesture is the "beat." Beats (sometimes called batons) are off-propositional gestures that have no content of their own. They are small, simple movements that do not evoke imagery but indicate an extranarrative comment. A beat might show that the material it accompanies was not part of the main narrative. For example, the utterance, "The runner—he's the one who you saw yesterday—continued in that direction," could be accompanied by a simple batonic gesture, such as a hand movement, marking the phrase "he's the one you saw yesterday" as a subpropositional, or extranarrative, element.

In contrast with emblems, iconix, metaphorix, and beats are closely tied to speech. They are closely tied in real-time to the utterance, never crossing clause boundaries. They parallel speech in development (Bates, Bretherton, Shore and McNew, 1983); in fact, during development, gesture and word use are driven by the same scripts. Iconix appear at about two-and-a-half years of age and take the form of fairly literal enactments. As the child's understanding of the world progresses to more and more abstract levels, gestures and speech reflect this together. Standard metaphoric gestures appear at 5–6 years, and more original metaphorix appear about age 9 (Stephens and Tuite, 1983), closely tied to conceptual development.

It is difficult to accept that metaphoric or iconic gestures are always bound to a linguistic context; in some situations gestures may actually override a linguistic message. Contradiction between spoken and gestural meaning has been a standard dramatic device in comedy for as long as there are written records of comedies, and is a device used as well in aboriginal societies. Aristophanes' parody of absent-minded intellectuals in his play *The Clouds* (423 B.C.) was built on such a device: Socrates is shown wandering, open-mouthed, contemplating the universe, just as a bird lands its droppings in his mouth. Socrates' words convey one thing, while his actions provide the contradiction. Parallel meanings: one linguistic, the other mimetic.

In the case of emblematic gesture, the parallel and independent course of gestural meaning is a common phenomenon. An example might be, "Global free trade obviously works to the advantage of the Third World," accompanied by the standard emblematic gesture for insanity (index finger circling temple). There are two parallel meanings in counterpoint, the one contradicting the other. Children learn to use and appreciate parallel mimetic-verbal puns and jokes very early; a simple pun might consist of making an angry face while speaking in a friendly voice. A more complex metaphoric example, which I have seen acted out many times by young boys, would be the following: the utterance, "Here's the world's greatest X (cyclist, tree climber, football player, whatever)," accompanied by a mime of incompetence at X's skill. The same skill, using parallel mimetic–linguistic channels, can be used for manipulative purposes, such as social ridicule.

The loss of gesture in aphasia is not a simple pattern of deficit but reinforces the independence of the mimetic and linguistic systems. Broca's aphasics—nonfluent but able to comprehend—continue to produce numerous and elaborate iconic gestures but, understandably in the absence of speech, few or no beats (Cicone et al., 1979). Pedelty (1985) claims that Broca's aphasics can use iconix only in a referential manner, losing the ability to mark relationships between items. Wernicke's aphasics—fluent but uncomprehending—seem to lose iconix and metaphorix altogether, while retaining beats. However, Pedelty has observed correct gestures accompanied by paraphasic errors in word choice or word construction. This suggests that gesture may sometimes be independent of the process of word finding. Pedelty has never observed the opposite—an incorrect gesture with correct speech—which reinforces his idea that mimetic representation is the more fundamental level of representation.

Considering the whole literature on linguistic gesturing, it is difficult to disagree with McNeill's conclusion that gesture and speech share a computational stage, or with Bates' conclusion that gesture and speech develop together and reflect the progress of developing thought. But nothing in McNeill's argument precludes the possibility that the gestural ability which normally accompanies speech might have preceded speech in evolution. Indeed, Butterworth and Hadar (1989) have asserted the global autonomy of gesture. Given the hierarchical organization of skill and knowledge, it would be expected that language would often dominate lower levels of skill. This might produce a real-time correlation between speech and gesture, but it does not imply an identity of mechanism.

Moreover, the data on aphasic gesticulation supports the distinctness of gesture and speech. In the aphasic we can see the underlying modular structure exposed. Speech production, including internal speech, can be completely destroyed without affecting the ability to produce and invent iconic and metaphoric gestures. This suggests that iconix and metaphorix are not absolutely tied to speech, any more than emblematic gestures are. Beats are more closely tied to speech, but this is hardly surprising, given their definition.

In conclusion, the mimetic brain can generate a rich variety of gestures. It does not follow, however, that all of the forms of gesture available to modern humans, usually used as a supplement to speech, were available to earlier human cultures. McNeill noted a lack of certain conduit gestures in the Turkana, and it may well be that the modern human repertoire of gestures is the result of millennia of gestural, and conceptual, accumulation. It does follow, however, that fairly elaborate systems of gestural symbolism might have preceded the speech adaptation, especially during the transition period.

Symbols and Mental Models

Episodic minds (as in apes) can use symbols when provided with them, and mimetic minds employ symbolic mimetic displays; each uses symbols in its own way. Modern humans, similarly, use symbols in our own way. The value of a symbol depends on the kind of mind putting it to use. Episodic minds create episodic models of the world; mimetic minds create mimetic models. Signs and symbols, given to such minds, possess no magical powers to change this. By extension, modern minds create the kinds of symbols that they do because their thought processes are different. The symbol-driven cultures of humans did not advance because they were suddenly the beneficiaries of symbols that unleashed hitherto unheard-of cognitive powers. On the contrary, humans must have *invented* their symbols because they needed them for the types of mental models they were creating.

The mental models of apes derive from event perception. Higher mammals could develop very complex models based on this data base—models of social situations and relationships, for example, that would involve the integration of numerous visual "scenes" into scenarios. Mimetic representation, as conceptualized in the last chapter, is also directly dependent on a high-level modeling process. But it must be able, in addition, to map models of event sets onto models of self-action—an achievement impossible for the episodic minds of apes.

Spontaneous mimetic representation must derive from a mental model; in an original mimetic representation there are no notations, no grammars, to help in the act of representing. Mimes must generate the representation directly from a perceptual model. They cannot proceed any other way; there are no other data to go on. For example, a fight might be mimed by retrieving an episodic memory of a fight and re-enacting it fairly literally; or by means of a more abstract, prototypal mime. Either way, the mimetic act is guided by perceptual metaphor. It is also understood by the perceiver by means of perceptual metaphor.

Iconic and metaphoric gestures also work directly from perceptual metaphors; they are perhaps not so literal as most mime, but their meaning still resides in a direct use of episodic representations. In McNeill's example of a Turkana speaker using a conduit metaphor to express how Europeans were "extracting" all his knowledge, the gesture (something flying out of the head) stands on its own, even though it was, in this instance, closely tied to its context in the sentence. In McNeill's theory the gesture acquires its specific meaning from its linguistic context; but the inherent meaning of the gesture is entirely mimetic: the gesture /something flying out of the head/ could be employed mimetically in any context where such a meaning was appropriate.

True arbitrary symbols, on the surface, do not appear to model in the same way; they seem only to signify, not to model. Words, or hand signs, or even written ideograms do not seem to be models of anything but only the objective carriers of information, arbitrary access devices for addressing certain memories. The model appears to exist elsewhere, at another level of representation. The symbol can thus be detached from its meaning, the lexicon and syntax detached from semantics, and the problem of reference ignored, or left to a "lower" level of intellect.

Carrying this curious but common logic to its conclusion, the invention of a simple language should require only that a lexicon be built, to connect the "knowledge" of the world, on level 1, to appro-

priate symbols on level 2. But this strategy runs into serious difficulty. Aphasics equipped with rules and symbols do not learn to use
them competently. And the "rules" apparently needed to connect human language to the real world turn out to be extraordinarily convoluted and awkward; see, for example, Jackendoff's (1983) heroic attempt to describe the rules regulating extralinguistic reference.

Perhaps for this reason, most linguistic and psychological theories
of meaning have tried to define semantics in isolation from the problem of reference. Three theoretical schools of thought have tried to do
this: semantic networks theory, meaning postulate theory, and lexical
decomposition theory. Johnson-Laird has effectively scuppered all of
them in his book *Mental Models* (1983).

It seems intuitively odd that psycholinguists would even have considered the possibility of constructing theories of meaning without
addressing the problem of reference, but the reason becomes clear
when the history of the field is considered. Lexicographers live in a
world of words; the "definition" of a word thus consists of other
words. Grammarians live in a world of words; grammars are verbal
descriptions of rules that are implicit in word use. In this context,
meanings are easily perceived as just another part of an entirely symbolic, self-contained system. Within the self-contained world of language, to find the meaning of a word in the mind one merely looks
up its definition, or searches through a mental filing system (network); or reduces it to a set of elementary symbols or semantic markers. In all three cases, there is no need to leave (that is, refer outside
of) the symbolic system.

The three approaches mentioned have been elaborated at various
times during the past twenty years. Lexical decomposition theory,
proposed by Katz and Fodor in 1965, held that when an utterance is
received as an input, it is decomposed into various universal and innate concepts. Thus, the statement "A man lifts a child," when decomposed, might be represented thus:

Cause [activity (human, adult, male) upwards (move) (human, not
adult)]

The more specific a statement, the more decomposition is required to
reduce it to universals; therefore it should take more time to comprehend some utterances than others. Experimental tests of lexical

decomposition theory have tested this by measuring the average processing time required by different types of utterances. The experiments yielded negative results; Fodor and his colleagues (1975, 1980)
did not find any evidence of increased processing time with increases
in the semantic complexity of an utterance. The theory has since
drifted out of favor.

Semantic network theory had its origin in the work of Collins and
Quillian (1972), who regarded words primarily as nodes in a hierarchical system of memory retrieval. Meaning networks were supposed
to gather assertions, stored in memory, about an entity; such networks were centered on the word representing the entity. Experimental data on retrieval time seemed to support the idea; the more
"nodes" to pass through to find a word, the longer it took to find it.
However, the initial support dwindled when it was realized that category size was a confound; moving up the semantic hierarchy from,
say, "tiger" to "feline" to "mammal," the number of exemplars in
each category increases. Increases in search times as the network is
traveled could be therefore due to the larger number of items scanned
in each category, rather than the number of nodes. Other problems
developed for the theory, even within the limits of its own limited
objectives. Johnson-Laird's key complaint, however, is that the theory
serves no semantic function; it merely provides a search mechanism
within the symbolic system.

Meaning-postulate theories were prosed by Kintsch (1974) and by
Fodor, Fodor, and Garrett (1975). The basic idea here is that utterances, to be understood, are translated into a form of internal speech
(or mentalese) made up of tokens that resemble the original surface
utterance but are not literally identical to it. The meaning of the utterance is tested, again entirely within the symbolic system, by means
of postulates that resemble the operations of formal logic. The internal consistency of the language system is thus testable, without external reference.

The problem with all three theories, according to Johnson-Laird, is
that they have nothing to say on the question of how language is
related to the world. They are primarily concerned with resolving
ambiguities and anomalies in sentences without reference to the outside world. Their logic is similar to that of mathematics, where it is

possible to state that the statements $a = c$ and $a = b$ imply that $b = c$, even if a, b, and c are left undefined.

The underlying assumption of this approach to semantics is the assumption of the "autonomy of intensions," or the idea that the intended meaning of an utterance can be processed independently of extralinguistic reference. This assumption is fundamentally unjustified, for a variety of reasons. For one thing, context is absolutely necessary for the resolution of lexical ambiguity, and ambiguity is the *rule*, not the exception. That is, very few words have good definitions, and the most frequently used words are generally the most ambiguous. One of Johnson-Laird's examples: "The plane banked just before landing, but then the pilot lost control. The strip on the field runs for only the barest of yards, and the plane just twisted out of the turn before shooting into the ground" (p. 233).

Virtually every word in this expression is ambiguous. Yet the expression is effortlessly comprehended; the most appropriate meaning of each word is understood with reference to its context. Since the utterance is unique, it would not be possible to understand it with reference to standardized sets of postulates, or interpretative contingencies; the only solution would be to construct a "mental model" of the message contained in the utterance. The construction of such a model would provide an efficient mechanism for dealing with inherent ambiguities. Ambiguity is often resolvable only through indirect reference, as in: "The ham sandwich is sitting at table number 5 and getting impatient" (p. 241). This statement is intelligible only if a model of the situation is constructed so as to "make sense" of the utterance. There are many such examples; geometric inferences in particular run into difficulty without reference to context, as in the following: A is on B's right, B is on C's right; therefore, A is on C's right. As Johnson-Laird points out, the conclusion would be true of a rectangular table but not a round one. Therefore access to the referential context, in the outside world, is the only way to resolve the issue; the internal logic of the language system does not contain the answer.

Johnson-Laird's theory of semantics explores a different idea, which is attractive because it tackles the problem of reference somewhat more directly. An utterance is understood in terms of a mental model.

The mental model is a theoretical construct; it is an attempt to understand the underlying structure and objectives of the processes of human thought, especially inferences. Models represent states of affairs, sequences of events, the appearances of things, social situations, and actions of varying degrees of complexity and abstraction. Mental models are necessary to draw inferences, to understand utterances, and so on.

The crucial question to ask about Johnson-Laird's theory of mental models, of course, is the same question he was asking about other theories of meaning: How do mental models represent the external world? Curiously, Johnson-Laird falls back on an evolutionary approach in trying to answer this question. He defines the problem nicely: the problem is how to establish the *intentional* correspondence between a symbolic expression and a state of affairs. Automatons, even the most complex automatons such as insects, do not have this problem, since they are incapable of constructing their own reference systems. Higher creatures, like mammals, are not mere automatons. They construct their own mental models. The origin of this modelling ability is in the evolution of perceptual systems that provide a model of the world, in which perceptual objects may be located. Mental models, or as Johnson-Laird puts it, "three dimensional kinematic models of the world," are still primarily perceptual in origin, even in humans, having emerged as a consequence of further evolutionary advances.

Mental models can be used to understand linguistic discourse. There are underlying cognitive devices that construct, interpret, and revise models; these can use algorithms that map propositions into the model. One remarkable implication of this proposal is that such algorithms would forego the need for specifying the semantic relations between words; there is no need for mental equivalents of dictionary entries, semantic networks, or meaning postulates.

Mental models are a somewhat effective device for dealing with the problem of extralinguistic reference. Extralinguistic reference starts early; a child points and utters a monosyllable such as "Boh," and the mother understands the intention by assessing the extralinguistic context. The child's skill in representing "what it means" is still primitive; but the mother, more skilled perhaps, but still probably imperfect in language use, understands the game the child is playing. She too has been trying to bend words to her own extralinguistic purposes most of her life.

A very clear example of extralinguistic reference is seen in pragmatic anaphora, which are in wide daily use by adults. Pragmatic anaphora are utterances that cannot be understood without nonlinguistic reference to the real world, as in: "I picked those up at the park yesterday." This sentence is unintelligible unless the word "those" is defined, and it can be defined only by looking where the speaker is pointing or somehow assessing the speaker's intent. Jackendoff (1983) describes at least seven different classes of anaphora, each referring to a different ontological category: thing, place, direction, action, event, manner, and amount. Mental models thus provide a referential basis for the "wh-" questions: why, which, where, when, who, what, and how much.

Ultimately, however, Johnson-Laird's proposals fail to break out of the self-contained symbolic world. His approach to model construction requires tokens and algorithms that must operate in a quasi-symbolic way, since his implementations of them are entirely symbolic. The mind must be capable of attaching tokens to aspects of events *without any outside help* (that is, without a programmer). But it is not obvious how this can be achieved. Where is the perceiver who "knows" the intention of the speaker from context and finds the appropriate tokens? Presumably, the presymbolic mind cannot read the tokens, while the symbolic mind, which can read the tokens, cannot read the perceptual model. There is a need for a level of mind that is competent in both ways of knowing—one that can "read" both the perceptual and the symbolic inputs and construct the most likely resolution, or model, of the situation, based on both inputs.

Where would the evolution of language and thought fit into this picture: Are mental models—or, in general, theories of this sort that emphasize formal structures in a computational framework—compatible with the evolutionary data on the emergence of hominid culture? Johnson-Laird does not address this question directly, and it is risky to bend such a distinctive theory to a different purpose. But the question must be asked. All higher mammals evidently employ, and construct, complex mental models in their dealings with the world. As we have seen in the discussion of episodic cultures, their perceptual models of the world, or their event perceptions, can be remarkably subtle and complex. Their limitations would seem to come in primarily on Johnson-Laird's higher conceptual levels; that is, they might appear incapable of constructing what he calls relational, metalinguistic, or set-theoretic models. This suggests that specific aspects of his system, perhaps certain recursive procedures, come late in evolution, allowing humans to construct these high-order models.

However, Johnson-Laird's framework is generally difficult to translate into biologically implementable terms. I have argued that mimetic modeling is virtually independent of linguistic modeling; yet a clear distinction between the linguistic level and the mimetic level is not easy to tease out of Johnson-Laird's magnificent edifice. At least, I have difficulty deciding exactly where in his version of mind/brain, filled with recursive procedures, propositional representations, and models, this distinction would fit in.

The most obvious place to look for a simple evolutionary break in the hierarchy of mental models is in his split between "perceptual" and "conceptual" models. However, the distinction between perceptual and conceptual models does not correspond perfectly to the linguistic-mimetic distinction. Mimetic representation is at an intermediate level; that is, perceptual models seem to correspond nicely to the realm of episodic cognition, while conceptual models correspond to the realm of symbolic thought. However, mimetic representation has some properties of both levels, while evidently missing some of the crucial properties that support symbolic thought.

There are many anomalies in the hierarchy, when mimetic representation is taken into account. Mime and metaphoric gestures are by-products of what appear to be metalinguistic models; they involve reference, and they assert meanings. They also contain relational elements; it is possible to express relations like "more than" or "higher than" with metaphoric gesture. Yet they are rather poor at creating simple situational models and apparently could not be used in isolation to express or construct highly abstract set-theoretic models.

Thus, the specific adaptation that led to symbolic language is not easy to single out in Johnson-Laird's paradigm. This is not a trivial difficulty, since in the present theory, the demarcation line between mimetic and linguistic representations looms very large indeed, involving much more than a simple step up from physical models to conceptual models. Mimetic representation laid the groundwork for language and symbolic thought, but lacked some critical element; the mental modeling apparatus was still incomplete.

It should be emphasized that this problem is not contingent upon accepting the evolutionary proposal set out in this book; it is inherent in the structure of modern human cognition. Mimetic representation is an isolable, parallel channel of representation that requires its own level of description regardless of the evolutionary scenario leading up to it. New mental models of mental models have to account for mimetic culture.

The principal conclusion to carry away from this brief discussion is that modeling is at center stage, especially in areas like discourse comprehension and syllogistic logic. Moreover, thought and language are so closely related as to be two sides of the same coin; there are many forms of thought that are literally unthinkable without language and other semiotic devices. Most importantly, where humans differ from apes and other mammals is not so much in their possession of signs and symbols but in the types of mental models they construct.

Linguistic Invention and Its Market

Johnson-Laird was concerned primarily with the process of constructing mental models in order to understand logic and thought; mental models are seen primarily as a device needed to understand the use of symbols. But the evolutionary approach taken here is the opposite; the modeling process would have first been needed not to decode a language that did not yet exist but rather to invent it in the first place. How was the mind able to invent symbols in the first place? And how were these preexisting mental models enhanced by the acquisition of linguistic capacity?

The need for an extralinguistic referential system reminds us of a more fundamental truth: the words and symbols of language must ultimately have originated outside of language. Symbols, like the countless new words introduced over the millennia of humankind's existence, were initially invented for *nonlinguistic* purposes. This had to be so, especially in their initial appearance in the collective lexicon. At one time, none of the words we possess existed, and literally none of the concepts expressible in English today existed in the lexicon. If Recorde's relatively minor invention of the equal sign within a highly developed symbolic repertoire of his time was a creative leap, then imagine the leap it required to have invented, for the first time in the history of evolving mind, a word to express the concept "equal." Or a symbol for any number of common objects, actions, feelings, and relations, as each received its first representation.

Linguistic invention is public and market-driven. The rapid modern growth of the number of available verbal symbols shows that symbols—new words, expressions, and notations—continue to emerge in large numbers and enter the public domain in an unending stream. Some of these become part of the accepted linguistic convention, whether of society in general or of a special group or class, and some are abandoned. Language is dynamic and collectively determined; it must have been this way from the start.

Each invention is a hard-won linguistic victory. Even street talk, which used to be unfairly regarded among the educated as the "vulgar" tongue, is enormously creative in generating new ways of expressing things, and vulgar expressions survive only if they capture, in a very competitive market, a notion the public wants expressed. Most importantly, each invention has to serve a useful representational purpose; that is, its usefulness is evaluated in terms of its public verifiability in terms of some publicly understood extralinguistic referent.

The mind needs, evaluates, and accepts various cognitive tools for its immediate purposes. Just as a carpenter evaluates and accepts a new tool, a mathematician, or a modern city dweller, might discover the need for a new symbolic device to express some important as-yet-uncaptured knowledge. This kind of invention must be driven by an underlying conceptual agenda. Once the appropriate symbols and use rules have been developed, the operations that the mind has been trying to encapsulate can be carried out symbolically. They become easier to learn, and easier to execute. Certain types of new operation may appear as well. The process of symbolic encapsulation of truly new ideas is our strongest clue to the true nature of language. For most modern humans, language is a given. A collective lexicon is our inheritance, as surely as any set of genes. The use of the lexicon and the use of the accepted grammatical conventions of a society allow us easy entrance to society's symbolically held collective knowledge. Anyone lacking this key is excluded. Anyone possessing it is included, if they can grasp the underlying model.

Ideas that are not within the purview of one's language are not accessible. Through the efforts of a few individuals, new ideas may be introduced and the collective system will grow, gradually. But for most people, ideational growth is limited to the constraints set by the existing language, and in most cases the existing language far exceeds the capacity of any single individual to grasp it in its entirety. The point is, the invention of language was a collective enterprise, one that depended heavily on the capacity for truly original linguistic invention, as opposed to the "innovative" uses attributed to children, which are not inventions in the same sense but rather experiments to verify existing use rules.

Chomsky (1965) coined the phrase "linguistic competence" to describe the child's rapid, innovative use of language; but the competence of the average language user, impressive though it may be, is far less than the competence required to invent truly new symbols. Some individuals have had a determining effect upon an entire culture; their efforts at linguistic invention created new words and phrases—and, perhaps more importantly, new grammatical conventions—that endured across the centuries: Spenser in the English language and Dante in Italian, for example. They and other great innovators were probably close to the kind of linguistic invention that led to the initial creation of language.

The same holds for grammar and other conventions of use, such as prosodic rules. Each rule, and each convention, must have had its original invention, followed by the collective acceptance and implementation of the new rule. Invention required the capacity to have conceptualized the need, and represented the convention, in a communicable way. Once in existence, new rules would have been easier to master than in their original invention. The notions of the subjunctive, the conditional, the idea of plurals and tenses, of conjugations and declensions, of active and passive voice: these all had, at some time, to be invented. They are obviously useful; they serve clear pragmatic and referential functions. They disambiguate and clarify otherwise ambiguous and vague utterances. Presumably their origins were closely related to their anticipated and perceived usefulness to the growing human cognitive agenda.

Symbolic invention within the context of an existing lexicon would become somewhat simpler, in the sense that many words can be invented from other words, and many linguistic conventions could be seen merely as improvement over existing, but qualitatively similar, conventions. But even within the context of highly evolved languages, the process of fundamental linguistic invention continues. Constant inventions and modifications of use rules are going on in all living languages, despite the efforts of the academies to freeze and prevent linguistic change. A new word is defined by rules of use, and the rules keep changing. Every new lexical invention requires a manual of use on its arrival. Word grammars are never given; they are won honorably, on the linguistic battlefield. Inventive individuals continue to pace the development of language, 50,000 years after modern language started its explosive growth. Users of a language vote for each new invention by adoption; when a really effective new expression comes along, its usefulness is affirmed by plebiscite.

The process of fundamental linguistic invention is close to the phylogenetic roots of language. The bond that ties language to extralinguistic knowledge can never be clearer than in the case of original linguistic invention; it is extralinguistic knowledge that recognizes the need, and perceives the utility, of new symbols. Language was something humanity needed for another cognitive agenda, an adaptation driven by something it could provide for the prelinguistic brain. The absurdity of dualistic arguments for isolating symbolically encoded knowledge from other kinds of knowledge becomes especially clear in an evolutionary context.

The Speech Adaptation

As observed earlier, the accelerated human invention of semiotic devices seems closely correlated in time with the arrival of a new type of vocal apparatus. This is no coincidence. The speech system has inherent advantages over other media that made it a necessary step in the evolution of human linguistic capacity and in the emergence of modern mind.

Speech was the primary form of human linguistic expression; writing and its relatives are very recent developments. It is important to remember from the start that speech is not the only medium of human language; sign language, written script, and Braille are other media that allow truly linguistic cognitive operations. Nevertheless, the speech adaptation was a necessary trigger for early linguistic invention and conceptual evolution. Like the earlier adoption of bipedalism, speech involved a complex pattern of change, of which vocal skills constituted only a part.

An overview of relevant cognitive changes in the pattern of adaptation accompanying speech should include (1) a high-speed phonetic vocomotor control device, consisting of a new, descended larynx with a more supple glottis, an altered suboral cavity and tongue, changed oral and lower facial musculature, and corresponding new sensory and motor pathways, and new cortical representation of vocal skill; (2) a new auditory adaptation, allowing the perception of "auditory objects" and "auditory events"; (3) a related device sometimes called the "articulatory loop," which involves a dedicated short-term phonological store for vocal and subvocal rehearsal; (4) a great increase in the vocal repertoire, to include tens of thousands of sound-units for use as symbols and hence an increase in specialized auditory memory and procedural phonological algorithms; (5) an enhanced capacity for the *definition* of event features, including what we call syntactic features; and (6) new cognitive skills on the integrative-modeling level, whereby the analysis of discourse and narrative, and formalized thought, could eventually be achieved under symbolic control. These last two features must have been on the leading edge of the adaptation; but they would have developed in concert with the first four, in an iterative loop of selection pressure originating at the cultural level.

The degree of cognitive change at this stage of human evolution is tremendous to contemplate—even after conceding that mimetic skill and a degree of symbolic gesture, along with many of the emotional and social structures characteristic of human society, were already in place. Yet this cognitive change must have happened, and it must have happened as part of a single adaptation, albeit an adaptation that evolved over 200,000 or more years, culminating in modern humans. Language and advanced conceptual development on the symbolic level are so closely interdependent in the brain that they appear to be inseparable; they must have taken the great leap together.

The High-Speed Phonological Device

The human speech adaptation involved a major change in the anatomy of the primate vocal apparatus. The result of that change was an increase in the range and variability of the sounds that hominids could produce, especially in the rate at which they could be produced and modulated and in the length of time over which phonation could be maintained. The best available discussion of the physical evolution of the speech apparatus may still be found in Lenneberg (1967), who illustrated the many anatomical features involved in the adaptation. Both Lenneberg and Lieberman (1984) have relied heavily on the unique anatomy of the vocal apparatus as evidence for a single biological speech adaptation.

The physical mechanism that enabled sustained, high-speed production involved a change in the voluntary control of respiration. Speech outputs tend to ride on long waves of expiration, and therefore the length of utterances tends to be influenced by breathing, and vice versa. Humans have a unique ability to override the reflexive loops controlling breathing, allowing very long expiratory phases, with short, rapid inspirations between utterances. Utterances thus cluster into what are called breath groups—a single expiratory phase that usually contains a coherent syntactic unit such as a sentence or phrase. The changes in respiration were apparently more neural than peripheral, although there were also changes to the shape of the rib cage and lungs. Their new respiratory apparatus gave humans the ability to break up their otherwise regular cycle of inspiration and expiration to sustain long, and sometimes irregular, utterances. An extreme form of this ability is evident in opera singers and classically trained actors, who can hold a note, or sustain an oration, for very long periods, breaking up and concealing the inspiratory phases of their breathing.

The facial musculature and vocal tract also underwent great change. The muscles of the lower face became more complex and mobile, especially those around the mouth. The tongue became more flexible, and the back of the tongue, in combination with the elevated soft palate, became able to alter its position in the vocal tract so as to vary the shape of the supralaryngeal cavity; this allowed humans to produce vowel sounds that no other primate could produce, particularly the so-called point vowel sounds, /i/, /a/, and /u/.

The laryngeal apparatus itself, however, was the major peripheral change enabling speech. The larynx descended so that there is a considerable vertical distance from the glottis to the soft palate (the upper part of the back of the throat). This created a right-angled turn in the vocal tract, above the larynx, and a larger cavity, whose shape can be changed to alter the frequency characteristics of the sounds produced by vibrations of the vocal cords. The glottis (the opening between the vocal cords in the larynx) became more supple and easier to vibrate during respiration.

The degree, and speed, of voluntary control over the larynx is phenomenal in humans: high-speed motion pictures of the larynx in action during speech reveal a process even more complex than the mouth movements that control the other end of the vocal apparatus. The tension on the vocal cords, and the size of the laryngeal opening, can be controlled by several sets of muscles. The fundamental frequency of sounds emitted by expiring air through the partially closed glottis is altered by the shape of the supralaryngeal cavity and the shape of the mouth opening. Sudden, or fast-rising, sound patterns can be created by rapidly closing and opening the vocal tract at various points.

The whole complex represents a distinct anatomical device that is largely absent in apes and, according to reconstructive evidence, in archaic hominids. The ability to sustain vocalization for minutes or hours at a time, to produce such a wide range of sounds, and to produce complex sequences of sound modulations at a very rapid rate was novel. The neural control of this apparatus was an integral part of the adaptation, and the brain areas representing the vocal apparatus—tongue, lips, soft palate, larynx, and respiratory cycles—were expanded, particularly in the cerebral cortex. Compared with apes, whose vocalizations are largely controlled from the subcortical limbic regions of the brain, humans have a greatly increased cortical representation of the vocal apparatus.

This new human speech device was a vertically integrated system. It involved not only changes in the various anatomical structures underlying the skill but also new sensory receptors, new motor pathways innervating the new arrangements of muscles, and new neural machinery capable of the coordination of several parallel channels of vocomotor activity at once. Speech sounds are the integrated product of the entire apparatus working as a unit, even though the separate parts—larynx, mouth, pharynx, tongue, and soft palate—each have separate innervations and must serve other, nonlinguistic purposes such as swallowing, smiling, and breathing. Therefore, the speech system requires a neural integrator, a single mechanism where all the various phonological devices available to the speaker come together for programming.

As outlined in Chapter 3, Broca's region is sometimes thought of as such a device, but evidence remains somewhat equivocal on the question of exactly what function the area serves. There seems little doubt that the neural control of phonology is normally on the left side of the brain and involves the inferior frontal and prefrontal areas of the cortex, as well as interconnected parts of the thalamus, basal ganglia, and cerebellum. Damage to these regions, and in fact damage to the entire left perisylvian area of the brain or to the circuitry interconnecting this region with other areas, can result in specific phonological dysfunctions.

As mentioned above, Lieberman (1984) claimed that some archaic humans possessed the vocal adaptation, while others (notably the Neanderthalers) did not, or at least they possessed it only in part. This claim should be evaluated in the light of the interconnectedness of the adaptation just described. The central neural apparatus, the new muscles, sensory receptors, and motor fibers of the lower face, the newly designed and descended larynx and its new innervation, the changed tongue and its innervation, and the central device that brings together all these anatomical features in a coordinated speech output must have evolved as a functional unit. This is not to say it evolved all at once; selection pressure might have been sustained over the whole transition period, that is, over several hundred thousand years. But it is unlikely that the new neural apparatus would develop in the absence of the new peripheral anatomy it is designed to control. Therefore, it seems improbable that hominids who lacked the anatomical features of the lower face and throat would have had the high-speed phonological device underlying human speech.

The phonological device needs considerable programming to achieve the range of control required for speech production. Levelt (1989), who has reviewed current theories of speech production, provides a useful summary of the features of the phonological system, illustrating the complexities of the system at this level. Once a message has been formulated, it must be translated into an articulatory program, or a *phonetic plan*. The plan usually runs ahead of the actual execution; that is, the formulated plan is "unpacked" in chunks and

then unfolded into overt speech. This process is highly automatic, at
least in adults, and the result is a high-speed pattern of articulation
which may run as high as 15 phonemes per second.

The phonetic plan is a detailed motor program, delivered word by
word, or preferably phrase by phrase. The speaker has to retrieve the
detailed phonetic plan as each phrase is initiated and then unpack the
phonetic plan for each phrase. The amount of time needed to assemble and unpack the detailed phonetic plan depends on the complexity of the motor unit and the number of such units involved in a
particular phrase. These phases are shown in Table 7.1.

The final stage is the execution of the motor commands by the
musculature. This involves setting goals for each small segment in
time, phoneme by phoneme. Successive segments of speech each have
oral–sensory features that have to be preplanned, down to the timing
of the onset and duration of each feature in the utterance. This is
done by setting motor goals for successive chunks of time and fine-

Table 7.1   Phases in speech motor control (after Levelt, 1989).

Stage 1: Assembling the program
This is the stage of phonological encoding, with a phonetic plan as output.
The phonetic plan is a detailed motor program, delivered phonological word
by phonological word. When the task requires, phonetic plans can be stored
in the Articulatory Buffer. The preferred units of storage are phonological
phrases.

Stage 2: Retrieving the motor program
When the speaker decides to start a prepared utterance, its motor units
(the phonetic plans for the phonological phrases) are retrieved from the Articulatory Buffer. The time needed to retrieve each unit depends on the
number of units in the buffer.

Stage 3: Unpacking the subprograms
Once retrieved, the phonetic plan for a phonological phrase has to be unpacked, making available the whole hierarchy of motor commands. The
more complex a motor unit, the more time unpacking takes.

Stage 4: Executing the motor commands
At this stage the motor commands are issued to the neuromotor circuits
and executed by the musculature. Syllables can be drawled to absorb retrieval latencies.

tuning the relative strengths of each motor unit involved at each moment in the utterance.

The assembling, unpacking, and execution of speech is done automatically. This automaticity can be seen in fine-grained chronometric experiments. In a particularly clear demonstration, Gracco and Abbs (1985) constructed an apparatus that unpredictably interfered with lip-closing while speakers were trying to produce the labial sound [b]. The speakers were able to compensate very rapidly (within 40 milliseconds) in response to the apparatus, by increasing the amount of force on their lips so that they could produce a sound close to the correct sound. This extremely short latency of response showed that the complex phonetic plan for the labial sound [b] had been altered too fast to be the product of a conscious feedback loop; conscious responses take 100 milliseconds or more to come into play (Evarts, 1973, 1979; Evarts and Tanji, 1976). In addition to automatic monitoring of speech output, there is also conscious monitoring of the phonetic plan during its execution. This can be seen in the slower corrections people make—for instance, the speaker may start to say "vertical" but speak only the first syllable thus "ve—no, I mean horizontal."

Many of the features seen in the articulation of speech segments are also found in other skilled motor behaviors. All conscious motor acts involve retrieving a plan, modifying for it the particular conditions of execution, and then unpacking it in a finely tuned temporal order under constant conscious supervision. Motor-learning theory has built these considerations into the concept of the "motor schema"; MacNeilage (1980) and a variety of other theorists in the field of speech control have studied these issues. The distinctiveness of speech at this level is found mostly in its speed and complexity, but it is a mistake to underestimate the speed and complexity of athletic skill, or manual skill. The major evolutionary fact is not that articulatory skill is unique in all of nature, but that it was completely new in the primate line and that it arrived as part of a specialized cognitive adaptation.

Synthesis of Auditory "Objects" and "Events"

As an integral part of the new speech apparatus there had to be changes to the auditory system. There were two reasons why this had to happen. First, feedback modulation of speech output requires a dedicated auditory mechanism for regulating sound production. Second, the decoding of speech inputs would place new demands on the auditory system, specifically on the ability to perceive words and phrases as auditory "objects" or "events" rather than simply as sequences of sound.

The importance of feedback in speech regulation is evident from the literature on training the deaf to speak. Deaf humans, prior to the nineteenth century, simply did not learn to speak. Auditory feedback in the natural human environment (that is, prior to the intervention of science) was absolutely necessary for learning speech. Disturbances of feedback—distortion or delay of the sound of one's own voice—disturb speech greatly, and medical conditions such as middle ear infections in children that chronically interfere with feedback can lead to phonological disorders. The timing of feedback is precise and rapid, as can be imagined from the rate at which humans can produce speech. A difference of one-tenth of a second in the rate of feedback will cause stammering and stuttering in normal people (Taylor, 1976).

Finely tuned feedback implies a specialized auditory path that maps sound input onto the speech output device. The auditory map appears to be specialized for the categories of sound important to speech, that is, to vowel and consonant sounds, which have peculiar characteristics when compared with other (nonspeech) sounds. Lieberman and others have proposed the existence of dedicated speech-sound analyzers, each of which specializes in a particular sound feature important to speech: for instance, voice-onset time, or place of articulation, or formant-frequency transition. These analyzers, in the secondary auditory cortex, should feed directly into the frontal speech-production areas.

This claim is very probably true; feedback systems in the body typically are quite specialized. Examples may be found in the peripheral visual control of eye movements, vestibular control of limb coordination, and stereoscopic control of manual skill. In each case, a specialized cortical area or subcortical nucleus (often both) brings specialized sensory information to bear on a specialized motor system. In the case of speech, phonological rules—the programs that control the serial formation, organization, and pronunciation of speech sounds—are probably vested in a feedback-controlled circuit going from the secondary auditory centers to the speech motor areas. Geschwind (1970) proposed such a circuit, flowing from secondary auditory regions of the cortex to Broca's area, via a large direct fiber tract known as the arcuate fasciculus. It is also likely that the basal ganglia are an important component of that circuit (Lieberman, Friedman, and Feldman, 1990).

The human auditory system, however, goes beyond providing feedback for fast, sophisticated sound modulation. It has a second major feature: speech sounds are "reified," that is, heard as if they were objects or events. A word, or a common phrase, stands out from the other sounds of the environment in much the same way that visual objects stand out. In resolving a word, the human auditory system achieves object constancy, much like vision and touch. Therefore, the term "auditory object" will be used to refer to this property in words. The brain can recognize tens of thousands of auditory objects; and longer utterances are perceived as events containing a number of objects in juxtaposition, much like visual events. The difference, however, is that auditory objects are synthesized primarily in time, rather than in space. Auditory objects possess the properties of other perceptual objects: they are constant under various transformations; they are difficult to describe precisely yet they possess a complex unity; and, once identified, they stand out from background information.

The perceptual skills needed for breaking down the speech stream into its constituent parts are uniquely powerful in humans. Many animals can hear as well as humans in terms of basic sensitivity and range, most nonhuman mammals apparently cannot segment and break down long human speech utterances into independent word units. This is perhaps because the words of speech are not clearly separated in the sound stream, and animals therefore hear the speech stream as a continuous, highly complex sound instead of a series of independent, identifiable auditory objects. To circumvent this limitation in most animals, trainers usually emphasize the first or most salient syllable of a command, and reduce the need to segment the sound stream into separable units.

Apes may be the sole exception to this rule, at least among mammals. Patterson (1980) used speech as well as manual signing while training Koko (a gorilla) to use sign language, and claimed Koko could eventually follow speech commands somewhat independently of the accompanying manual signs. However, Koko's speech comprehension was restricted to predictable contexts in which there were many other clues as to what was being said, and there was no evidence that she had an independent grasp of spoken words. However, Savage-Rumbaugh and her colleagues (1990) have recently produced convincing evidence that some apes (bonobos) can segment speech into meaningful word units. Their subjects learned to recognize many individual English words by about 4 years of age and also learned to understand simple spoken sentences.

Importantly, they responded appropriately to changes in the position of a word in a sentence: for instance, they could tell the difference between "the snake bites the elephant" and "the elephant bites the snake," showing that the lexical items were perceived as interchangeable units in their own right. It is not yet known whether this ability is special to bonobos or shared with other apes; nor is it known how far it can develop. But the presence of this perceptual skill in bonobos, who are thought to be close to humanity's apelike ancestors, appears to be an evolutionary antecedent to the exceptional auditory segmentation skills of humans.

Auditory-object constancy should not be confused with the notion of categorical perception, which originated with the Haskins laboratories some thirty years ago (Studdert-Kennedy, 1975). The idea of categorical perception was originally applied to the phonetic components of speech, which had a curious property: gradual variations along phonetic dimensions were perceived not as graduated changes but as sudden switches from one category of sound into another. For instance, the gradual physical alteration of the consonant sound "ba" into the sound "da," which generates a number of intermediate stimuli, is perceived as a sudden switch. Given, say, five intermediate stimuli, the observer might hear the sequence, ba, ba, ba, ba, da, da, da. There would be no intermediate perceptions.

But this categorical ability is not unique to humans; chinchillas can also achieve it, as can a variety of other animals (Kuhl and Miller, 1975; Trehub et al., 1981). Moreover, it is not unique to hearing; many visual stimuli are perceived categorically. In fact, categorical perception is now seen more as the rule than the exception (see Har-

nad, 1987, for a thorough review). Given its ubiquity, categorical perception of stop-consonants cannot be held up as evidence for a unique
auditory speech adaptation.

The idea of auditory-object constancy, and its extension in complex
auditory event perception, goes well beyond the categorical perception
of features to the synthesis of complex objects. Words, and more complex auditory events such as sentences and longer utterances, take on
the perceptual characteristics usually attributed to three-dimensional
visual and tactile objects. This is quite independent of their meaning;
a meaningless speech utterance may be heard as a series of auditory
objects, although it is less accurately perceived than a familiar sound.
But most environmental sounds are aspects of events, rather than
events in themselves; words stand out as events on their own. Some
nonverbal sounds are highly salient and recognizable items—for instance, a bird call or a crackling fire—and come close to being auditory
objects, but they do not constitute a class of sound objects so infinitely
variable as speech. Words are at least as recognizable, and as variable,
as faces, another class of perceptions for which humans have a very
large memory capacity. This suggests a specialized storage device.

Phenomenologically, speech perception involves the clear perception of much longer streams of sound, as though there were an auditory buffer that allows the simultaneous analysis of fairly substantial
segments of the sound stream at once, or at least a sliding window of
sound several seconds in width. The tendency of the system to analyze large chunks of sound as a unit can be seen in a variety of so-
called top-down effects found in psycholinguistics. For instance, the
semantic context of words greatly influences our perception of them.
Words are better understood when presented in intelligible sentences
(Miller, Heise and Lichten, 1951), and when isolated from context
they become harder to identify (Pollack and Pickett, 1964). Word perception, in turn, influences the way syllables and phonemes are heard.
The best example is found in the phonemic restoration effect (Warren,
1970) in which subjects tend to perceptually "fill in" missing sounds
in a speech stream, as though whole words were called up on partial
information, much the way half-obscured visual objects are perceived
accurately by filling in missing detail. These effects show that the
listener is able to hold a number of previously heard words in mem-

ory while trying to decode new inputs. In effect, expectations are at work in speech perception, as in other kinds of event perception.

The ability of humans to discriminate and remember long utterances raises the issue of auditory memory. Part of the speech adaptation seems to involve two enhanced aspects of auditory memory. The long-term verbal memory store consists of a lexicon of thousands of words carried by every speaker; and the short-term aspect is the so-called "articulatory loop."

The Articulatory Loop

Baddeley (1986) has proposed a model of working memory that has an interesting property, from an evolutionary standpoint: it has two semiautonomous subsystems, one auditory–verbal, the other visual–spatial, both feeding into a central-executive system. These working memory systems correspond to the two great subsystems of human thought, which, between them, process the vast majority of semiotic representations. Both subsystems appear to have properties that are uniquely human. The visual–spatial subsystem will be discussed in the next chapter. The auditory subsystem, which Baddeley (1986) has called the articulatory loop, is properly seen as a necessary aspect of the speech adaptation. It is really the mnemonic aspect of the phonological system; that is, in possessing the ability to produce words and sentences, the brain needs to be able to hold a significant amount of material available for enough time to process larger chunks of sound and to "rehearse" material.

The articulatory loop works on a phonological code; that is, the sounds of words themselves are important to its function. This is evident in the phonological similarity effect, first shown by Conrad (1964)—the systematic phonetic errors made by subjects trying to memorize lists of consonants. Speakers evidently relied on literal reproduction of the sounds in the list rather than on a more abstract semantic record of what the items were. Similarly, Baddeley (1966) showed that lists of similar-sounding words are harder to memorize, in the short term, than different-sounding words, regardless of their meaning. This reinforces the impression that the short-term memory system recalls the items literally, according to their sound. Murray (1967) contributed to the evidence for this by demonstrating the value of articulation in strengthening short-term verbal memorization; suppression of articulation reduces recall of auditory–verbal material. These effects break down for long-term memory storage; that is, it is the meaning of what was said that determines long-term memories, not literal phonetic replays. But in the short term, phonological storage predominates.

The span of the articulatory loop appears to be quite short—Baddeley estimates slightly under two seconds—but has proven hard to quantify, perhaps because it varies somewhat with different recall conditions. Word duration is crucial in memory span; even when the number of syllables is held equal, sets of long-syllable words (harpoon, Friday) are harder to remember than short-syllable words (bishop, wicket). Baddeley interpreted this to mean that there was an absolute time constraint on the articulatory loop, which would show up whether memory span was measured in number of syllables or total spoken duration. Words are held in a temporary auditory store, then channeled into a rehearsal loop, which can maintain the trace for several seconds.

The loop can function subvocally, but the traces are weaker in this case; this is what Baddeley calls the "inner voice." The inner voice has the same properties as the articulatory loop and supports inner speech, or verbal thought. The articulatory loop is the basis of verbal rehearsal and thus an integral part of the speech adaptation. Vocal rehearsal is vital to language acquisition and the development of thought.

The first three aspects of the speech adaptation—the new vocomotor device, the new auditory phonetic analyzers, and the articulatory loop—constitute a self-contained system. The system represents a considerable advance on prosodic vocomotor control, which is attributed, in this evolutionary scenario, to an earlier, purely mimetic adaptation. The advance is evident in the increased range and complexity of the sound repertoire that can be remembered and reproduced; instead of the two dozen or so sounds in the possession of the ape, humans can utilize tens of thousands of morphemes. It is also evident in the speed, or rate of speech, factor: prosody, the archaic mimetic precursor of phonetic speech, is a very slow-moving skill compared with phonology.

Phonology was apparently such a radical change that completely new neuroanatomy was called for. Instead of replacing or elaborating upon the archaic vocomotor apparatus, phonological skills were built upon a new, parallel vocal apparatus and did not supplant the older adaptation, which still served for emotional vocalizations and their modulation. The neurological dissociability of both aprosodia and emotional speech from phonological disorders (Ross, 1981) shows that these aspects of voice control have remained separate in the modern human brain. In normal speech, phonetic and prosodic elements are often (not always) coordinated; but the elements are nevertheless dissociable.

The phonological system possessed great advantages over previously available media for symbolic invention. Distinctive utterances could be generated in virtually unlimited numbers, and they could be remembered and rehearsed. This feature allowed the symbolic repertoire to expand rapidly into a very large body of easily remembered, retrievable items. Moreover, phonological invention and communication did not interfere with other major ongoing brain activities, like locomotion, skilled manual behavior, orientation in space, and visual perception. Thus, phonological invention could go on in parallel with the rest of behavior. It remains the ideal medium in which a primarily visual–motor organism, living in close proximity to his conspecifics, can develop a system of symbols to use concurrently with other mental activities.

The Lexicon

The phonological system, in itself, is nothing but an elaborate vocomotor skill system. It invents, recognizes, rehearses, and recalls utterances at a fast rate and in large numbers of permutations and combinations. It acts as though it is rule-driven on two levels, which can be roughly labeled phonological/morphological and semantic/syntactic. The phonological and morphological "rules" for constructing words in a language seem to be able to function somewhat independently of semantic and syntactic rules, suggesting that they are stored and executed separately from the latter. This can be seen in certain cases of pure Wernicke's aphasia, where patients lose the higher control functions of language but continue to construct phonologically legal utterances. But the latter is a limited residual ability, since the correct phonology of many words depends upon their meaning. And the phonological apparatus, complex and skilled as it may be, cannot generate meaningful, grammatical, language on its own; for this it must have a linkage to the real world, that is, a referential system.

The linkage to reference invests the utterances of the phonological system with meaning and syntactic structure. The primary symbolic device in this reference system is the "word," and collectively words are often referred to as the "lexicon." A lexicon, in ancient Greek or Arabic, was a dictionary, a list of words and their definitions. An alternate term is "vocabulary," which is also an expression of Greek origin for a list of definitions, or a dictionary. But the parallel with written dictionaries is deceiving. The phonology of a word provides a memory address, or the equivalent of its dictionary entry, and a means of its reproduction. But the "meaning" of a word in the brain, as we have seen, cannot be a definition entirely in terms of other words, tokens, or symbolic devices, on whatever level. Because it does not need to provide a system of reference for its contents, a dictionary makes a poor metaphor for the lexicon. Ultimately, all dictionaries and symbolic devices, including the products of computational algorithms, depend upon human lexical entries for their interpretation.

The idea that the lexicon is at the center of the language system is sometimes called the *lexical hypothesis*. The lexical hypothesis holds that lexical entries are unifying devices that tie together grammar, meaning, phonology, and specific morphological rules. Levelt (1989) places the lexicon at the heart of the speech system, mediating grammar and phonology, and mediating the conceptual message and its verbal surface structure. In his view, illustrated in Figure 7.1, each lexical entry is the center of a complex network that contains four classes of information: (1) meaning, or reference; (2) syntactic properties, which include the category of the entry (for example, verb) and the types of arguments attachable to it (transitive); (3) morphological details, concerning things like affixes for plurals or tenses, and (4) form specifications, or phonological information, which includes syllabic structure and accents. Hudson (1984) has proposed a related theory of "word-grammars" that adopts a similar position with regard to the central role of the lexicon in syntax. In his proposal, the internal structure of the word-system determines the syntactic structure of a sentence, so that the rules of grammar are inherent in the lexical entries themselves.

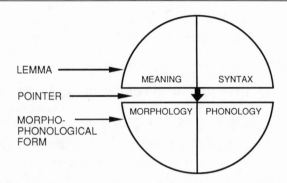

Figure 7.1 A lexical entry system (after Levelt, 1989) consists of an elaborate system of knowledge, usually centered on a word. For any lexical entry there are two major parts: morphophonological form, including the rules for pronunciation as well as modifications for plurals and tenses; and the lemma, including syntax (the rules for using the word in a sentence) and meaning (the word's various semantic referents). The lemma controls grammatical encoding, while the lower part of the diagram controls phonological encoding, and ultimately pronunciation. The lemma "points" to its corresponding form, that is, it can find the form address in the brain to produce the intended word.

In some formulations, the meaning and syntax information are called the "lemma" (Kempen and Huijbers, 1983) of a lexical entry, while the form information, items 3 and 4 on Levelt's list, are a separate part of the lexical entry. The form information aids in the decoding of phonological inputs, while the lemma encompasses both reference and syntax. The dissociability of lemmas from word-form information is an essential point: a lexical entry can harness the phonological system quite independently of its central role in the formation of meaningful utterances, in which role the lemma plays a key part.

The lexicon must be regarded as part of a representational system that is separate from the mimetic system. Phonetic symbols would have been somewhat separate from the start, partly because they originated in a completely different perceptual–motor module. But the accelerated evolution of speech, when it occurred, involved more than a phonological adaptation; it involved a lexical explosion, a capacity for symbolic invention on a grand scale, and a whole new repertoire of reproducible and rehearsable acts. Perhaps it was the detachment of speech from most mundane sensorimotor activities that allowed it to extend its reach beyond the concrete. Precisely because it was a new system, unencumbered by the demands placed on the visuomotor mimetic system, it could grow in capacity, in parallel, without interfering with pre-existing cognitive functions.

Undoubtedly, the words of speech constitute the brain's "mother" lexicon. One strong indication of this is its universality; speech, in the absence of medical abnormality, is truly a species-universal skill in humans. Writing, Braille, sign languages such as ASL, and various other symbolic systems are modern inventions and are far from species-universal. Speech is also the only modality of language production that is self-triggering in development, another testimony to the special, and primary, biological role of speech in language. But given the variety of symbolic media that can supplement speech, the mechanism of forming a lexicon cannot be a restricted part of the phonological "speech machine." Unlike, for instance, bird songs, which are restricted to one sensorimotor system, words are inherently supramodal constructs. Even though words in the first instance might have been vocalizations, they must have been inherently general-purpose constructs from the start.

If we follow Levelt or Hudson, the lexicon would have emerged specifically to link together a unified language system, as the central part of the language adaptation. But this appears to be putting the cart before the horse. In view of what was said previously in this chapter about mythic culture and the earliest uses of language, the lexicon must have been a by-product of evolving narrative skill, a necessary device in the formulation of commentaries upon experience. Thus, the lexicon was really a secondary result of the attempt to impose meaning, by the "upper" end of the language adaptation, upon specific phonological and morphological codes, through a process of social consensus.

Language could not have developed, especially in its early stages, without fundamental changes in the mental models held by individuals. The process of *defining the world* is really one of bringing these models under symbolic control. But this process also results in different kinds of models, in which the event structure of the world has been differentiated and the components made independently accessible in memory. In this sense, the models so constructed encompass the words used in their definition, and the words are an integral part of their definition.

Once the mind starts to construct a verbally encoded mental "world" of its own, the products of this operation—thoughts and words—cannot be dissociated from one another. Thus, Brother John (Lecours and Joanette, 1980), our archetype of the mind stripped of its words, cannot think linguistically *at all*; the models and their words are so closely intertwined that, in the absence of words, the whole system simply shuts down. There is no surviving "language of thought" from which the words have become disconnected. No symbols, no symbolic thought, no complex symbolic models.

But what about anomias? Patients with anomias cannot find the precise word but often seem to have retained their mental models. In searching for a word, they may access the category of the word but not the word itself, and their mimetic representations show they often have a good sense of what it was they were trying to say. But anomia does not throw into question what was said about the linkage between thought and the lexicon—on the contrary. Most anomias are specific failures of memory, not a general dysfunction of language; most of the system is still up and running, and failures to find a word can be circumvented in various ways.

The lexicon contains a variety of entries, and the standards for classifying something as an entry in the lexicon are debatable. For instance, is the lexicon strictly limited to "legal" core words or does it include within it various stock phrases and partial words as well? Levelt is willing to admit idioms, such as the expression "red herring," as lexical entries. This seems a reasonable position to take, since the meanings of idioms cannot be inferred from their component words. In daily discourse, the lexicon does not appear to be limited to words and conceivably might contain entries even larger than idioms, such as larger stock phrases and even proverbs. In practice, the lexicon is a collection of utterances of variable length, each of which has a set of specific symbolic functions and use rules attached to it. Most of the time, words appear to be the functional units of the lexicon, but longer and shorter units may also be included in the lexicon when they are treated like words.

How many lexicons are there? This is another red herring; in some neuropsychological theories there are as many as four different lexicons: (1) the auditory-input lexicon; (2) the visual-input lexicon; (3) the vocal-output lexicon; (4) the graphemic (written-output) lexicon. This is based on the study of modern, literate humans, of course; in preliterate societies (2) and (4) are completely absent. One might add Brailled lexicons (both input and output) for the blind; sign-language lexicons (both input and output) for the deaf; and in the case of some reading systems, ideographically or hieroglyphically encoded lexicons. For instance, in Japanese there are kana symbols, which are alphabetic, and kanji, which are ideographic (see Paradis, Hagiwara, and Hildebrandt, 1985). In a multilingual polymath, in theory, one could establish all of these, in a number of languages. Thus, one could have, in one brain, a dozen lexicons or more, all independent in the sense that they may all contain different items, with different rule systems governing their use. For instance, the same brain may have many more vocabulary items available for reading than for writing, or for one language as opposed to another.

Note that most of these different "lexicons" are defined in terms of the perceptual modality or motor system of use. But some lexicons are only labeled in terms of their specific content—for instance, lexicons for different languages, including special languages for mathematics, computing, or music. Given the number of different ways lexicons can be constructed, it is clear that the construction of symbols is not in principle restricted to speech and involves a general-purpose capacity that extends beyond the vocal–auditory pathway. In other words, no matter how many lexicons, real or potential, one might postulate, there is one superordinate process supervising lexical invention, and it is rooted somewhere in the higher end of the system. Speech is easy, natural, and efficient, and tends to be the first path to lexical invention. But in the absence of a phonological system, other media may be substituted, because symbolic invention is only an aspect of the linguistic modeling process.

In speech, each word system has a phonological address, and the same address may be used by a variety of word systems, producing apparent ambiguities in speech. But this is not a problem if the dictionary model of the lexicon is abandoned in favor of an embedding model; the currently dominant mental model determines the temporary meaning of a specific utterance. Words are virtually never used outside of an embedding cognitive context. The notion that words are "embedded" in mental models is not new; Wittgenstein (1922) appreciated this aspect of language and perhaps expressed the complexities of word functions better than anyone else. In the *Tractatus* he repeatedly stressed the difficulties involved in defining words precisely, even while the same "undefinable" words are used successfully in thought and discourse: "Man possesses the ability to construct languages capable of expressing every sense, without having any idea how each word has meaning or what its meaning is—just as people speak without knowing how the individual sounds are produced" (4.002). In fact, the process of thinking and formulating propositions seems to go on without any precise definition of the component words: "The meanings of simple signs (words) must be explained to us if we are to understand them. With propositions, however, we make ourselves understood" (4.026).

In other words, symbolic thought is *primary*; it is the driving force, the invisible engine, behind word use. This idea is not a minor point, or a diversion, from the more serious business of understanding the lexicon: items in the lexicon cannot always be identified in terms of their surface form. Their identification resides in their meaning in specific instances, that is, in their current use. Wittgenstein gives this example: "In the proposition 'Green is green'—where the first word is the proper name of a person and the last an adjective—these words do not merely have different meanings: they are *different symbols*" (3.323). These two words are different symbols, not because they differ in some aspect of their input value, or phonetic properties, but because they are embedded in different models.

The same physical symbol can be ambiguous, and in fact usually is ambiguous in itself. Lexical ambiguity and indefinability is not unusual or restricted to the writings of philosophers; it is the norm in word use, particularly in street talk, where metaphor reigns and words continually evolve in their uses. As Johnson-Laird reiterated, following up Wittgenstein's conclusion, humans formulate unambiguous propositions, which are easily understood, from inherently ambiguous material. It is a mistake to assume that most words, especially common words, are definable; they usually are not.

So why is the lexicon of interest? Why not focus exclusively on the modeling process itself? For one thing, the lexicon is the most easily describable access route to linguistic knowledge; we have no alternative but to use words as addresses for the complexities contained therein. For another, during the process of learning language, children acquire individual words as thought tools, and it is these that are the most direct and visible evidence of their progress. They acquire a vocabulary and use it in an indeterminate number of situations. They also acquire thought skills and knowledge, but these are less easily analyzed and measured. Moreover, the words in the lexicon, difficult as they are to define, are key symbolic units and the principal tools available to the thinker.

The lexicon is probably best regarded simply as a compendium of available devices, a glimpse of the mind seen from a single viewpoint: from the lexical switchboard, from the central meeting place of myriads of word systems. The complexity of these systems cannot be overestimated: a human being is constantly modeling the world and storing the result; modeling again and storing the result. During a lifetime, a modern educated human in English-speaking society will utilize 30,000–40,000 words in production and, according to Oldfield's (1963) estimates, 75,000 in the passive or receptive lexicon. These are drawn out of the pool of half a million or more words available in modern English.

The cumulative memory load involved must also be taken into account. Assuming each speaker lives for 70 years or more as an active speaker, tending to use language more or less continuously during waking hours during that 70-year period, a person will produce at least 25 million utterances, of which the vast majority will be in sentence form or in longer discourses, embedded in a more complex thematic or narrative context. Conceding that there will be redundancies in these utterances, on every level, there is a tremendous cumulative load on semantic memory that psychologists and linguists still do not have the slightest idea how to assess. The richness of the semantic network contained within a single brain at the end of such a life has so far defied any formalisms or experimental methods. In light of such complexity, the lexicon may appear a poor thing, and only a part of the system, but it provides a valuable access route to the system, one of the very few available.

Narrative Skill and Myth

Bruner (1986) distinguishes two major modes of thought, the narrative and the paradigmatic. Narrative imagination constructs stories and historical accounts of events. Paradigmatic imagination seeks logical truth. Narrative skill develops early and naturally in children, whereas the logical–scientific skills that support paradigmatic thought emerge only after systematic education. The difference between these modes of thought run very deep, even to the definition of truth employed by each. In modern culture, the narrative mode still predominates in the arts, while the paradigmatic predominates in the sciences. There is little doubt that narrative thought developed earlier in human history than scientific and logical thought. The latter depends on certain symbolic tools that will be discussed in the next chapter; but narrative imagination can be supported in a purely oral, or preliterate, tradition. Aboriginal hunter–gatherer cultures, in their possession of elaborate mythical accounts of reality and in their daily uses of language, show a predominantly narrative mode of thinking.

Narrative skill is the basic driving force behind language use, particularly speech: the ability to describe and define events and objects lies at the heart of language acquisition. Group narrative skills lead to a collective version of reality; the narrative is almost always public. Thus, the adaptive pressure driving the expansion of symbolic capacity, the usefulness of symbolic invention, and the value of a high-speed speech mechanism with a huge memory capacity all depend upon the ability of the mind to harness these abilities toward the reconstruction of reality or, in the minds of some philosophers such as Goodman (1976) toward its construction in the first place.

Bruner classified narrative skill as a form of thinking, rather than as an aspect of language. But it might be seen more simply as *the natural product of language itself*. Language, in a preliterate society lacking the apparatus of the modern information-state, is basically for telling stories. Language is used to exchange information about the daily activities of the members of the group, to recount past events, and to some extent to arrive at collective decisions. Narrative is so fundamental that it appears to have been fully developed, at least in its pattern of daily use, in the Upper Paleolithic. A gathering of modern postindustrial Westerners around the family table, exchanging anecdotes and accounts of recent events, does not look much different from a similar gathering in a Stone age setting. Talk flows freely, almost entirely in the narrative mode. Stories are told and disputed; and a collective version of recent events is gradually hammered out as the meal progresses. The narrative mode is basic, perhaps *the* basic product of language.

The supreme product of the narrative mode, in smaller preliterate societies, is the myth. The myth is the authoritative version, the debated, disputed, filtered product of generations of narrative interchange about reality. In conquering a rival society, the first act of the conquerors is to impose their myth on the conquered. And the strongest instinct of the conquered is to resist this pressure; the loss of one's myth involves a profoundly disorienting loss of identity. The myth stands at the top of the cognitive pyramid in such a society; it not only regulates behavior and enshrines knowledge, but it also constrains the perception of reality and channels the thought skills of its adherents. And those who preserve and regulate myth—priests and shamans—hold positions of great power in the collective cognitive hierarchy.

The importance of this role is seen clearly if the place of myth in the architecture of mind is considered. Myth is the inevitable outcome of narrative skill and the supreme organizing force in Upper Paleolithic society. Judging from the continuing role of mimetic representations, as seen in ritual, song, dance, and games, the construction of narrative and myth did not displace mimetic cognition but rather complemented it perfectly. Myth had its special place in the cognitive armamentarium; it stood at the very center, in a controlling role. Mimetic skill, so crucial in the cultural achievements of early *Homo*, was still important but relegated to a less central role.

Linguistic Innovation and Diffusion

One feature of spoken language that is very well documented is its extremely rapid and continued growth and development. Comparative linguists have documented the evolution of various language groups, and the one consistent rule is change and experimentation. All of the modern European languages, whose immediate roots are well understood, are products of the last millennium: English took its modern form in the thirteenth century, French in the twelfth, Spanish in the fourteenth, Portuguese in the sixteenth. We have excellent documentation on the rate of change of written languages and considerable evidence on the spread and evolution of unwritten ones as well. For instance, the languages of Pacific Islands spread out south of Malaysia have been thoroughly studied (Renfrew, 1987), and their rate of linguistic change has been estimated.

Recent studies of the Indo-European linguistic family (Renfrew, 1987; Gamkrelidze and Ivanov, 1990), which includes most languages of Western Europe, Eastern Europe, the Mediterranean, and India, suggests that they all derived from a common ancestor about 8,000 years ago, spreading across Eurasia with agriculture. Thus, languages as diverse as English and Sanskrit differentiated from a common tongue only eight millennia ago. Given this evidence on the common origins of very diverse tongues, it is conceivable that all known human languages, including such diverse types of language as the many agglutinative tongues found around the world, descended from a common ancestor in less than a hundred thousand years. The linguistic heritage of *Homo sapiens sapiens* is a heritage of incredibly rapid, unceasing change and innovation.

Rapid, continuous innovation is thus an inherent property of language. This serves to justify, in retrospect, the emphasis I have placed on symbolic invention as the central underlying linguistic skill. This inventive, unifying, imaginative force must have had an explosive impact on mimetic culture, which had moved very slowly for more than 1 million years, changing in relatively minor ways. Some aspect of language had a multiplier effect on every cognitive capacity that had gone before. Words somehow enabled a new level of cognitive discourse, in a realm of their own.

An Overview of the Language Adaptation as a Unity

Viewed from an evolutionary perspective, the language system brought with it not only a new vocal apparatus but a *wholly new system for representing reality*. The entire propositional storage system and the whole of what is known as verbal semantic memory were products of language, and form part of *one vertically integrated adaptation*, ultimately unified under a "linguistic controller." The vertically integrated language system is illustrated in Figure 7.2. Note that the linguistic controller (L) encapsulates the mimetic controller; that is, access is one-way, with the linguistic system able to model the outputs of the mimetic system but not vice versa. It also has direct access to episodic event representations.

The linguistic controller is a representational process whose products are narrative models. Narrative thought is the normal, automatic activity of the linguistic controller; it produces oral–verbal commen-

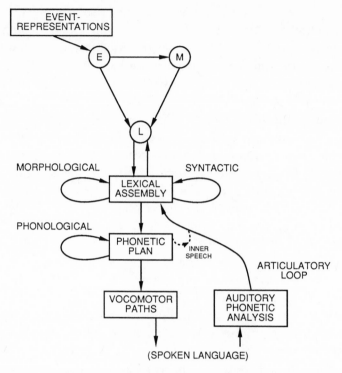

Figure 7.2 The linguistic controller (L) at the peak of a vertically integrated speech system. L constructs narrative models of inputs from episodic memory (E) and the mimetic controller (M); in turn, it drives the lexical assembly process. Lexical assembly involves not only the correct sequencing of lexical entries but also their correct form. The phonetic plan maps the assembled utterance onto neuromotor paths and, ultimately, the vocal musculature. Inner speech constitutes the silent monitoring of a phonetic plan. The articulatory loop constitutes temporary storage of a speech input.

taries on experiences just as effortlessly as the mimetic system produces representations in terms of action patterns. It is a *modeling device*, and the driving force behind linguistic invention. The controller operates the rest of the system: specifically, it directs the process of lexical assembly, whereby appropriate lexical entries are assembled into complex sentences, according to syntactic and morphological constraints. Accordingly, the core of the language adaptation resides at the top of the system, not below.

Lexical assembly must occur at the word-form, or morphological, level; it is the generic form of a word that is initially selected for a given utterance. Only after morphological constraints (such as affixes on particular words) and syntactic constraints (on appropriate word order) are taken into account can the specific utterance be assembled. Once assembled, a phonetic plan is devised, employing appropriate phonological constraints; prosodic inputs are also integrated into the utterance at this level, presumably through a parallel vocomimetic pathway (not shown) that is monitored by (L) but driven directly from (M). In the same manner, gestures can interact with spoken language by parallel pathways; the mimetic controller independently provides the "mimetic envelope" for an utterance, but (L) is the ultimate governor of any synchronization between the two representational systems. Note that the semantic content of a word is altered by its context and is specific to the particular operative mental model resident in (L).

What Is Localized?

Just as in the case of mimesis, the language adaptation had to involve many different parts of the brain; one cannot hope to place the whole of such a complex adaptation in any single brain area. The phonological device, auditory object constancy, the articulatory loop, and even the lexical entry system may be tracked to a number of different specific regions of the brain: these include the cerebellum, premotor cortex, supplementary motor cortex, thalamus, basal ganglia, superior temporal lobe, hippocampus, angular gyrus, supramarginal gyrus, insular and opercular regions, and inferior frontal cortex. Lesions in any of these areas may affect language in a variety of ways, and yet it is not clear that any of these lesions is destroying the linguistic controller. We cannot even conclusively limit language control to the left side of the brain; as shown in Chapter 3, the right side seems to play a complementary role to the left in a variety of metalinguistic tasks. In other words, when the language adaptation evolved, most brain regions changed to some degree. Once again we are looking at mosaic evolution driven at level four—the programming of a highly complex anatomical change by forces largely ensconced at the sociocultural level of selection pressure.

Pierre Marie thought he knew the location of the linguistic controller: it must be, he thought, in Wernicke's region, in the first temporal gyrus. The temptation to assign this role to Wernicke's region was understandably great, since the most crippling aphasias seemed usually to involve this area. Yet there are tremendous difficulties with this position. The greatest problem is the variability of symptoms that result from its destruction. Take the following two cases, drawn almost at random from the many single case studies in the literature. Case number one was reported by Badecker and Caramazza (1985): a 54-year-old male with an extensive posterior perisylvian lesion extending from the first temporal gyrus to the supramarginal gyrus, with some involvement of the internal capsule (deep motor paths) and the insular part of the temporal lobe. His primary symptom was "agrammatism," a symptom that involves difficulty with the production of complex sentences and discourse. His comprehension was preserved, and he could find the appropriate word in a fill-in-the-blanks test; but he could not produce a sentence without missing prepositions, omitting auxiliary verbs, or substituting inappropriate words. He seemed unable to retrieve and assemble complex sequences of words, whereas he had no difficulty finding words one at a time.

Case number two is a famous one reported by Marshall and Newcombe (1973) in their classification of dyslexia. A 45-year-old male with a left temporal-parietal (perisylvian) lesion that 75 percent overlapped the lesion in the previous case, the main difference being that it spared the internal capsule and extended down to the angular gyrus instead of up to the supramarginal gyrus. The main symptom here was a specific impairment of reading and writing; there was no agrammatism and no difficulty with oral language. In both cases, the first temporal gyrus on the left was severely damaged, and in neither case was there evidence of the devastating comprehension deficit that is supposed, according to the classical theory, to follow damage to Wernicke's region. Expressive language was seriously impaired in both cases, but in drastically different ways: spoken grammar was impaired in one, and writing in the other. Oral comprehension was intact in both. Such cases are not exceptional; a lesion to either Wernicke's or Broca's area might result in any number of neurolinguistic symptoms, and in some cases none at all (Mohr, 1976).

Computerized tomographic (CT) brain scans have confirmed these clinical observations. Selnes, Knopman, Niccum, and Rubens (1983) measured auditory comprehension in 39 stroke patients, classifying them into four groups according to the degree of initial impairment and subsequent recovery; CT scans were used to localize lesions. They found that the physical lesion most common in impaired patients was roughly in, or close to, Wernicke's area: the superior temporal gyrus and the supramarginal gyrus. However, 22 percent of the *least* impaired patients, including some who had no measurable language deficit, also had lesions in Wernicke's area! Moreover, when the same research group looked at patients with severe deficits in articulation and speech, the CT scans showed lesions primarily in the face area of the motor cortex rather than in Broca's region. In fact, there was no significant difference in the incidence of Broca's area lesions between the most fluent and least fluent patients! Thus, we are faced with the reality of a significant number of patients with lesions in the classical speech areas, and yet with none of the predicted aphasic symptoms. Such inconsistencies in the underlying anatomy raise serious questions about the viability of the classic model, and introduce the distinct possibility that cortical language localization is not as fixed across individuals as the model predicts. This would apply especially to the highest levels of linguistic and metalinguistic control.

Another difficulty with the classic model is that it does not account for complementary evolution of many other structures in the emergence of language. Mythic culture, like all major hominid innovations before it, was a complete pattern of cultural adaptation, including some very complex anatomical adaptations. The variety of structures involved in such a major change is staggering. Changes occurred to most areas of the brain, as well as to many peripheral nerves and receptor surfaces. There was major muscular and skeletal redesign, including the face, body mass, cranial shape, respiration, and posture; there was a revolution in social structure; and there was a great change in the fundamental survival strategies of the human race. The entire nervous system had to adjust to its new selection pressures and changing conditions; it was not a simple matter of acquiring a new "language system" with a cleanly isolated cerebral region attached to a modified vocal tract.

Through all of this, however, the mimetic system remained functionally distinct from language; and this is the most compelling reason for predicting broadly different cortical representations for the two systems. The big difference at the top of the two great cognitive hierarchies, mimetic and linguistic, seems to be shown in the broad demarcation between the "perisylvian brain" of language, and the dorsally located association areas probably underlying mimetic representation and reproductive memory. If local cortical expansion reflects the relative growth of these two systems (admittedly a very big assumption), then the association cortex of *erectus* should have expanded (relative to habilines and australopithecines) more dorsally and less in the temporal–perisylvian region, while the sapient expansion (relative to *erectus*) should have had a much larger temporal aspect. Further comparative examination of fossil skulls may resolve this issue.

Regarding the inevitable issue of laterality: phonology is highly lateralized, probably because it is an *axial* skill; that is, it harnesses vocal muscle-groups that straddle the central vertical axis of the body. Thus, it requires precise, high-speed integration of the two sides of the vocal tract and mouth. Cortical motor representation is generally crossed, with the right brain controlling the left side of the mouth, and vice versa; this might impede high-speed coordination of the two halves of the vocal apparatus. Integration of the two sides might be achieved more efficiently from one locus, rather than two. In this regard it is interesting that split-brain patients do not show any asymmetry in speech motor control, even immediately after the operation. If any aspect of their vocomotor apparatus was bilaterally controlled at the cortical level, and the callosum was completely severed, they ought to speak normally from only one side of their mouth, much like some hemiplegic stroke patients. But they don't; they articulate symmetrically, apparently controlling both sides of the vocomotor apparatus from the isolated left hemisphere.* This is powerful testimony to how strongly lateralized phonology has become in the modern human brain.

---

*The critical neural structures on the left might be subcortical, as well as cortical; this applies especially to the basal ganglia. As reviewed in Chapter 3, recent radiological evidence suggests that most expressive aphasias involve damage to these areas, which sit just below the inferior frontal region and are often injured simultaneously with the surface cortex. The basal ganglia are closely interconnected with frontal and prefrontal cortex, and broadly involved with various forms of skilled motor control. Injury to the right basal ganglia region apparently produces no expressive aphasia; thus, the lateralization of phonology extends to subcortical structures as well.

If the phonological apparatus is usually lateralized, then following the hierarchical, lexically driven model espoused in this chapter we must conclude that the phonological (and morphological) aspects of lexical entries are lateralized. This in turn implies that lexical access could be severely impaired by left-sided lesions, even if the linguistic controller, and thus the semantic content of lexical entries, had been left unscathed. If sentence construction is lexically driven, as Levelt has argued, then apparent "agrammatisms" could also result from impaired lexical access, due entirely to the remote effects of partial disconnection and destruction of the phonological apparatus. Anomias could be another product of impaired lexical access, explainable without necessarily implicating the highest levels of the system. Paragrammatisms, paraphasias, verbal amnesias—a whole range of apparently distinct symptomatology might similarly be due to the disruption and removal of critical addresses and related elements of the lexical entry system.

But studies of "signing aphasia" in the deaf suggest that the lateralization of expressive language in the deaf brain extends beyond phonology, to include formal sign languages. Poizner, Klima, and Bellugi (1987) also found that those left-sided cortical regions involved in signing aphasia in the deaf were somewhat different from those supporting phonology in the hearing population. Specifically, the parietal lobes appeared to play a more important role, and the temporal lobes a somewhat different role, in the deaf population. It is not yet known whether this is a result of a general redeployment of temporal cortex in the deaf or evidence for a fundamentally different cortical organization for sign language that would prove valid in both hearing and deaf people.

The left lateralization of signing in the deaf has been typically explained at the modular level, and the language module, at least as far as it is made visible in aphasia, appears to be localized on the left, regardless of modality of expression. Poizner and colleagues have drawn an even broader conclusion, that the representation of intrapersonal space, and hence of skilled movement, generally resides in the left hemisphere. However, this is moving somewhat ahead of this part of our evolutionary scenario, since formal sign language is a very recent invention. Sign language is really more similar to ideographic reading and writing than to speech, and neither sign language nor writing were part of the linguistic picture in oral–mythic cultures. Thus, the localization of these skills is not relevant at this stage. Both topics are addressed again in Chapters 8 and 9.

Damage that was restricted to the linguistic controller would result in quite a distinct syndrome: spoken language would remain fluent and word finding would not be impaired. There would be no reason to expect difficulty with repetition or with limited comprehension and production of single words and short, highly overpracticed stock sentences and phrases. Mimetic uses of vocalization would likewise remain unaffected. But narrative and discourse-level uses of language would break down; symbolically based thought would be selectively impaired; and the comprehension of linguistically encoded ideas would be destroyed. Moreover, linguistic invention would be lost. In short, damage to the linguistic controller would produce symptomatic evidence that narrative skill, the dominant form of representation driving language invention and use, was impaired.

Is there an aphasia syndrome (call in discourse-level aphasia) corresponding to the above? There would have to be a reassessment of Wernicke-style aphasias, whereby those whose symptoms were explainable without recourse to the controller were separated from the population and those remaining were reclassified as deficient *primarily* in the highest (narrative/discourse) aspects of language. The closest thing in the exiting literature to discourse-level aphasia might be the deficit in the recall of stories shown by some left temporal-lobe epileptics (Milner, 1965). These patients have difficulty recalling the details of narrative material, even very short narratives. They have difficulty attributing agency and action; if asked about the content of a story, they have difficulty even with major aspects of the plot and the major characters. The deficit is made worse (although the epileptic symptoms are usually relieved somewhat) by neurosurgery of the diseased areas. This is a very specific cognitive loss, not attributable to aphasia in the normal sense, or to dementia, intellectual loss, or sensory loss.

Frisk and Milner (1990a, 1990b) have shown further that this deficit is not due to a reduction in immediate working memory capacity, or in the rate at which verbal information can be processed. The deficit is characterized by loss of narrative information after a short delay (even overlearned material is lost in 20 minutes) and is made worse in proportion to the amount of left hippocampal tissue removed in the surgery. The specificity of this deficit points in the general direction of discourse-level aphasia, but also illustrates the difficulty of distinguishing between the mnemonic and linguistic aspects of discourse comprehension and expression. Further work will be needed, especially on the question of whether these patients suffer an equivalent and parallel deficit in the self-directed organization and production of discourse.

Until more careful neuropsychological investigation of these questions, we cannot say where the linguistic controller is, or even whether it is exclusively on the left. Perhaps the highest levels of both the mimetic and linguistic systems are bilaterally organized. (Why not? It is not clear that high-speed lateralized systems would retain any advantage at this slower-moving level of control.)

We cannot even say to what degree language is localized in similar ways in different individuals. The information-flow pattern in linguistic modeling is crucial here. The linguistic controller receives its inputs from a variety of sources, and Geschwind's (1965) idea that many symptoms may result from disconnection might apply: interrupt the flow of information in a complex system and the results are somewhat unpredictable. Fortunately, for the type of model presented here, localization is not a critical issue. Neuropsychological information was critically important in establishing the functional independence of cognitive subsystems but not in locating them in specific neural structures. Nevertheless, functional localization, or at least a localization *principle*, must remain a long-term goal in this field.

Conclusion: Mythic Culture and the Uses of Language

Mythic culture tended rapidly toward the integration of knowledge. The scattered, concrete repertoire of mimetic culture came under the governance of integrative myth. The importance of myth is that it signaled the first attempts at symbolic models of the human universe, and the first attempts at coherent historical reconstruction of the past. By definition, history, and most narrative, is a reconstructive act; it is an attempt to piece together a large number of episodes so as to give a place and a meaning to the smaller-scale events the myth encompasses. The intellectual gulf between an episodic culture and a mythic culture is huge and obvious; but the intellectual leap from mimetic representations to an integrative system of myth, however primitive, was also formidable.

Mythic integration was contingent on symbolic invention and on the deployment of a more efficient symbol-making apparatus. The phonological adaptation, with its articulatory buffer memory, provided this. Once the mechanism was in place for developing and rehearsing narrative commentaries on events, an expansion of semantic and propositional memory was inevitable and would have formed an integral part of the same iterative process of evolutionary change. At the same time, a major role in human attentional control was assumed by the language system. The rehearsal loops of the verbal system allowed rapid access and self-cueing of memory. Language thus provided a much improved means of conscious, volitional manipulation of the mental modeling process.

The social consequences of mythic integration were evident at the cultural level: narratives gave events contextual meaning for individuals. In Paleolithic cultures, and in aboriginal cultures in general, the entire scenario of human life gains its perceived importance from myth; decisions are influenced by myth; and the place of every object, animal, plant, and social custom is set in myth. Myth governs the collective mind. This remains essentially true today, even in modern postindustrial cultures, at least in the realm of social values.

Symbolic invention on a grand scale allowed the inherent structure of episodic events to be articulated. Symbolic devices, particularly the lexicon, enabled and triggered mythic invention, by which events could be mentally restructured, interrelated, and reshaped in the mind's eye. The human mind had come full circle, starting as the concrete, environmentally bound representational apparatus of episodic culture and eventually becoming a device capable of imposing an interpretation of the world from above, that is, from its collective, shared, mythic creations.

Third Transition: External Symbolic Storage and Theoretic Culture

From Mythic to Theoretic Governance

Episodic, mimetic, and mythic culture are all broad, unifying concepts that express the dominant cognitive quality of the individual mind in relation to society. The two previous transitions represented major qualitative breaks with the cognitive past. A third cognitive transition, which will be developed in this chapter, likewise is signaled by a major break with the previous cultural pattern—that is, a break with the dominance of spoken language and narrative styles of thought.

From the start, I have made the simplifying assumption that each cognitive adaptation in human evolutionary history has been retained as a fully functional vestige. The simplest working hypothesis, by far, is that, when we acquired the apparatus required for mime and speech, in that order, we retained the knowledge structures, and the cultural consequences, of previous adaptations. A corollary is that evidence of fundamental change would not be found within expansions of existing representational systems; evidence should rather be sought in new systems and new classes of cognitive output.

I have also used a structural criterion for establishing previous cognitive transitions: they were accompanied by basic architectural change, meaning that a different superordinate organization was imposed on cognition. Thus, we might be justified in postulating a new level of cognitive functioning if it could be documented that there were truly new representational systems in the human brain and if there was a new overall architecture to human cognition.

But before attempting that documentation, it might be useful to review, diagrammatically, the changes in the architecture of cognition that occurred during the first two transitions (Figure 8.1). For the purposes of this summary diagram, the details of each subsystem are left out, and each new representational system in human cognitive evolution is represented simply by a circle. Each level of development is shown as a complete architectonic configuration, that is, an arrangement of representational modules. These modules are functional, and not necessarily anatomical, entities.

Thus at level I, the starting point, representational structure is very simple; the mind had only one way of representing reality—as event perceptions. Therefore, the whole range of complex event representations held by chimpanzees, and by extension australopithecines, is enclosed within a circle, labeled "episodic" (E). The transition from level I (episodic) to level II (mimetic) is also simple to conceive in this way; the entire range of mimetic knowledge, from games and tool-making skills to group ritual and standardized gesture, is enclosed

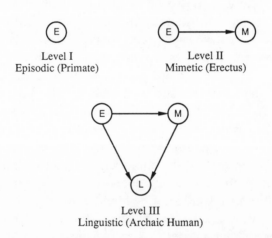

Figure 8.1   Emergence of distinctively human representational systems. The entire repertoire of primate episodic representations is enclosed within the circle marked E. The mimetic system (M) came with the first transition in human cognitive evolution; the linguistic (L) system came with the second. Newer representational systems can model the outputs of archaic and vestigial ones; thus L encapsulates both E and M, while M encapsulates E.

within another circle (M). Mimetic representations are metaphoric, rehearsable self-representations built upon episodic knowledge; therefore, by definition, episodic outputs are available to the mimetic representational process. An arrow indicates this access route. This relationship is asymmetrical: episodic systems cannot model the outputs of the mimetic representational system; hence, the arrow is unidirectional.

Note that the new representations emerge at the top of the system; cognitive structures that formerly served as the "central processor" of mind are superseded, and encapsulated, by other structures. This is clearly the case after the first transition; episodic representations could not have been informationally encapsulated in the primate. In apes, episodic event representations are at the cognitive pinnacle; they are supramodal, and it is reasonable to assume that they can access outputs from all sensory modules and thus define the forms of their conscious experience. But, after becoming embedded in a matrix of higher representations at level II, the episodic mind would have become encapsulated, gradually surrounded by more powerful methods of representing reality, while it continued to produce its traditional outputs. Fodor would have little choice but to place episodic experience in his "central processor" category in apes. But in humans, the central processor—that is, the highest, unencapsulated structure— seems to have moved elsewhere. The integrative machinery of episodic experience was superseded by higher mimetic structures: *sic transit gloria mundi.*

Mimetic skill was at the core of the cognitive style of *erectus*, and it could access all that went before. Therefore, in its time, it must have been unencapsulated; but it too was bypassed and relegated to the cognitive provinces, when the linguistic system emerged at level III, after the second transition. Although mimetic representations continued to access episodic experience, they could not model linguistic content. Thus, they suffered the same fate as episodic mind and were encircled by a more powerful representational apparatus. Nevertheless, mimesis continues to exist as an independent representational system: for someone lacking language, mimesis remains the highest, or governing, way of representing reality and presumably dominates the forms of conscious experience. But a linguistically able mind will relegate mimesis to a secondary role.

The architectonic configuration of level III, where speech is introduced into the structure, is more complex, since speech—that is, narrative skill—can encompass *both* lower levels but not vice versa: thus two more asymmetrical relationships appear. Speakers can assign descriptions to episodic material—events and things—and they can construct commentaries built on mimetic representations, for instance, on a rehearsed skill sequence, or an event re-enactment, or a group ritual, or a pattern of gestural communication; but speech is not transparent to the mimetic or episodic mind. In the diagram, speech and narrative skill are enclosed within a circle labeled "linguistic" (L), which encompasses all of the oral–narrative knowledge held by an individual. It includes a collection of stored commentaries on episodic and mimetic material, which amounts to a superordinate class of models held by the linguistic system.

The speech system thus has a special status at the top of the hierarchy; it can access, independently, both episodic and mimetic outputs and formulate linguistic descriptions and encapsulations of these; and this relationship is unidirectional. Neither episodic nor mimetic mind can, by definition, comprehend speech outputs per se; it follows that, when skilled mimetic acts are synchronized with speech, semantic control of this synchronization must rest with the language system; the governing system is oral–linguistic.

This was the situation in oral–mythic culture. The main source of evidence in favor of a further evolutionary transition in human cognition would lie in evidence of a fundamental change in this overall architecture, and therefore architectural change will be reviewed repeatedly in this chapter as a history of modern symbolic representational systems is developed. The evidence for this change might be found in the anthropological and historical record, as in the case of previous transitions; but there is a great deal more evidence to consider, almost too much. The third transition must have been recent; the key question is whether there are aspects of modern cognition that were absent from oral–mythic culture, that is, from simple hunting–gathering cultures.

Three crucial cognitive phenomena appear to have been underdeveloped, or virtually absent, in oral–mythic culture. These phenomena are *graphic invention, external memory,* and *theory construction.* Graphic invention is the first clue to what has happened in recent cultural transformations; accordingly, the next section will examine the emergence of new methods of graphic representation over the past few thousand years and the parallel development of thought skills. On a simple level, the invention of graphic representations signaled a shift in the relative importance of the two major distal–perceptual modalities, from auditory to visual representation. On another more important level, it signaled the invention of entirely new classes of symbols from those used in mimetic and oral linguistic communication. Visual symbols have become the dominant form of representation in modern society.

The second major clue is likely to be found in the realm of memory. Whereas oral–mythic cultures rely heavily on individual biological memory, modern cultures rely much more on external memory devices, mostly on various classes of graphic symbols, from pictures and graphs to ideograms and writing. Thus, the shift is from *internal* to *external* memory storage devices. As the pattern of memory use shifts toward the external symbolic store, the architecture of the individual mind must change in a fundamental way, just as the architecture of a computer changes if it becomes part of a larger network. This architectural change will be discussed in a later part of this chapter.

But the most important cultural product of human cognition is less obvious, and much more dominant in terms of cognitive governance: it is a relatively new kind of thought product known as *theory*. As we have seen, Bruner (1986) pointed out that there are two broadly different modes of thinking evident in modern humans. One is sometimes called narrative thought, and the other is variously called analytic, paradigmatic, or logicoscientific. In modern culture, narrative thought is dominant in the literary arts, while analytic thought predominates in science, law, and government. The narrative, or mythic, dimension of modern culture has been expressed in print, but it is well to keep in mind that in its inception, mythic thought did not depend upon print or visual symbolism; it was an extension, in its basic form, of the oral narrative.

The major products of analytic thought, on the other hand, are generally absent from purely mythic cultures. A partial list of features that are absent include: formal arguments, systematic taxonomies, induction, deduction, verification, differentiation, quantification, idealization, and formal methods of measurement. Argument, discovery, proof, and theoretical synthesis are part of the legacy of this kind of thought. The highest product of analytic thought, and its governing construct, is the formal *theory*, an integrative device that is much more than a symbolic invention; it is a system of thought and argument that predicts and explains. Successful theories often convey power.

Combining the three variables just mentioned, the governing cognitive structures of the most recent human cultures must be very different from those of simple mythic cultures. They exist mostly outside of the individual mind, in external symbolic memory representations, which are dependent upon visuographic invention, and they culminate in governing theories. The aspects of culture that fall under the governance of theories are here labeled "theoretic." The third transition, from mythic to theoretic culture, was different from the preceding two, in its hardware: whereas the first two transitions were dependent upon new *biological* hardware, specifically upon changes in the nervous system, the third transition was dependent on an equivalent change in *technological* hardware, specifically, on external memory devices. Theoretic culture was from its inception externally encoded; and its construction involved an entirely new superstructure of cognitive mechanisms external to the individual biological memory. As in previous transitions, earlier adaptations were retained; thus, theoretic culture gradually encompassed the episodic, mimetic, and mythic dimensions of mind and indeed extended each of them into new realms. The last part of this chapter will focus on theoretic development.

What was truly new in the third transition was not so much the nature of basic visuocognitive operations as the very fact of plugging into, and becoming a part of, an external symbolic system. Reading, for example, is a very distinctive mode of knowing, one that raises disturbing questions about the true locus of human memory. Moreover, theoretic culture broke with the metaphoric style of meaning in oral–mythic culture. Where narrative and myth attribute *significances*, theory is not concerned with significance in the same sense at all. Rather than modeling events by infusing them with meaning and linking them by analogy, theory dissects, analyzes, states laws and formulas, establishes principles and taxonomies, and determines procedures for the verification and analysis of information. It depends for its advanced development on specialized memory devices, languages, and grammars.

The first step in any new area of theory development is always antimythic: things and events must be stripped of their previous mythic significances before they can be subjected to what we call "objective" theoretic analysis. In fact, the meaning of "objectivity" is precisely this: a process of demythologization. Before the human body could be dissected and catalogued, it had to be demythologized. Before ritual or religion could be subjected to "objective" scholarly study, they had to be demythologized. Before nature could be classified and placed into a theoretical framework, it too had to be demythologized. Nothing illustrates the transition from mythic to theoretic culture better than this agonizing process of demythologization, which is still going on, thousands of years after it began. The switch from a predominantly narrative mode of thought to a predominantly analytic or theoretic mode apparently requires a wrenching cultural transformation.

Mythic culture, in the purest sense of the word, extended to include all Upper Paleolithic, Mesolithic, and Neolithic societies. It continues today in many traditions, and its vestiges are still highly visible in some sectors of postindustrial civilization. Its exact outer boundary cannot easily be drawn. Theoretic culture grew from within and has gradually encompassed mythic culture. It has been developing for several millennia, and has become the dominant thought form of postindustrial society.

The three central cognitive innovations underlying theoretic culture will be discussed in the following sections, which deal respectively with visuographic invention, external memory, and theoretic thought.

Visuographic Invention

The critical innovation underlying theoretic culture is visuographic invention, or the symbolic use of graphic devices. Judging from available archaeological evidence, it took sapient humans thousands of years to develop the first methods of visual symbolic representation. Visuographic invention ultimately provided three new visual symbolic paths. The historical order of emergence of these three paths is important, because it demonstrates the independence of the three visual paths and speaks to the original underlying function of each. Thus, the history of visuographic invention has to be reviewed in some detail.*

Visuographic invention developed within a predominantly mythic cultural environment, but it did not unfold all at once. Very early in human prehistory, progress was slow. There was very little use of graphic devices prior to 40,000 years ago, when they proliferated more or less simultaneously in Africa, Europe, Australia, and Asia; only much later did they appear in the Americas. The development of visuographic symbols then went through several apparently orderly stages, independently following a roughly similar course in a number of locations.

Prior to the late Upper Paleolithic, some antecedent developments pointed the way to intentional visual symbolism. Body painting probably dates back to the earliest archaic sapiens, as testified by a large collection of yellow, brown, red, and purple pieces of ochre found at Terra Amata, near Nice, France, in 1966, that dated back to about 300,000 years ago. The inhabitants of this site had good basic constructive abilities; they built simple huts out of branches, and one such hut was rebuilt at least eleven times on the same site (Hadingham, 1979). They also left other traces in the sand that testified to an impressive range of skills: impressions of wooden bowls, outlines of skins used to carpet the hut floor, remnants of stone tools. All of these traces suggest the beginnings of craft, but not necessarily symbolic uses of visual representations.

One of the earliest artifacts with purposeful visuographic markings is a 200,000-year-old engraved rib at Pech de l'Azé, Dordogne, France (Marshack, 1972), which has an interconnected series of carefully engraved double arcs. But this object is an isolated example from that period. In any case, the engraving on the Pech de l'Azé rib does not have any recognizable referent, and it is possible that it, like early body painting, might have been done without any focused representational purpose.

---

*The reader is referred in particular to a two-volume history of writing by Diringer (1948, 1962) and to Marshack's research (1972) on early pictorial representations.

Another very early precursor to visuographic invention was the purposeful arrangement of objects, usually in a ritualistic setting. Very old artifacts have been discovered in spatial arrays that suggest a symbolic purpose; for instance, circles made of bears' skulls, arrangements of mammoth tusks, and graves decorated with flowers. Some of the skull arrangements are associated with Peking Man and are dated back more than 400,000 years; apparent grave decorations have been found in Neanderthal grave sites dated back over 60,000 years. But the arrangement of significant objects falls far short of using the visual modality for either pictorial or linguistic representation. Thus, for several hundreds of thousands of years, from the late *Homo erectus* period to Neanderthal culture, hominids showed no further progress, and no evidence of systematic visuographic symbolic invention.

The earliest cultures of biologically modern humans (*Homo sapiens sapiens*) did not initially achieve more than their predecessors, despite the high development of spoken language. Among documented hunting–gathering cultures, the most universal visuographic feature seems to be self-adornment: body painting, ritual scarring of the face and torso, elaborate hair arrangements, and refinements of costume. Costume in particular took on vital symbolic functions in ritual and religion, probably drawing on the visually creative resources of the society more than any other medium. Costume, and especially face masks, could be regarded as a true graphic invention, one of the most universal visual symbolic devices, which have been documented in all human societies, even the poorest and most isolated. Thus, costuming the body and masking the face might be regarded as the most ancient graphic (quasi-symbolic) inventions, but they are still an essentially mimetic use of graphic skill.

Artifacts from the late Upper Paleolithic period, however, signal a change in the social role, as well as in the style and purpose, of vision and of graphic invention. Unlike body decoration or the purposive arrangement of significant objects, these new graphic inventions were pictorial in nature; that is, they were either two- or three-dimensional representations of recognizable perceptual objects, usually animals. Starting around 40,000 years ago, there was a proliferation of engraved bones and carved ivory: the carvings were highly skilled two- and three-dimensional representations, mostly of contemporary animals. The first evidence of truly advanced painting and drawing skills dates back to about 25,000 years ago, in hundreds of illustrated limestone caves of the Ice Age, of which the best known are at Altamira and Lascaux. Clay sculptures and figurines and a wide variety of trading tokens were common 15,000 years ago. The earliest evidence of writing dates back about 6,000 years, to the emergence of large city-states; and the idea of the phonetic alphabet is less than 4,000 years old.

The newness of graphic symbolic invention might be surprising, given the pre-existence of fully developed spoken languages and of various forms of mimetic representation, including ritual and gesture, that employ visual signs extensively. Apes, and by extension all hominids, can perceive and use complex visual symbols, and the motor skills used in toolmaking would have sufficed for basic visual symbolic invention. Therefore, once again, as in the case of spoken language, the brunt of the evolutionary burden must be shifted to the highest cognitive aspects of symbolic invention. Before visual symbols could be invented, the modeling intellect had to perceive a need for them. In principle, this is similar to the argument for the invention of the oral lexicon; symbolic invention is generally driven by the demands of mental modeling.

Visuographic invention was thus not primarily a visual–perceptual innovation, despite its name. Rather, it evolved as a set of high-level production systems that operated according to conventions and rules. From the start, there was no single "correct" or "perfect" way to represent the world visually. The symbolic inventions underlying drawing and sculpting techniques, and conventions that led to geometric representations of space, were not straightforward perceptual or motor innovations as much as they were guidelines for the manufacture of symbolic artifacts.

There are three broadly different modes of visual symbolic invention, which might be called pictorial, ideographic, and phonological. They are not mutually exclusive; for example, pictographs can be used to represent linguistic events directly. But writing generally does not depend upon perceptual isomorphisms and metaphors, whereas the pictorial mode does. The pictorial mode is primarily dependent upon visual norms and does not necessarily engage linguistic structures. But the phonological mode depends upon its linkages to the

speech system; and the ideographic mode falls somewhere in between. The three modes are quite distinct in the type of thinking they
allow.

All visuographic inventions are recent, and even the development
of pictorial representations came long after spoken language. Thus, it
would be wrong to argue that, somehow, the mere possession of spoken language automatically led to graphic invention. Spoken language usually has not been accompanied by a writing system, even
within recorded history. Harris (1986) has observed that, of the many
thousands of languages spoken at different places and times by humans, fewer than one in ten have ever evolved an indigenous written
form. Of those that have, the number that have yielded a significant
body of literature barely exceeds one hundred. Writing was not only
a late development, it was a very rare one.

Highly evolved spoken languages, and fully developed oral–mythic
traditions, were not necessarily accompanied by a sophisticated pictorial tradition either. Although well-developed pictorial representations are often found in some preliterate societies, there are many
preliterate societies in the historical record that lack any tradition of
sophisticated visual representation, particularly of two-dimensional
representation. Moreover, the possession of advanced graphic skills
did not lead automatically to literacy; the exquisite drawings and
paintings in the caves at Lascaux are just as refined as the visual products of many early literate civilizations, such as those of Mesopotamia
in the fourth millennium B.C. But the Lascaux artists lived 10,000 to
20,000 years before the invention of writing, while their Mesopotamian equivalents were working within an emerging literate culture.
Thus, pictorial skills can be dissociated from both oral and written
language, on the basis of their respective histories. The pictorial aspect of the visuographic system was the first to evolve.

Pictorial Representation

Visuographic invention was limited to body decoration, grave decoration, object arrangement, and other simple symbolic applications
until the late Upper Paleolithic. Then, relatively suddenly, the situation changed. The most dramatic evidence of this change is found in
the Ice Age caves of southern Europe, which contain thousands of
sculpted figurines and carved ivory and bone objects and numerous

paintings, some of great beauty. In a relatively short period of time, graphic art had been elevated to a very high plane, one that was not to be surpassed for at least 10,000 years. Writing was nowhere in sight yet. During this period, the visual symbolic world was limited strictly to pictorial products, in two and three dimensions.

Pictorial representations emerged to serve a perceived need and immediately implied the existence of certain capacities. They appeared in the context of an existing oral–mythic culture and can only be understood in terms of that environment; therefore some exploration of what is known of that environment is necessary. The earliest pictorial products were undoubtedly simple pictographs, but surviving pictographs are generally too crude to be taken as evidence of a real visuosymbolic breakthrough. The earliest evidence of a qualitatively new, highly sophisticated pictorial skill in humans is found in the artifacts of several hundred Ice Age caves in Europe. The painted images and sculptures found in these caves are well known and have prompted much speculation about their proper interpretation. Their real significance to the many artists who produced them over several millennia will never be known with certainty. But some things about the cave paintings can be said with confidence.

One is that most of the elaborately decorated caves were not used for living but were ceremonial centers. The driving motive of early pictorial representation thus seems to have been at the very core of oral–mythic culture: ritual images with great religious significance. Evidence for this statement derives from two sources: the locations of the caves and the subjects represented in the images. Many of the caves were, and still remain, virtually inaccessible; they were far underground, dark and damp, and reached through endless dangerous passages and dead ends. These facts suggest that they were not used for regular habitation but rather for a special purpose. The remoteness and difficulty of access in some cases were extreme. Hadingham (1979) describes the forbidding entrance to a remote cavern at Montespan, in the French Pyrenees:

> For over two kilometers the stream known as Houantou near the village of Montespan flows underground through a series of galleries which require the visitor to wade in a crouching position for much of the time in order to avoid impact with the low and irregular ceiling . . . the dangerous entrance at Montespan is entirely filled with water. Through

this siphon Norbert Casteret swam with great daring in 1923, and emerged to find a gallery filled with Paleolithic engravings and the remains of sculptured animals. (p. 185)

Another example is at Niaux, also in the French Pyrenees, where the cave complex is so vast that the visitor has to walk through unlit corridors for half an hour before reaching a pitch-black circular hall known as the Salon Noir, located a kilometer inside the complex, where large images of hunted animals—ibex, bison, and so on—are painted on the walls. These would necessarily have been painted, and displayed, to firelight. The interior is too remote and uncomfortable to serve as regular living quarters, and there are no artifacts to suggest it was ever used for this purpose. The generally accepted significance of the locale of such caves is that they were used primarily for ceremonial or ritualistic purposes. They were the Paleolithic equivalent of high art. Their remote locations made them secret and special, while the paintings and sculptures on the walls became the religious icons of those societies.

The second set of certainties about Paleolithic cave art surround its major *themas*. The themas of paintings or sculptures are their subject matter, usually cast in mythic or narrative terms. Their subject matter often goes beyond simply reproducing or generating a visual image, to the expression of narrative themata. Broadly speaking, the two major themas were hunting and fertility, the first reflected in thousands of images of large game animals, some of which have spears or arrows buried in their hides, and the second reflected in countless drawings and sculptures of the female figure, usually with great emphasis placed upon the breasts and reproductive organs.

Most of the figures seem to be painted or sculpted in isolation, but some more complex compositions have also been found. The most famous of these is at Lascaux, France, in what appears to be a ceremonial pit whose ceilings are covered with hundreds of multicolored images. On one side is a three-part composition consisting of a disemboweled bison, apparently still alive and moving; a man falling backwards, whose head is in the shape of a bird and whose penis is erect; and a rhinoceros with six dots under its tail. A large number of small stone lamps were discovered on the floor nearby. The significance of the composition has not been deciphered, but its basic format is familiar to the modern eye: juxtaposition of visual metaphors, possibly recounting a story that had become embellished and enshrined as a myth.

Fertility and hunting were, and continue to be, the two great mythic themas of hunter–gatherer societies. This is reflected in their oral–mythic traditions. Beliefs in animal powers, magical identification, and fertility rituals have been documented widely in various aboriginal cultures (Frazer, 1890; Campbell, 1959). These myths are entirely consistent with the first widespread use of visual symbolic representations. It is surely no coincidence that the first recorded society to exploit the graphic medium on such a high level used its new skill for a very serious cognitive objective—to explore and develop the mythic ideas that were already the governing cognitive constructs of human society. Graphic art was initially subsidiary to existing mythic cognitive structures; but this should not surprise us. The power of this new method of representation and its usefulness in enhancing the expression and preservation of myth were evident to the creators and keepers of those myths. The first uses of visuographic inventions were thus completely congruent with the forms of oral–mythic culture.

A third certainty is that the knowledge involved in constructing these images was considerable; this speaks to the underlying skill structure suggested by pictorial invention. One could argue that the cave painters were even more skilled than many modern artists, since they had to know a great deal about materials and their use and had to fashion their own paints and tools from raw materials. Moreover, they had to work under extremely difficult conditions, in low-ceilinged caverns without natural light. Many hollowed-out blackened stones have been discovered, which seem to have served as hand-held lamps. In the cave of Tito Bustillo in northern Spain, archaeologists discovered 11,000-year-old painting materials on the floor of the cave underneath a magnificent polychrome frieze of horses. Pieces of coloring—mostly ochre—lay in the barnacle shells in which they had been mixed. A variety of flint and bone tools were scattered around; presumably they were used to apply the colors and make the engravings.

The paintings and sculptures themselves were obviously the products of a long and highly developed tradition. The evidence for this lies partly in the long time span—several thousand years—between the newest and oldest of the ceremonial caves, but also in the level of skill and knowledge of materials manifest in their contents; such knowledge would have taken time to acquire. The images created by these artists were generally realistic, with a very modern feeling for animal anatomy, and the use of color and line shows a sophisticated awareness of the methodology of two-dimensional reduction of three-dimensional figures. The sculptures were also realistic in style, at least in the earliest period, and remarkably well-proportioned. There are also some recurrent and apparently abstract painted symbols, as well as sculptures that deliberately distorted the human figure, especially the male penis and the female genitalia. Many lesser artifacts were also discovered in the same regions—carved antlers and tusks and clay figurines—some of very high quality, which confirms the presence of great graphic skill in those societies.

As Goodman (1968) has pointed out, the pictorial image is essentially symbolic; it is a *referential* device for the user of the image. But where, in the representational architecture of the oral–mythic mind, is the reference system for visual–pictorial images? Some pictorial images can be accurately perceived by apes: a gorilla can respond appropriately to certain quite subtle pictorial images; thus pictorial images, on one level, must possess episodic reference. Thus far we know of no animals other than apes that can interpret drawn images, that is, see them in terms of their referents; however, this question probably has not been explored as thoroughly as it might.

If the reference system for interpreting pictorial images is, at least in part, episodic, the *invention* of pictorial representations is another matter. There would appear to be at least two additional components at work here: mimetic manufacturing skill and linguistic conceptual skill. The same mimetic skill that enabled humans to achieve tool-making would have sufficed for the constructional aspect of the manufacture of pictorial art, while speech would have aided in devising and maintaining the complex sequences required in their manufacture. Such skills are also relatively rare; they are not species-universal features like speech.

This having been said, it might appear that pictorial representations did not really lead to any fundamental change, and that the first graphic images were simply a cooperative product of the three preexisting systems of human thought and representation. This is probably true, at least in the initial stages of their emergence. In fact, since no new biological capacities are proposed here, it could not have been otherwise. But the accompanying diagram (Figure 8.2) suggests that pictorial invention might have signaled the start of a new cognitive structure. In the diagram, the first three steps leading to modern representational architecture are shown, with the addition of a new structure, indicated by a circle in the center of the diagram marked V/S, which stands for visuosymbolic codes. The line connecting this to episodic knowledge is labeled as the pictorial path. V/S at this stage is simply a visuosymbolic input module that contains the interpretative codes and conventions for "reading" complex manufactured visual images. It necessarily has a strong linkage to episodic representations, since pictorial images, whether realistic or somewhat fanciful, have at least a metaphoric link to episodic experience. Most importantly, the first pictorial images themselves were also external representations. They existed outside of the individual, rather than in visual memory. Therefore a technological bridge was under construction that would eventually connect the biological individual with an external memory architecture.

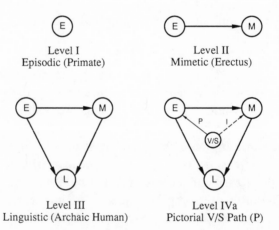

Level I
Episodic (Primate)

Level II
Mimetic (Erectus)

Level III
Linguistic (Archaic Human)

Level IVa
Pictorial V/S Path (P)

Figure 8.2   Emergence of the first visuosymbolic path, the pictorial (P) path, in the late Paleolithic, with the invention of crafted, permanent visual images. Note the nascent (still implicit) ideographic (I) path. Visuosymbolic (V/S) codes are interpretative strategies for processing crafted visual images.

Cuneiforms, Lists, and Numbers

Pictorial representation, in two or three dimensions, was the only visuosymbolic path available to humans for at least 15,000 years. During that time, pictorial skill was elaborated in some societies, but it rarely, if ever, improved on the standards set by the cave artists. The next step in visuographic invention was rudimentary writing. The invention of writing followed multiple parallel evolutions in different places, and its early history is highly complex. Writing eventually resulted in the establishment of two new visuosymbolic paths, but their emergence was covert, and therefore the argument for their existence requires a more detailed examination of their historical origins.

Writing had its origins in the most mundane of daily dealings: trade. The great majority of early written documents are records of transactions. The earliest documented artifacts are over 100,000 cuneiform tablets found in the ruins of ancient Mesopotamia, the oldest dated back to Uruk about 5,000 years ago. Cuneiform writing, named for its wedgeshaped characters, was a very successful system, persisting in use for 3,000 years, until the early Roman period. The oldest cuneiform artifacts are clay tablets on which there are usually three kinds of marks: pictures of items traded, numbers, and personal seals of identity. The items traded in quantity included animals (chickens, oxen, and so on) and food items (oil, wheat, barley). Some of these early symbols were pictorial in derivation. For example, oxen were indicated by a full-length drawing that gradually became standardized into the head of an ox. These standardized signs were eventually put on the tablets using a kind of script, which was made up of wedge-shaped impressions (cuneiforms) pressed into the clay; some of the earliest cuneiforms resembled their pictorial predecessors.

The emergence of cuneiform writing systems appears to have been very slow. Some of the earliest written Sumerian symbols can be traced to a primitive accounting system developed in Mesopotamia about 8,500 B.C. and used for several millennia (Schmandt-Besserat, 1978). Clay tokens of about twenty different types—spheres, semi-spheres, discs, and cones, each bearing incisions and punch marks— could be used both to count items and to identify them. Some of the tokens contained schematic animal forms, later modified in the Sumerian writing system. Thus, some of the symbols later found pressed into cuneiform tablets were derived not from direct attempts to model animal or objects but from pictographs of tokens, which themselves were already stylized pictorial representations, three-times removed from the original.

The reason for resorting to two-dimensional accounting systems seems to have been completely pragmatic. Shipments of goods were accompanied by a clay vessel containing the accounting tokens. Gradually the custom developed of impressing on the outside of the sealed clay vessel a list of its contents. The list was pressed into the clay in the form of two-dimensional copies of the three-dimensional tokens. Eventually, the tokens themselves became redundant and were replaced by tablets containing lists in cuneiform script (Schmandt-Besserat, 1978).

Although some of the Sumerian signs can be accounted for in this manner, others cannot, and have no obvious predecessors in pictorial or token form. Were the early accounting tokens themselves a form of writing? The answer to this question will depend on one's definition of writing. Bloomfield (1933) defined writing as any set of characters fixed to represent words. Cohen (1958) placed a further requirement on the definition: the characters have to be at least indirectly phonetic, as in a rebus system.* Gelb (1963) went even further and insisted that only the earliest syllabaries qualified as true writing systems because they were truly phonographic; that is, they directly represented the sounds of spoken language. Clearly, linguists have found it difficult to agree on exactly which feature of writing marks the essential break-point before which writing did not exist and after which it did. Harris (1986) proposed a welcome antithesis to the phonetic definition of writing: namely, that writing cannot be reduced to

---

*A rebus system uses a visual image to elicit sounds. For example, the pronoun "I" might be elicited by a picture of an eye. All early writing systems used the rebus principle to a degree. A rebus image is only *indirectly* phonetic, since the pictorial image must first be identified correctly in order to know what sound to produce, and there are usually several possible visual interpretations. In a *directly* phonetic system, like an alphabet, the primary purpose of the symbol is the evocation of sounds, and the reader has fewer degrees of freedom regarding which sounds are considered to be acceptable interpretations of the image. Phonograms often have no pictorial identity. For example, the Western alphabetic image "I" has no inherent pictorial message.

a system for reproducing the sounds of spoken language, but rather that it is a system of representing ideas in its own right. Writing, like spoken language, was used to represent ideas directly, and not merely to produce a graphic echo of speech.

The essential independence of writing is especially evident in systems of counting. Early counting systems usually possessed symbols for unity, ten, and larger units such as one hundred (in Egypt), or sixty (in the Assyrian script). Larger numbers were built out of these basic symbols. The number signs were logograms, like modern Arabic numbers: that is, the sign had no phonetic specification (the number "3" can be read as "three," "trois," "tres," or "drei," and so on; it is not a phonetic character; however, the word "three" can only be pronounced one way).

Thus, number signs were among the first true visual symbols; they were totally arbitrary, bearing no perceptual relationship to what they represented. Usually, the number one was represented simply by a slash or a dot, and these were accumulated to indicate higher numbers, until reaching a base limit, which was usually 6 or 10. Higher base numbers had different symbols—for instance, a small circle indicated units of 10 in the earliest Sumerian system, while an inverted triangle indicated 60. By combining symbols, it was possible to express powers or multiples of numbers; thus, a large inverted triangle containing a small circle represented 60 times 10, or 600. A large circle represented 60 squared, or 3,600; a large circle containing a small circle was 3,600 times 10, or 36,000; and so on. In this way, fairly large numbers, as large as were needed for trade at that time, could be represented. As the pictographs were succeeded by cuneiforms, the circles evolved into squares made of straight lines and wedges, but the same visual embedding principle continued.

The earliest texts from Uruk were mostly concerned with the economic transactions of the temples: livestock, food, and textiles were the main items traded. The texts did not contain any grammatical elements and used a principle of simply applying one sign to each concept or category. There was, at first, no phonetic principle in the construction of visual words; that is, the sign had no phonetic purpose: it was only a picture assigned to a concept by convention. Thus, at this stage, writing could be broadly classed as ideographic. A significant percentage of the texts were lexical lists, used to train scribes, which remained unchanged for at least 600 years, illustrating the conservative nature of the training system (Walker, 1987).

However, despite the absence of grammatical markers, and the lack of sentence structure, there were other principles at work that were effective in communicating the ideational intent of the writer. One of these purely visual conventions was the grouping of units of information into boxes. Instead of stretching sentences out linearly, in the same word order used in speaking, meanings were presented in *visual clusters*. Lists were not yet as orderly as lists later became; a tablet was sectioned off by the writer into squared sections, each containing an item. The order apparently was not absolutely fixed at the earliest stage of cuneiform writing, and the reader was left with the task of reconstructing the message in each box, or visual cluster. Just as tokens in the earlier Mesopotamian economy were grouped into a single container for each set of transactions, the tablets themselves served to designate the grouping into which a given set of transactions fell.

The first cuneiform writing system was not yet able to express grammar, but already something new had been gained over spoken language. Part of the gain was in the transportability and permanence of records; but another important part was in the ability to arrange virtually endless lists of items. The *list* is a peculiarly visual institution. The usefulness of oral listing is very limited, owing to memory limitations; orally memorized lists tend to tie up working memory, preventing further processing of the list. In contrast, visual lists can be arranged in various ways, and juxtaposed to simplify the later treatment of the information they contain. List arrangement can facilitate the sorting, summarizing, and classifying of items and can reveal patterns otherwise not discernible. With the invention of visual lists, the newly created state could acquire, analyze, and digest the information it needed to function.

After about a thousand years of use, cuneiform evolved into a syllabic script; the syllabic uses of cuneiforms became evident in Ur about 2800 B.C. The first syllabaries were rather loosely constructed collections of symbols, testifying to their ad hoc invention. There was no unifying principle or organization to the system. The smallest sound elements were usually monosyllabic words, and these visual syllables could be combined to form other words. In some cases, a syllable was represented by only one sign, usually a commonly used sign; for instance, the Sumerian word for barley was "she" (as in shepherd), and thus the cuneiform sign for barley could be used, reliably, to form other words that contained the same sound. But in other cases, a syllable could be represented by many different visual signs; for instance, Walker (1987) gives the example of the syllable "gu," which could be written in Sumerian in at least fourteen different ways. The opposite also held: a given visual sign might convey various sounds; for instance, the visual sign for mouth (a head with the teeth marked) could be pronounced "ka" or "gu," depending on whether the writer meant to convey "mouth" or "shouting"; semantic context was the main cue for deciding which. Thus, the linkage between visual sign and spoken language was not simple; in fact, it was secondary to the economic functions served by writing, which were distinctly dependent on its uniquely visual characteristics.

Cuneiforms acquired grammatical conventions about the same time that they became partially phonetic. They also became linear: whereas the earliest scripts had been read in loosely clustered boxes or rectangles, later cuneiforms were turned around ninety degrees, and written from left to right in straight lines, starting to imitate the spoken order of words. The progression was thus from a primarily visual medium, inventing completely new representations like lists of numbers, to a medium which, increasingly, tried to map the narrative products of the language system. To help the reader resolve ambiguities, scribes invented category markers: for instance, wooden objects were given a certain prefix, while stone objects had a different one; thus, if the same symbol could have applied to both a stone and a wooden item, the ambiguity was resolved by the prefix.

Sumerian was eventually replaced by the Akkadian language, a semitic tongue, and by 1800 B.C. the major use of cuneiforms was in Akkadian. By this time there were 600 distinct cuneiform signs in use, and the list of grammatical conventions, category markers, and possible interpretations relevant to deciphering a text would bewilder a modern alphabetic reader. Literacy was an elite skill in ancient Mesopotamia, and the reasons are obvious when the complexity of the system is taken into consideration. Despite (or perhaps because of) the complexity of the system, cuneiforms eventually were used to record legal and historical texts of some grammatical sophistication, such as the Code of Hammurabi.

The major lesson to be taken from considering cuneiform writing is that it was a crafted, hard-won symbolic invention whose fundamental cognitive function changed over several millennia. Early cuneiforms were largely pictorial in style and lacked syntactic structure. Later cuneiforms were partially phonetic and contained some grammatical markers. The latest examples of cuneiform script, dating to the Roman era, were mostly phonetic and grammatically more sophisticated.

Another lesson is that many of the conventions used in the earliest writing were uniquely *visual:* they had no fixed auditory equivalent. The visual arrangements of sentences and lists have no equivalences in sound; sound is not enclosed into pages, columns, tablets, or into lines or boxes; spoken words are not spaced apart. There are pauses, of course, in spoken discourse, but the pauses do not necessarily correspond to the spatial separations used in writing, and cannot communicate the clustering information common to written articles. Moreover, the prosodic information enveloping phonetic utterances was generally left out of visual symbolic systems from the start.

The distinctness inherent to the operation of "reading" is perhaps better revealed in these primitive systems than it is in modern alphabetic scripts, where the illusion can be created that reading consists mainly of translating visual signs into their phonetic equivalents. In assessing the content of a cuneiform text the reader is trying primarily to gain access to the ideas behind the symbols—to guess at, or reconstruct, the mental model that drove the scribe to write the message in a particular way. Pictographs were not very effective at doing this because they left so many avenues open to interpretation; grammatical markers, category markers, and phonetic signs were introduced to allow more complex messages to be written down in a less ambiguous manner. In effect, scribes were evolving a multidimensional system of notations that served to express their mental models.

Visual symbols had immediate advantages over speech. Lists of transactions and numbers were much better expressed in writing than in speech. Lists of genealogies, and other historical sequences, were also much clearer in written form, and devices such as astronomical almanacs, which were produced in cuneiform in the later periods of its use, simply could not be formulated or expressed in spoken language. Note that, for the most part, these advantages are quite independent of the grammatical or phonetic sophistication of the script. Large numbers are also an inherently visual symbolic invention, and they were perhaps the first major contribution of nonpictorial visual symbols. Looking at the early Mesopotamian cuneiform records, the number system is by far the most sophisticated representational symbolic invention on the tablets.

Harris (1986) pinpointed the invention of writing to the invention of "slotting" systems for accounting, a technique that eventually encompassed most of writing. The earlier token accounting systems had a serious limitation: the recordkeeper had to recount every item when a total had to be calculated; this is an "iterative" strategy; that is, it requires the repetition of the same operation, over and over. So long as numbers remained small, this was no problem; but when larger-scale trading developed, involving thousands of items, the limitations of such a system would have been crippling. The great invention, as Harris called it, was the shift from a token-iterative strategy to what he calls an "emblem-slotting" or slot-listing system for recording numbers. A minimal slot-list contains two signs, one of which fills a slot reserved for the class of item, and the other a complementary slot reserved for indicating a total. In his example, a token-iterative system would list 60 sheep in the form, "sheep, sheep, sheep . . ." or "sheep, another, another, another . . ."; while a slot-list system would contain a message of the form, "sheep, sixty," where the item-class slot was filled by the word "sheep" and the number slot by the word "sixty."

Slotting is a semiological principle used in oral language, and it imposes a superordinate structure on word use, regardless of the specific message. Slotting necessarily involves the isolation and differentiation of properties of events and objects from their expression in independent signs. Number signs are a specific category of sign, intended for a specific slot; a number cannot be placed in a slot intended for an object or an action or a feature, such as color. Thus, in inventing a graphic representation system for numbers, the scribe was not trying simply to draw the sounds already assigned to numbers in spoken language; the scribe was inventing completely new, and urgently needed, numerical concepts and placing them into a slot-listing system with an implicit linguistic structure.

The readers of early cuneiform literary texts were performing a complex task quite different from the reading of modern English. There were many different processes going on in parallel: direct pictorial representation, visual metaphors, categorical emblems, phonetic signs, spatial juxtaposition and enclosure, directional conventions, and the use of order information. All of these processes served to help the reader construct models of the scribe's intended message. The writing system assumed a great deal of knowledge on the reader's part. From the start, the interpretation of visual symbols was a multichannel effort, highly fluid and complex, and not a simple linear–mechanical process.

Hieroglyphs and Ideographs

The principle of applying several parallel channels of high-level processing to visual symbolic inputs is even more evident in the other great systems of ancient writing: hieroglyphs and ideographs. Egyptian hieroglyphic writing was probably the first system of writing that could convey virtually any linguistic message. Like cuneiform, Egyptian writing started out as a series of standardized drawings of familiar items, such as birds, the sun, water, wheat, kneeling priests, human faces, and so on. At this early stage, the drawings were, by our standard, still pictorial but much more sophisticated (as drawings) than early cuneiforms. Images, standardized supplementary signals such as lines, boxes, and markers, and phonetic cues were combined to convey a message.

Egyptian hieroglyphs were phonetic long before cuneiforms became phonograms; and while later cuneiforms adopted a syllabic phonetic strategy, with each syllable containing both consonants and vowels, hieroglyphs were solely consonantal; vowels were not represented (Davies, 1987). The relation between the written hieroglyphic symbol and spoken language was very complex. There were function markers, order rules, gender conventions, and other grammatical aspects of hieroglyphics that make them a puzzle for anyone who is used to seeing written language only from the viewpoint of modern English. Many of the rule relationships were encoded in a strictly visual, or graphic, mode and did not map onto the sound stream. The same principle of visual independence applies here as in cuneiform reading, although the specifics are different.

For the reader learning to read hieroglyphics out loud, the phonetic linkage often follows the rebus principle: from visual sign, to meaning, and then to sound. Thus, the sun ('ra') is portrayed as a small circle within a larger circle; to pronounce the sound corresponding to the symbol, there is no direct grapheme-to-phoneme correspondence, and the phoneme in this case can only be known after the meaning of the symbol is understood. And while in this particular example the image appears to have a metaphoric relationship to the actual physical sun, this is illusory; a circle could be made to represent almost anything, even within the range of so-called mimetic possibilities: for instance, a mouth, an eye, a hole, the self.

Thus, even in the simplest case imaginable, hieroglyphic conventions were arbitrary; their rules were complex; and most importantly, their meaning often preceded, and did not follow from, phonetic decoding. There were exceptions to this, but the point stands. This indicates the visual autonomy of the hieroglyphic sign, at least on one level. Another way in which the visuographic dimension dominated was in the representation of certain relationships and rules. There were markers for male and female endings and for connectives or modifiers. The marker for plural was a set of three parallel lines, usually placed under the sign for the pluralized word. There were also determinatives placed in sentences that were not read aloud: for example, if a sign was intended to be read as a logogram, a determinative line was placed underneath it. Sometimes determinatives were quasi-pictorial: for instance, the word "sh" could mean either "scribe" or "writing"; to resolve this ambiguity the symbol for "sh" could be written with an adjacent logogram that indicated either a man or a book-roll, the former indicating to the reader that the symbol was intended to designate a scribe and the latter the act of writing. The determinatives indicated the meaning assigned to the attached symbols and thus affected their pronunciation. The reader had to infer the meaning of the message from a variety of visual cues, and the correct phonetics were transparent only after having done this.

In some ways, this visual dominance is even clearer in Mayan writing, which appears to have been a more primitive form of hieroglyphic than Egyptian. Mayan writing had less standardization of graphic convention, and the artist was given considerable leeway in how to assemble the ideograms and phonetic markers of a given message, which were often incorporated into a larger pictorial representation

(Stuart and Houston, 1989). In Mayan hieroglyphs we can see how the early inventors of graphic traditions must have struggled with the problem of representing events and complex ideas visually. Ancient Mayans employed logograms and syllable signs in what appear to be a much less rigidly structured order than Egyptian, leaving the scribe a great deal of freedom—and the reader, presumably, a great deal of ambiguity, although the latter was undoubtedly reduced by extensive training (Stuart and Houston, 1989).

Spoken language is only one of several vehicles with which to communicate ideas, and the task of the early scribe was not necessarily to replicate sound in graphic form. Rather, it was to convey, in visual symbols, the *ideas* normally communicated and formulated with language. If an idea could be symbolized with metaphoric devices, such devices were used; writing was, after all, an attempt to represent the *message* visually, not the sounds associated with a narrative version of the same message.

Could hieroglyphic writing simply have emerged out of drawing? There is a traditional school of thought that holds this to be the case; one of the key pieces of evidence used to support this view is the Rosetta Stone, discovered by Napoleon's soldiers in 1799. The stone dates back to about 200 B.C. and was critical evidence in allowing the deciphering of ancient Egyptian. Its importance was due to the fact that it contains three scripts relaying the same message: Egyptian hieroglyphics, the later demotic Egyptian script, and Greek. At first glance, undeciphered hieroglyphics look simply like pictures, whereas the demotic and Greek scripts look like writing. However, once the hieroglyphics were deciphered, it became clear not only that hieroglyphs were a true form of writing but that the demotic Egyptian script was a shorthand version of hieroglyphs for use in the bureaucracy. It was less obviously pictorial and looked more like modern handwriting; nevertheless, its origins were hieroglyphic, not alphabetic. The Greek script, on the other hand, was the ancestor of the modern phonetic alphabet. Thus, the apparent trend of the symbolic evolution shown on the Rosetta Stone moved from a pictorial syllabary, to a quasi-pictorial syllabary, to an alphabetic writing system, the latter being the culmination of centuries of refinement.

This suggests a direct line of evolution leading in several steps from pictorial representations to writing. On one level it is hard to argue with that view: the earliest symbols were obviously more pictorial in nature than the later ones. However, Harris (1986) questioned the value of this apparent evolutionary trend in understanding the origins of writing; he called this widely held view the "evolutionary fallacy." How many lines of intermediary scripts would it be necessary to add above the hieroglyphic, Harris asked, in order to describe the "gradual" evolution of writing from drawing? His point was, of course, that there could be no gradual transition from drawing to writing. Hieroglyphs were already a fully developed writing system; they could express virtually anything that could be said or thought. Whether or not they were as efficient as the alphabet, they were just as expressive; and their distance from the pictorial mode was just as great.

The strongest evidence for the autonomy of visual symbols came from the existence of elaborate, purely ideographic writing systems. Chinese writing survives today as a living example of the third type of major writing system to emerge before the alphabet. Chinese writing, and its relatives in the Sino-Tibetan family of languages, operates on quite a different principle from either cuneiforms or hieroglyphs. Unlike the latter, traditional Chinese writing is not truly phonetic; the writing system is made up of thousands of ideograms, or logograms, each of which represents a concept or word. Ideograms have no direct phonetic implications; therefore the same written sentence can be read in various languages and dialects without the need for translation, in much the same way that an accountant's balance sheet can be read in English, French, and Japanese, without translation. Ideographic characters are used in modern Japanese as well; they are known as "kanji," and are derived from the ancient Chinese characters. (Japanese writing also uses phonographic characters, known as "kana.")

The existence of highly successful systems of ideographic writing is a serious challenge to those who would define "true" writing as inherently phonetic. Chinese civilization has continuously employed this writing system, to the exclusion of others, for thousands of years. Although the complexity of the system has prevented mass literacy, ideograms are able to express virtually any idea, whether literary or scientific. Chinese culture passed through millennia of development, pioneered in advanced scientific and literary inventions, and built the largest bureaucracy in human history, all without any kind of phonetic writing system. Nothing could better verify the independence of written from spoken language.

Early writing altered the overall architecture of representational systems; Figure 8.3 reviews the previous stages of development and shows additional paths that were added by establishing early systems of writing. The underlying structures—episodic, mimetic, and linguistic—are still in place, as is the pictorial path. But the ideographic path is now fully developed, and the weak phonetic path is under construction. More significantly, there is a new element called the *external memory field*, or EXMF, which is essentially a cognitive

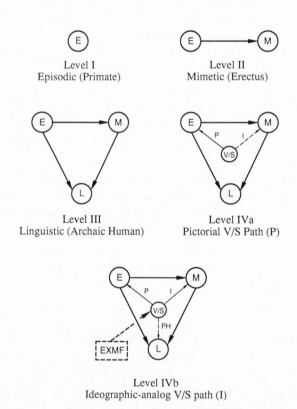

Figure 8.3   Second, or ideographic (I), visuosymbolic (V/S) path, a Bronze Age innovation that arrived with early hieroglyphic, cuneiform, and ideographic writing. Note the nascent, indirect phonological (PH) path. The external memory field (EXMF) is established, as an external working memory for processing and refining visual symbols.

workspace external to biological memory.* In early writing, there was extensive use of the EXMF, especially in the processing of lists. The individual writes down an image or idea, processes the image visually, alters the image and adds symbolically encoded messages, reprocesses the image visually, and so on. The EXMF will be discussed later in this chapter.

The Phonetic Alphabet

The alphabet is now in such wide use that it has become the criterion by which all other writing systems are judged. It is traditionally touted as the most important step in visual symbolic invention. This must be conceded, if only because of its wide and rapid diffusion. While cuneiforms, hieroglyphs, and ideographic writing allowed the accumulation of written records, they imposed a tremendous burden on human memory. The invention of the phonetic alphabet reduced the memory load imposed by reading skill and allowed a much wider diffusion of literacy. Instead of remembering six or seven hundred hieroglyphs or cuneiforms, or several thousand ideograms, the reader of the early Greek alphabet could achieve literacy with about two dozen phonograms, or letters. The main advantage of the alphabet was the economy and precision with which it allowed the reader to map visual displays of symbols onto spoken language. The complex visual–ideographic route could thus (supposedly) be bypassed.

The alphabet was the product of several thousand years of experimentation with writing in the Middle East and Mediterranean basin. Many of the symbols of the Western alphabet have a complex, but traceable, history; for example, the Greek letter alpha has been traced to the first symbol of the Old Hebrew syllabary, aleph, which came from the North Semitic word for oxhead and ultimately from the Egyptian hieroglyph of an ox's head, turned on its side (Diringer, 1948). Various Semitic scripts developed uniconsonantal syllabaries in the second millennium B.C.; these contained no vowel sounds but in other respects were similar to the alphabet in principle (Gelb, 1963; Diringer, 1962). Between 1600 and 1300 B.C. various consonantal syl-

---

*The EXMF usually consists of a temporary array of visual symbols immediately available to the user. The symbols are durable and may be arranged and modified in various ways, to enable reflection and further visual processing. The section starting on page 325 deals extensively with this subject.

labaries continued to evolve in the Middle East and differentiated into the first forms of the Arabic, Hebrew, Aramaic, and Phoenician alphabets.

It was the Phoenician alphabet, containing 22 consonantal characters, that was transmitted to the Greeks through commercial contact sometime between 1100 and 700 B.C. Once the Greeks possessed it, the alphabet was applied for the first time to an Indo-European language. Since Indo-European languages place great importance on fine distinctions in vowel sounds, the Greeks found it necessary to specify a number of vowel sounds in their alphabet; they combined these new vowel symbols with the Phoenician consonants, plus three new consonants: theta, phi, and psi (Gelb, 1963). The result provided the basis for all later Indo-European alphabets and was, in the opinion of most writers, the first truly phonetic script.

Judging from the measurable success of writing systems—ease of use, speed, and popularity—the alphabet is by far the most successful method of writing ever invented. Speech is a universal trait among humans, whereas visual language is not; it follows that visual language should be easier to learn if directly harnessed to an ability everyone starts with—speech. However, this begs the main question; while the ease-of-acquisition argument must be conceded, it is a secondary issue. The more important issues are (1) does an alphabetic writing system really *depend* upon the link to speech for its expressive power, and (2) is an alphabetic system necessarily any less "visual" than an ideographic system?

On first examination, the answers to both questions appear to be in the affirmative. Alphabetic writing systems do not appear to use ideograms. But this is because of an explicit rule of exclusion; in fact, ideograms abound, even in modern Western society, and for some messages they are recognized as more effective than (alphabetic) writing. For instance, in English-speaking societies a cancelled circle is an ideogram for "do not enter." Hundreds of such ideograms can be recognized by any Western adult. Logograms are also widely used especially in number systems; and many mathematical and logical symbols, including brackets, sum signs, horizontal lines, and slashes are used in a nonphonetic manner. Rebus systems are still used, especially in children's puzzles. Thus, all the earlier visual symbolic devices are still around; but aside from mathematical symbols, they are not generally defined as valid forms of writing.

The phonetic alphabet appears to have increased our dependence on the linkage to sound; it short-circuited the thousands of direct associations that had otherwise to be established between the visual and auditory forms of words. These associations were so numerous, and the rules for their use so complex, that reading was out of reach of, and thus of little direct value to, the ordinary citizen. Syllabaries were the first step toward a more economical visual–auditory mapping device, but they remained cumbersome and ambiguous. A typical syllabary has a single sign for one or more syllables; in Egyptian hieroglyphics, there were uniconsonantal, biconsonantal, and triconsonantal signs; the sound "m" was represented in one uniconsonantal sign and at least twelve other bi- and triconsonantal signs, such as "ms," "mr," "hm," and "ndm." This multiplied the number of possible syllabic signs and led to duplication of the number of characters needed to represent each basic sound in the language. To complicate the picture even further, hieroglyphs also used logograms and determinatives in the writing system proper.

The alphabetic principle bypassed this problem by eliminating logograms altogether (except for numbers) and also eliminating visual determinative signs. Everything written down could thus be read aloud, and conversely anything that could be said could be written down. Although alphabets are not usually phonetically perfect, in that there are still some phonological ambiguities in most systems, the number of interpretative ambiguities is enormously reduced in alphabetic writing. The act of reading is thus directly plugged in to a specific auditory complement of the visual record. Also, by linking reading so closely to sounds, the power of auditory memory is brought into play in a more direct manner than in ideographic systems; items in the written record tend often to be remembered in their acoustic forms.

Independent Visual and Phonetic Paths

All of this might appear to argue against the autonomy of writing from speech, at least in a society with an alphabetic writing system. It might seem that writing traded its autonomy as a uniquely visual language modality in return for an efficient marriage with speech. However, things are not so simple. When ideographic writing was invented, the path of association was first and foremost between a visual symbol (which was not necessarily a pictorial symbol) and a concept or mental model. This was a direct path: the modeling intellect was using the symbols as a vehicle for its reconstructions of conceptual reality. The speech system had to construct a narrative version, in parallel, of the model under construction. But there was no single "correct" oral version; so long as the expressed idea corresponded to the concepts expressed in the ideographic script, it was an acceptable reading.

When there is more phonetic information in the script, as there is in hieroglyphs, the modeling process is nevertheless similar. The reader has to take into account ideograms, determinatives, phonograms, and a variety of higher-order contextual cues that reside in the content of the text itself. The resulting mental model can be vocalized, but now there are more constraints on the form of the vocalization than in the ideographic case. Extending this logic to alphabetic systems, the reader might still be employing a parallel strategy, in which the visual path to the modeling process is still somewhat autonomous from the auditory. As in earlier writing systems, the reader might be simultaneously modeling the world represented by the visual input and vocalizing this model; but now there are even more constraints on the form of that vocalization, imposed by the phonetic cues in the text. In other words, readers of alphabets might only appear to be using the phonological route, since anytime they are asked, they can correctly "read" the phonetic information in the text.

But, it may be objected, this is more than mere appearance; how would an independent "visual" modeling of text be possible, if the text is made up exclusively of phonograms? Precisely; but the text is *not* made up exclusively of phonograms. This assumption, that alphabetic writing consists only of phonetic cues, appears to be the chief source of difficulty. It is not a solid assumption; an alphabetically written word may be a phonogram, but it can also simultaneously serve as an ideogram or logogram. Frequently used words, in particular, are recognized so rapidly that there appears to be no time to perform grapheme-to-phoneme mapping; highly trained speed-readers can take in whole phrases and short sentences as fast as single words. Alphabetic reading thus utilizes rapid, direct links between visual words and their conceptual referents; this is especially true when reading handwritten script, where the letters may be unintelligible by themselves, while familiar words are easily recognized. Unlike ideographic writing, however, a parallel phonetic path would allow a reader to simultaneously reconstruct an accurate spoken version of the same message.

What about infrequent words, unusual textual constructions, and long, convoluted sentences? These surely would resist any attempts to reconstruct them ideographically. It may be the case that rarer words are often retrieved more easily by means of an phonetic route; but this is hard to accept as the norm, since reading vocabularies often exceed listening vocabularies and often contain different lexical items, especially in second languages. Unusual or convoluted textual constructions, such as the passive–negative transformations so beloved of Chomskian psycholinguists, supposedly cannot be "understood" ideographically from alphabetic script, because the alphabet lacks appropriate ideographic grammar markers. But there must be nonphonetic, purely visual cues even in alphabetic English.

This is more than speculation; it is the only possible interpretation that can be placed on the reading skills of the congenitally deaf. Most deaf readers have no phonetic training, and many have no skill at signing, yet their reading abilities are considerable (Conrad, 1972). If alphabets required the reader to follow a phonological route, congenitally deaf people should not be able to learn to read at all, at least with alphabetic scripts. But the deaf can learn to read, albeit with difficulty; as one might expect, those who are most profoundly deaf do not make "phonological" errors in reading but rather tend to make "visual" errors, that is, errors that confuse the visual forms of words or letters. And subjects with some residual hearing are more like hearing subjects, making some phonological mistakes. Conrad (1972) concluded that since the profoundly deaf can learn to read, speech-based codes are not absolutely necessary for (alphabetic) reading.

Although the profoundly deaf can read, they do not read with the speed and facility of hearing people. Does this indicate some inherent limitation on nonphonetic reading? Conrad felt that this was not necessarily the case; after all, alphabetic writing was designed as an optimal system for *hearing* people. This means that alphabetically written words tend to have a convenient phonetic shape, but they are highly confusable visually. Thus, they are probably not an efficient system for a logographic–ideographic representation of ideas. This puts the deaf person at a disadvantage, not so much because phonological codes are necessary for reading but because alphabets do not make good ideograms, having been designed specifically to do something else.

But what *is* the alphabet to a profoundly deaf reader, if not a set of standard visual building blocks for logograms, and possibly larger-scale ideograms? It certainly is not a set of phonetic clues tapping a nonexistent speech system. As Conrad pointed out, there is a possible mediating role for manual signing, in those deaf readers who sign, and there is some experimental evidence to suggest that deaf readers make subtle hand movements while reading. But in the latter case the written words would still have to serve as logograms for the manual signs, since the individual letters would not map neatly onto manual signs the way phonemes do. Even finger spelling, specifically designed to map onto alphabetic letters, is necessarily nonphonetic to the congenitally deaf. Therefore, to them, the referents of deaf finger spelling are other arbitrary visual symbols that lack a phonetic code; the entire system is ideographic.

Although the fact of nonphonological reading of alphabetic script is clearest in deaf readers, it is present in varying degrees in experienced hearing readers. There is neuropsychological evidence for the independence of visual and phonological reading. Studies of the dyslexias are particularly relevant in this regard; reading can break down in two dramatically different ways. The most common pattern of acquired dyslexia is so-called "surface" dyslexia, in which the reader loses, partially, the ability to make appropriate grapheme-to-phoneme conversions, as in reading "just" for "guest", or "bik" for "bike" (Marshall and Newcombe, 1973, 1980). There are also errors of stress, as in reading "beggin'" for "begin," and regularization errors, as in reading "decease" for "disease" (Shallice and Warrington, 1980). The patient remains able to read individual letters and finds irregular words especially difficult.

The second major classification of dyslexia described by Marshall and Newcombe is called "deep" dyslexia and is in some ways the complement of surface dyslexia, as the name implies. The key characteristic of this disorder from our viewpoint is the existence of *semantic errors* in the patient's reading: for example, reading "muddy" as "quagmire," or "symphony" as "orchestra" (Patterson and Besner, 1984). This testifies to the existence of a nonphonological route for reading; to read "muddy" as "quagmire" is to access the meaning of the word, while having no idea as to its pronunciation. Since the phonology of a semantic error is completely inappropriate, the access route to meaning, which produced a semantically related word, must have been purely visual. Deep dyslexia is a very complex neurological syndrome and is usually produced by a stroke and accompanied by Broca's aphasia; thus, it is not an isolated set of symptoms and is far from a pure disorder. Moreover, there are several (at times confusing) lines of theory regarding the significance of deep dyslexia (Shallice, 1988), and a discussion of these is not in order here. Nevertheless, the existence of reliable semantic errors is an important piece of evidence to carry away from this literature.

Marshall and Newcombe's original interpretation of deep and surface dyslexia was to affirm two independent routes in accessing the meaning of a written word: the first operates via the phonological system and the second via some parallel visual path. These two paths correspond roughly to our phonetic and ideographic paths. Deep dyslexia involves damage to the first path, and surface dyslexia to the second. Thus, the deep dyslexic makes semantic errors because there is no phonological correction on the pronunciation of the word. The word's meaning is accessed through a purely visual route, like an ideogram, but the patient cannot use the visual information to guide pronunciation: therefore "orchestra" sounds as credible as "symphony" (an inexperienced reader of ideograms might easily make a comparable error).

Surface dyslexics are thought to make errors of pronunciation because the correct phonological reading of a word often depends on independent visual recognition of the form of the word. For example, the "ough" ending in English is phonetically ambiguous, depending on recognition of a specific word. Thus, the surface dyslexic knows the phonological rules and often misapplies them because the word cannot be recognized visually. Confirmation of the independent visual route comes from studies of Japanese dyslexics, who can lose their ability to read phonetic *kana* independently from their ability to read ideographic *kanji* characters (Paradis, Hagiwara, and Hildebrandt, 1985; Sasanuma, 1985). This independent visual route must operate on both ideographic and logographic principles.

Although the dyslexia literature has become more complicated since Marshall and Newcombe's theory, dyslexias cannot be understood without proposing at least two parallel routes to word (and, by implication, phrase and possibly sentence) recognition (Shallice, 1988). And to make any sense at all of the dyslexias, one of those routes must always be purely visual. Note that, for the most part, acquired dyslexias are observed in patients whose early experience and education were normal and who suffered impairment *after* language and reading were both well established. The existence of an ideographic reading route in these patients confirms that it emerges in normal development and is not restricted to early brain injury or congenital deafness.

By implication, normal alphabetic readers read words by at least two parallel routes, just like readers of ideographs and hieroglyphs. While they can utilize the visual route to resolve ambiguous phonology, they can use the phonological route to narrow the semantic ambiguities in the text. Presumably both routes access meaning, and, extrapolating from the properties of ideographic and hieroglyphic reading, the visual route is probably rather diffuse, allowing a variety of specific interpretations to be placed on the word signs. In fact, Conrad's complaint about the alphabet probably applies here; the alphabet is not designed specifically for visual ideographic representation and may yield less information on this front than writing systems designed to use the visual route exclusively (a corollary is that it might be possible to improve on the alphabet with a writing system that is perfectly phonetic, while allowing for greater ideographic power in the written signs.)

The visual route is a powerful aid in resolving phonetically ambiguous written messages, even with alphabets. Carroll (1972) gives the following example: the knowledge that the meaning of the statement "the sons raise meat" differs from that of the statement "the sun's rays meet" must reside entirely in the visual display, not in the phonemically transformed text. Carroll, in discussing Martin's (1972) review of nonalphabetic writing systems, concluded that the principles of teaching reading are not all that different for alphabetic and nonalphabetic systems. Most nonalphabetic systems evolve some quasi-phonological conventions to help the reader know the correct pronunciation of a sign and to reduce the number of symbols needed for daily use; and alphabetic systems are not as phonetically unambiguous as we sometimes assume.

Alphabetic writing led to a new visuosymbolic path and an enormous expansion in the number of codes to be held in the V/S store. The architectural diagram has to be changed again, to accommodate these new developments (Figure 8.4). Level IVa had already appeared

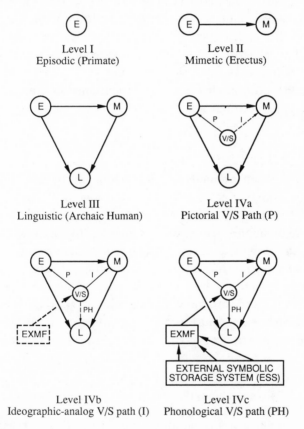

Figure 8.4 Third, or phonological (PH), visuosymbolic (V/S) path, established with the invention of the alphabet in the 1st millennium B.C. With the possibility of written storage of long narratives in the ESS, linguistically based thought gradually incorporated an external iterative loop through the external memory field (EXMF).

with the pictorial path. Level IVb introduced ideographic skill, which initially built a bridge between mimetic representations and language, leading to the earliest systems of writing. At this point humans were starting to construct interactive narrative commentaries based on lists, diagrams, and numbers and on other ideographic symbols, and a weak start on a phonological path was begun; this is indicated by a dotted line.

But it was with the addition of a direct phonological symbol route that the external symbolic storage system (ESS) really came into its own, shown as Level IVc.* Once this route was established, narrative commentaries could be externally stored and refined. The external memory field became a more complex and much more powerful device, since it could harness all three innate representational systems, and the larger, permanent external storage system expanded rapidly. The architectonic configuration shown at level IVc is several orders of magnitude more complex, reflecting the tremendous distance humans had traveled from their ancestors, in terms of overall cognitive structure.

Other Kinds of Visuosymbolic Invention

The rise of vision as a modality of symbolic invention thus progressed gradually from the early pictorial inventions of mythic culture to a variety of writing systems, culminating in the alphabet, which provided a two-channel strategy for reading. But visual symbolism did not stop developing at this point. The introduction of the new visual mechanisms of the third transition, which constitute a variety of access routes to external memory, changed cognitive architecture completely, since there were now mental vehicles not only for externalizing working memory but also for pooling the outputs of the three innate representational systems. The governing *biological* system was

---

*The ESS is distinguished from the EXMF on the basis of its availability and permanence. The term ESS applies to all memory items stored in some relatively permanent external format, whether or not they are immediately available to the user. The EXMF is a temporary arrangement of some of the material in the ESS, for the use of one person. Thus, I may have a whole library of material available for a project, but I can remove only a few items and arrange them for my immediate needs; the former is part of the ESS, while the latter constitutes my EXMF for the moment.

still the oral–narrative system; but it was embedded in a larger, partly external, structure.

Other uses of visual symbolism were eventually developed. Musical notation was one of the most important, but there were also new symbolic notational systems developed for representing various ideas: geographic maps, military plans, geometric concepts, astronomical lists, calendars and clocks, architectural drawings, and a variety of complex visual metaphors employed in graphic art. More recently, representational notations have developed for dance choreography; visual directorial scripts for filmmaking; charts of organizational structure; systems and conventions for advanced engineering drawing, and so on. Elaborate methods of scientific graphing and illustration have emerged, and the technique of computer graphics has allowed the visualization of complex relationships in ways never before possible. The visuosymbolic route, that cluster of visual codes (V/S) for the decoding of external symbolic memory devices, has grown enormously in size since its beginnings.

Goodman (1968) has considered the symbolic nature of various graphic and notational systems in the arts, including musical scores, painters' sketches, paintings, scripts, architects' drawings, and literary products such as poems and novels. All of the above serve a *referential* function for the artist's thought, and although the specific nature of the thought process may be radically different for various art forms, the common factor is a dependence on symbols. This is equally true of the scientist; and Goodman sees the two types of thinking, art and science, as closely connected. The differences between the types of symbols employed by these two broad categories of human symbol use reflect their specific thought strategies; but both share a common feature in that they are primarily symbol-based cognitive activities. In the present theoretical context, Goodman's idea can be rephrased, for emphasis: Science and art are both dependent upon external memory devices and thus upon the vast numbers of visuosymbolic codes and conventions stored in the modern brain.

One additional visuosymbolic invention needs further discussion here. The invention of formal sign languages, such as ASL, came fairly recently and used some of the features of ideographic writing. ASL communicates and represents in the visual modality and could be regarded as a form of impermanent writing. It employs some of the techniques of hieroglyphs and cuneiforms, such as determinatives and spatial–grammatical rules. A skilled ASL user can define and utilize several regions of space concurrently, effectively drawing symbols on air and flashing manual signs in successive bursts of activity. ASL should not be confused with gesture, any more than ideographic writing should be confused with gesture. It employs grammar, has a developed lexicon, and is able to represent highly abstract ideas (see Klima and Bellugi, 1979).

Finger spelling, which supplements ASL, is superficially equivalent to the phonetic alphabet, but this is misleading: it obviously cannot build on a direct phonetic route in a deaf person. It relies on existing convention for its usefulness. However, if the alphabetic convention had not already been in existence when ASL was invented, some purely ideographic method of indicating names probably would have been invented. Perhaps a totemic or clan marker, as in early pictographic and ideographic symbols, would have been used. The recency of formal sign language, and its resemblance to some forms of writing, argues in favor of its classification with ideographic writing and other modern visuosymbolic inventions, rather than with oral narrative skills and speech.

External Memory Devices: A Hardware Change

Visuosymbolic invention is inherently a method of external memory storage. As long as future recipients possess the "code" for a given set of graphic symbols, the knowledge stored in the symbols is available, transmitted culturally across time and space. This change, in the terms of modern information technology, constitutes a *hardware* change, albeit a nonbiological hardware change. A distinction should be made between memory as contained within the individual and memory as part of a collective, external storage system. The first is biologically based, that is, it resides in the brain, so we will refer to it as *biological memory*. The second kind of memory may reside in a number of different external stores, including visual and electronic storage systems, as well as culturally transmitted memories that reside in other individuals. The key feature is that it is external to the biological memory of a given person. Therefore we will refer to it as *external memory*.

External memory media—especially written records but also many other forms of symbolic storage—are a major factor in human intellectual endeavor; few people would dispute this. But do they deserve to be treated as an integral part of cognitive architecture? In the traditional view of psychologists, the mind has clear biological boundaries. "External storage" might be seen as just another term for the culture or civilization within which the individual exists. The individual picks and chooses, acquires skills and knowledge from society, but nevertheless exists as an easily identifiable unit within that society. In this view, while society influences memory and thought, memory and thought occur only in the individual mind or brain and therefore are to be regarded as attributes of the individual, and studied in the individual. Consequently, the proper approach to the study of memory, language, and thought is to study the individual. The literature of experimental psychology testifies to our adherence to this tradition; it is composed mostly of studies of individual performance or competence and the variables that influence it.

But external memory is not simply coextensive with culture in general, or with civilization. Culture and civilization are broader concepts, including material products, such as technologies and cities, and many aspects of human life that are not cognitive. External memory is best defined in functional terms: it is the *exact* external analog of internal, or biological memory, namely, a storage and retrieval system that allows humans to accumulate experience and knowledge. We do not possess any ready theoretical frameworks in psychology from which to view external memory. Fortunately, there is an excellent point of comparison in the field of computing science: networks.

Computing scientists are able to specify exactly what hardware a given computer possesses and what software. The "cognitive capacity" of a specific computer can be described in terms of its hardware features (such as memory size, central processing unit speed, and peripheral devices) and its software features (such as its operating system and available programming support). In detailing the two types of features, the machine would have been described as thoroughly as we know how. A comparable approach to the human individual would be to proceed in roughly the same manner, as is commonly the case in psychology: study the hardware (the nervous system) and describe the software (the skills, language, and knowledge carried by the individual).

However, the metaphor should be carried to its logical conclusion. If a computer is embedded in a network of computers, that is, if it interacts with a "society" of other computers, it does not necessarily retain the same "cognitive capacity." That is, the powers of the network must also be taken into account when defining and explaining what a computer can do. This is because networking involves a structural change; the processor and memory of the computer are now part of a larger network computational resource. As part of a network, the computer can now delegate computations beyond its own internal capacity. It can also assign priorities within the larger system. It can store its outputs anywhere in the network, that is, in external memory sources. It can even be assigned the role of controller of the system. The point is, in a true network the resources of the system are shared, and the system functions as a unit larger than any of its individual components.

Another relevant analogy might be the addition of magnetic tape or disk storage to a computer. The RAM (Random-Access Memory) of a computer is somewhat analogous to biological memory inasmuch as it lives and dies with the machine. A computer with no hard external memory medium has to rely on its own internal RAM capacity entirely. The only way to expand its capabilities without adding external memory would be to expand RAM memory; but this could not go on indefinitely, and would be extremely wasteful. In biological terms, a similar expansion of individual memory could not go on without running into the physical limitations imposed by factors such as energy requirements and metabolic functions. Just as the invention of hard storage was an economical and flexible, in fact essential, feature in expanding the powers of computers, external storage was an extremely important and efficient innovation in the history of human cognition.

Historically, culture is conventionally regarded as software, not hardware. There is some surface validity to this view. Learning a particular language will not alter basic memory skills or extend short-term memory and appears to leave basic cognitive skills unaffected.

Similarly, many other aspects of culture that might be classified as software do not bear greatly on basic cognition. But some cultural artifacts are in fact hardware changes and not software changes. Visuographic inventions of all kinds should be classified as hardware changes. They have produced a collective memory bank as real, in terms of hardware, as any external memory device for a computer. And because they are not bound by time and space, they transport the very limited capabilities of the individual into realms that would not be remotely reachable by an isolated mind.

Individuals in possession of reading, writing, and other visuographic skills thus become somewhat like computers with networking capabilities; they are equipped to interface, to plug into whatever network becomes available. And once plugged in, their skills are determined by both the network and their own biological inheritance. Humans without such skills are isolated from the external memory system, somewhat like a computer that lacks the input/output devices needed to link up with a network. Network codes are collectively held by specified groups of people; those who possess the code, and the right of access, share a common source of representations and the knowledge encoded therein. Therefore, they share a common memory system; and as the data base in that system expands far beyond the mastery of any single individual, the system becomes by far the greatest determining factor in the cognitions of individuals.

The memory system, once collectivized into the external symbolic storage system, becomes virtually unlimited in capacity and much more robust and precise. Thought moves from the relatively informal narrative ramblings of the isolated mind to the collective arena, and ideas thus accumulate over the centuries until they acquire the precision of continuously refined exterior devices, of which the prime example is modern science. But science, ubiquitous though it is at present, is atypical in historical terms. Human cultural products have usually been stored in less obviously systematic forms: religions, rituals, oral literary traditions, carvings, songs—in fact, in any cultural device that allows some form of enduring externalized memory, with rules and routes of access. The products of this vast externalized culture have gradually become available to more people, who are limited only by their capacity to copy (understand) them.

Individuals connected to a cultural network can access an exterior memory bank, read its codes and contents, store new contributions in permanent form, and interact with other individuals who employ the same codes and access routes. If cultural technology is not easily seen as hardware, it is probably because our links with our own culture cannot be described in obviously physical terms. Human minds float freely, without any apparent physical tie-in, either temporary or permanent, to cultural devices. This is more apparent than real, however. Our collective memory devices are undoubtedly physical in nature. A written tablet or book or papyrus is a physical artifact. So is a totem pole, a mask, a costume, a poster, a traffic sign, or an affidavit. So, more obviously, is an electronic memory device. Our links with these devices are physical: they respond to and emit energy, whether electromagnetic or mechanical. The hardware links of computers to networks will undoubtedly become more subtle as technology progresses, and even now they too do not always require a direct physical attachment: satellite linkages can be regarded as hardware devices even though the components may be thousands of miles apart from one another.

External memory is a critical feature of modern human cognition, if we are trying to build an evolutionary bridge from Neolithic to modern cognitive capabilities or a structural bridge from mythic to theoretic culture. The brain may not have changed recently in its genetic makeup, but its link to an accumulating external memory network affords it cognitive powers that would not have been possible in isolation. This is more than a metaphor; each time the brain carries out an operation in concert with the external symbolic storage system, it becomes part of a network. Its memory structure is temporarily altered; and the locus of cognitive control changes.

The Locus of Memory

In a network, memory can reside anywhere in the system. For example, a program running a computer may not "reside" there; or it may reside there wholly or in part. In terms of the execution of the algorithm, these things do not matter; the system functions as a unified system, for whatever period it is defined as a system. The roles assumed by individual components in the system depend entirely on two things: system demands and the limitations of each component in the network. The key to control, or power, in the network, for an individual component, depends on the level of access to certain crucial aspects of the operating system and on preset priorities. Any component that cannot handle key aspects of either the operating system or the programming language, or that cannot execute long enough or complex enough programs, is automatically limited in the role it plays and eliminated from assuming a central role in the system. (There may be ways to circumvent this, but in principle this statement holds true.)

Given a compatible network, the powers of an element may become that of the entire network. For example, a single computer might be positioned to delegate priorities and assign roles to others in the system. It could drive many peripheral devices, delegate very large or complex computations to larger but less well-located machines, "piggy-back" excessively large operations on other devices, and so on. And simply by reassigning priorities, it may be relegated to a lesser role, even reduced to the level of a terminal peripheral to the network.

Individual humans, utilizing their biological memories, may interact with their collective memory apparatus in approximately similar ways. Of course, it is historically more accurate to reverse the metaphor and say that, in constructing computers, we have unreflectively emulated this feature of human society. The locus of memory is everywhere or anywhere in the system at any given moment, depending upon the flow of information, access, and priorities. The biologically encapsulated mind might be called a *monad*, after Giordano Bruno (1584) and Liebnitz (1714). Monads may form temporary connections with specific nodes in the external symbolic system, disconnecting and reconnecting at will in various configurations. A node may be defined simply as a particular location within a network; thus a specific scientific paper in a clearly defined area of the ESS might be considered a node, whereas the entire collection of journal and reference works in the area constitutes an information network. The annual report of a specific corporation might constitute another node, which could form part of several cross-referenced networks; for instance, it might be part of various networks of annual reports organized by region, economic sector, corporate size, growth rate, and so on. Figure 8.5 illustrates this point; when several monads are connected to the same node, they share a memory resource, temporarily.

Given the enormity of modern information technology, and the

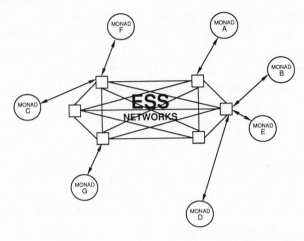

Figure 8.5 Schematic of temporary linkage between biological memory systems (monads) and external symbolic storage (ESS) networks. Patterns of linkage change continuously; connected monads may be temporarily networked. Monads sharing an ESS subsystem share a major source of memory material.

amount of information collected and disseminated daily in society, monads can hold only a very small fraction of what the ESS contains. The major locus of stored knowledge is *out there*, not within the bounds of biological memory. Biological memories carry around the code, rather than a great deal of specific information. Monads confronted with a symbolic information environment are freed from the obligation to depend wholly on biological memory; but the price of this freedom is interpretative baggage. The reason other mammals cannot appreciate the symbolic environment of humans is not that they cannot hear, see, remember, and occasionally use symbols; it is that they cannot interpret them. They lack the various levels of semantic reference systems, and the encoding strategies, that are essential to the extensive use of symbols by humans.

Properties of Exograms

Single entries in the ESS might be called *exograms*, after Lashley's (1950) term engram, which referred to a single entry in the biological memory system. An exogram is simply an external memory record of an idea. In both engrams and exograms there are retrievable traces of past experience that can be used to determine future interpretations of the world. Similarly, exograms and engrams are both interpretable only by the individual mind, which must provide the referential basis for understanding the memory record. Exograms, however, are inherently very different from engrams. Whereas engrams are built-in devices, genetically limited to the format and capacity of the human central nervous system, exograms are virtually unlimited in both format and capacity. Engrams are impermanent, at best lasting only as long as the life of a single individual; exograms can be made permanent, outlasting individuals and, at times, entire civilizations. Unlike engrams, systems of exogram storage are infinitely expandable, lending themselves to virtually any system of access, cross-indexing, cataloging, and organization. They also lend themselves to maximum flexibility and can be reformatted over and over in a variety of media and custom-designed for various purposes. Thus, a cognitive system containing exograms will have very different memory properties from a purely biological system (Table 8.1).

The most important feature of exograms as storage devices is their capacity for continuous refinement. Exograms are *crafted*; that is,

Table 8.1.   Some properties of engrams and exograms.

Engrams	Exograms
internal memory record	external memory record
fixed physical medium	virtually unlimited media
constrained format	unconstrained and reformattable
impermanent	may be permanent
large but limited capacity	virtually unlimited
limited size of single entries	virtually unlimited
not easily refined	unlimited iterative refinement
retrieval paths constrained	retrieval paths unconstrained
limited perceptual access in audition, virtually none in vision	unlimited perceptual access, especially in vision; spatial structure useful as an organizational device

they are symbolic inventions that have undergone a process of iterative examination, testing, and improvement. To a degree this may
also be said of biological memory—for instance, a procedural skill like
violin playing, or acting, or rhetoric may be rehearsed and improved,
and individuals may employ mnemonic devices—but the conceptual
products of cognition themselves cannot undergo extensive refinement in a purely oral tradition. Only in elaborate exographic systems,
such as written histories or mathematical or physical theorems, can
the products of thinking be frozen in time, held up to scrutiny at some
future date, altered, and re-entered into storage, in a repetitive, iterative process of improvement.

Exograms also have unique *retrieval* features. Their position in the
retrieval hierarchy, and the cues which access them, are a question of
design and are entirely arbitrary, a matter of convention. This is not
generally true of biological memory; episodic and procedural skills
have highly constrained retrieval paths, time-constrained in the former case and structurally constrained (by the performance sequence
itself) in the latter. Semantic memory has a highly fluid structure in
modern humans, but as observed above, it is likely that its present
internal structure is at least partially due to the influence of the external symbolic storage system, whereby the culture tends to restructure the individual's biological memory in its own image. In aboriginal human society, semantic structure might not have been so
multidimensional and fluid (to my knowledge this question has never
been investigated). In any case, even the semantic structure of biological memory is tremendously constrained compared with the ESS.

Because of the largely visual format of exograms, they can be clustered in ways that facilitate *perceptual* synthesis and analysis, capi-
talizing on the ability of human vision to achieve complex spatial
synthesis. Lists have already been mentioned as an example of a distinctly visual form of information clustering, but visual clustering of
scriptic and graphic images goes far beyond simple listing. Visual processing of ESS clustering patterns presents the brain with a massive
scanning problem; that is, clusters have to be processed fixation by
fixation, with each fixation presenting only a partial picture. Symbolic images have to be scanned in a certain order, or at least assembled in the correct order after scanning.

Complex visuospatial scanning and searching strategies underlie

reading and graphic interpretation in general. Bouma and de Voogd
(1974), Kolers, Wrolstead and Bouma (1979), Rayner (1978, 1979),
and many others have demonstrated that successive eye fixations are
controlled by various complex cues. For instance, the act of accurately
scanning a page, following landmarks like line endings and spaces, is
controlled mostly with extra-foveal spatial cues. Eye-movement scanning sequences are also controlled by the meaning (or meaningless-
ness) of the sentence or image unfolding in the viewer's brain. In fact,
the decoding of meaning is sometimes rapid enough to affect the duration of single fixations (Rayner, 1979). Eye fixation sequences are
not simple or linear; viewers hop back and forth, fixating on various
parts of the image, reviewing some areas several times, skipping
through others, varying tremendously in the amount of time they
fixate upon a given part. And the only possible way to make sense of
the sequence of fixations presented to the visual brain is to place the
contents of each fixation into the correct overall spatial, temporal, and
semantic juxtaposition. This complex visuospatial scanning framework extends beyond the control of eye movements; for instance,
Mewhort (1974) has shown that the alphanumeric images underlying
written words are scanned serially in the brain, even when eye movements are not made.

But the ability to scan visual images sequentially is common to
many higher mammals. Why is the symbol scanning of humans any
different? The uniqueness of symbolic scanning, and especially reading, seems to lie in the fact that the highly overpracticed and stylized
routines supporting human symbol scanning are driven by arbitrary,
often complex, interpretative conventions, whereas natural and pictorial image scanning is determined more by innate stimulus features.
The rules by which visuographic stimuli are scanned and analyzed
represent yet another additional burden on biological memory imposed by elaborate exographic systems (Kennedy, 1984). This burden
extends beyond reading. Twyman (1979) classified the many branches
of modern graphic symbols into linear and nonlinear categories: the
linear includes verbal and mathematical symbols, which are typically
presented in a fixed linear sequence; related quasi-linear items include
lists, branching diagrams such as family trees, and matrixes, including tables. Nonlinear images include pictorial and schematic images
(for example, maps). Cartoons and comics incorporate both linear and

nonlinear features into a single medium. The spatial arrangement of symbols thus demands a variety of visual scanning strategies and requires a complex of visual interpretative skills that had no equivalent in earlier human societies. These strategies are an essential part of the visuosymbolic (V/S) coding modules.

The overall spatial properties of a cluster of symbols, and the relation of these spatial properties to the contents of the image or text, can also determine the interpretation of the meanings of symbols. The ultimate use of this sort of display is found in mathematical formulas. In these a variety of abstract entities can be interrelated in a complex manner and reduced to a single, graspable visual display, the precise form of which is important to its comprehensibility. In poetry, spatial juxtaposition is particularly critical; the impact of a line of poetry can be greatly affected by its spatial position relative to other elements of the poem. Spatial juxtaposition effects are largely unexplored, and they are unique to exograms. Where the engrams of biological memory have to work within the format restrictions imposed by genetic structure, the spatial conventions and formats of exograms are virtually unrestrained.

Although chunking and batching of information occurs in biological memory (Miller, 1956), such chunking is mostly in terms of rates, durations, and simple spatial clustering and is dependent upon the learning of larger meaningful groupings of items. Ironically, most psychological studies of visual chunking have utilized "prechunked" symbolic stimuli—that is, printed items gleaned from the external symbolic storage system. But in a preliterate world, the only likely sources of prechunked information would have been in the temporal organization of spoken language and gesture. In the ESS, chunking and batching involves much spatial flexibility, and not only in the batching of words and phrases. It is possible to organize longer entries, such as historical accounts, novels, and so on, into smaller, digestible chunks such as chapters, sections, and paragraphs; or to highlight features in separate tables and figures to enhance comprehension; or to organize pictorial images into easily digestible batches, as in early ideographic writing. Biological memory cannot lend itself easily to this type of organization, and this undoubtedly imposed a serious limitation on human thinking prior to the growth of the ESS. Moreover, linguistic properties such as iteration and recursion become truly unlimited in the ESS. Speech may be recursive, but due to the limits on biological memory, its recursiveness is not infinite in practice; written records, on the other hand, are truly unlimited in this regard.

Perhaps the most powerful feature of the ESS is its unlimited size. The number of items stored in collective human experience has grown exponentially with the development of the ESS, both because the encoded knowledge of the past can be better preserved and because the process of producing ESS entries has resulted in a huge industry for generating, inventing, and mass-producing exograms, much like the toolmaking technology of earlier times. Just as the number of tools in the archeological record increased in the late Upper Paleolithic, exogram technology has exploded in the last few hundred years. Many of the images included by Twyman (1979) in his classification of graphic designs were inventions of the eighteenth century or later; this is particularly true of combinatorial media that juxtapose image and word: catalogs, comics, newspapers, magazines, illustrated dictionaries, illustrated textbooks, scientific papers, cartoons, and poster art. Although they had precedents in hieroglyphic texts and illuminated manuscripts, the integration of image and written text is much more widespread in modern media. This serves to emphasize that the modern reader carries around much more than a fixed encoding strategy for linear script and a visual lexicon. The modern reader has developed a comprehensive strategy for dealing with, simultaneously and in various combinations, script, ideograms, and pictorial stimuli.

Changes in the Role of Biological Memory

The existence of exograms eventually changed the role of biological memory in several ways. Human memory had, from its inception, expanded the range of primate memory. The earliest form of hominid culture, mimetic culture, depended on an expansion in the self-representational systems of the brain and created the initial base for semantic memory storage, which consisted initially of representational action scenarios reflected in mime, gesture, craft, and skill. With the evolution of speech and narrative ability, there were even greater increases in the load on biological memory, adding not only the storage networks for phonological rules and the lexicon in its entirety but also a very large store of narrative conceptual knowledge.

Thus, the first two evolutionary transitions would have greatly increased the load on biological memory. However, the final step in this tremendous cognitive expansion might have *reduced* the load on some aspects of biological memory, by gradually shifting many storage tasks onto the newly developed ESS. At the very least, the existence of the ESS must have forced a great change in priorities and memory organization. While the reliance on auditory memory may have been diminished, a complex new memory apparatus was required for visuographic invention. An entirely new memory structure was needed for reading and writing, including highly specialized skills, such as scanning and analyzing the written image. There was a rapid increase in the number of logograms, ideograms, alphabetic items, words, and graphic images in use; these had to be encoded in biological memory. Whatever capacity had been freed up by reducing human dependence on auditory memory was probably quickly used in the great expansion of visual memory.

The expandability of the ESS also meant a great increase in the number of symbolically encoded things that could be known. This trend became particularly evident after urbanization and literacy, but it was already apparent in ancient Egypt and Mesopotamia, at least among the literate elite. Compared with the monotony and redundancy of the hunting–gathering lifestyle, these early centers of graphic invention exploded with symbolically encoded things to be mastered. Large state libraries were already a reality in ancient Babylon, and by the time of the Greeks ESS products had been systematically collected and stored in several world centers of learning. At this point in human history, standardized formal education of children was needed for the first time, primarily to master the increasing load on visual–symbolic memory. In fact, *formal education was invented mostly to facilitate use of the ESS.* Scribes were trained extensively; fully 15 percent of the more than 100,000 cuneiform tablets in existence are lexical lists for training scribes (Walker, 1987).

The expansion of the ESS can be seen not only in the total number of entries but in the type of entry as well. The ESS introduced a completely new dimension into the human cognitive picture: large-scale memory management, including selectivity and the setting of priorities. How does human biological memory deal with its inherent capacity limits, when confronted with an expanding, virtually limitless ESS? It has first and foremost to learn how to search, locate, and choose items from the potential sets of items available. But how can it do this? The archaic strategies of biological memory simply do not apply here; there are too many possible routes, too many possible methods of classifying and structuring items. Temporal codes and phonological codes, the two retrieval workhorses of episodic and oral memory, are largely irrelevant to managing the ESS. In the ESS, it doesn't much matter what items sound like, or when, or in what order or juxtaposition, they were acquired. Once items are entered into the ESS, information about time of acquisition or physical contiguity may or may not be retained and may or may not be made relevant to their retrieval.

One requirement for successful use of the ESS is a *map* of its contents. Thus, biological memory must contain information about the structure and access routes of the ESS, as well as its retrieval codes. These skills are necessary, above and beyond the decoding skills needed to understand ESS entries, in order to find the appropriate information when needed. A significant percentage of the knowledge contained in the literate brain falls into this category; this is particularly true of scholars and administrators, whose familiarity with their "field" or the "levers of power" usually translates into the possession of a good cognitive map of the ESS entries relevant to some highly specific pursuit, for instance medieval Iberian history or Pacific Rim banking policy. The skilled user of the ESS does not try to carry around too much fine detail in biological memory but has learned to be very adept in accessing and utilizing the relevant parts of the external symbolic storage system. This has been true since the earliest great bureaucracies: a Chinese bureaucrat of the Han dynasty (roughly contemporary with Imperial Rome) spent a lifetime training for, and immersed in, the ESS.

Narrative memories—the elaborate semantic constructs that were enabled by speech—would have been a more immediately useful aspect of biological memory and must have served as the basis for interpreting and constructing ESS structure in some fields. In certain areas, therefore, biological memory exported its own organization; narrative versions of events are organized according to thematic codes, and classifications of themas are still useful in accessing, interpreting, and structuring certain areas of the ESS, particularly historical and literary items. But even here, the ESS product is far more refined and detailed than anything in biological memory, and there are many more items to recognize, analyze, store, and comprehend. The problem of selective biological storage from unlimited archives of ESS items thus possibly extends even more to the literary realm than to science.

When confronted with the modern electronic expansion of the ESS, the problem of accessing the system takes on a different complexion. It is fair to say that humans have not developed a strategy for dealing with the magnitude of the twentieth-century ESS. Children have to be trained for many years just to be able to master parts of some highly specialized, narrow area of the ESS. There is no longer any question of extensive crossing of linguistic or disciplinary lines; survival demands great specialization. But what are the children taught? Many of the skills being taught in schools now are really memory-management skills; maps, flags, and pointers are three appropriate computer-related terms that throw light on what students must learn. The map of the ESS is its overall structure; by exposing students to a wide variety of fields at the entry level, they become acquainted with the general geography of ESS-stored knowledge. Flags are priority markers that alert the biological memory to the relative importance of various items and access routes. Pointers are necessary direction markers that signal the addresses of relevant items at appropriate times. In addition, students have to be able to find a variety of ESS entry addresses, and therefore they must master a huge number of category labels. Biological memory is now necessarily cluttered with such items; however, for the most part they are empty of content and are concerned primarily with memory management: where to find things, how to scan them, how to assign their importance or priority.

The controlling structures of the modern mind must depend heavily, in modern humans, on this sort of information. The allocation of selective attention in an extensive ESS environment is ultimately a function of flags, pointers, and priorities, rather than simple front-end perceptual or motor selectivity. Much of ESS-oriented biological memory, then, is necessarily filled with directional information, address categories, and a general framework, or map, of that part of the ESS to which the individual has access.

The implicit assumption of modern education is that ESS devices will be a regular feature of most cognitive operations and that biological memory must be equipped primarily to deal with them, rather than to carry around a great deal of finely detailed information. This reflects a real change in the importance of the ESS. The mnemonic arts and rote learning, once a major part of education (Spence, 1984), have receded into the background as the reliance on biological memory for storage has faded. In earlier human societies many fewer items were in external memory, and thus fewer ESS-related skills had to be mastered. For example, in ancient Egypt, very few people expected to be able to read hieroglyphics—Davies (1987) estimates less than 10 percent of the population—and for those who could, the only ESS items encoded in a form they could comprehend were inscriptions and sculptures in ceremonial buildings and systems of numeration used in trade. By our standards, even professional scribes would have had very few inscriptions and manuscripts to read and remember. Biological memory would have exceeded in size and influence, by far, the useful contents of the ESS (except for lists of trade and government records such as censuses).

In such an environment, the ESS was a secondary player, the servant of a fundamentally oral–mythic society, dominated by ritual and tradition. In modern society this has changed; a great deal of public life takes place through ESS media, whether printed or electronic. Even illiterate people depend upon exographic channels of experience, provided by television or some other medium. There is much less dependence on biological memory. Even episodic experience depends upon electronic media. The great events—D-day, the first moon landing—are vicariously experienced and interpreted through the ESS. In modern society, it is not only the technological and bureaucratic aspects of life that are ESS-dominated; public life itself is lived through the ESS, whether in the print arena of periodicals or the newsrooms of network television. The biological memories of individuals are supplemented by photographs, school yearbooks, diaries, and innumerable symbolic artifacts collected over a lifetime. Even personal identities are stored, refreshed, elaborated, and sustained with the aid of ESS entries.

One of the things sought by Marcel Proust in his massive attempt at reexamining his own life (it was, actually, a recasting of his life in his own biological memory, with the mediation of ESS devices) was a return to the "truth," or things as they really happened. Proust was aware, in that psychoanalytic era, of the tricks played by biological memory and, perhaps naively, thought that truth might be found in the careful construction of externally encoded accounts of his biological recall. The irony is that Proust ended up much farther from the "natural state" he sought and much more dependent on the ESS. In fact, his written account *became* his memory for the events; even though his book did not exist in his head, it influenced, and probably reorganized, his biological memories. It is unlikely that he could have recalled, verbatim, a single page of *Remembrance of Things Past* from biological memory; his own life was now visible to him only through the lens of his crafted words, a device external to him that eventually outlived him. He could not have gotten further away from "natural truth."

The process of reading is interesting to contemplate in this context. Reading is a cognitive state whereby the biological mind is brought temporarily under the complete dominance of an ESS device. The mind is literally "played" by the book, moved into a state crafted by the author. Disconnect the reader from a book, and detail is immediately forgotten; reconnect, and the state quickly reforms itself. In very long novels, the amount of detail is staggering and impossible to recall once the book is put down; but immediately upon starting to reread it, the mind-state induced by the book is reestablished and the reader's awareness is reconfigured, following the intentions of the author; the ESS is in control.

This is even more true of film. In this case, the pictorial aspect of the medium is effectively more powerful than normal episodic experience, because, unlike most of the naturally occurring events of life, film is contrived and crafted to a focus. The viewer's mind comes temporarily under the control of the director; and, just as in the case of the novel, the recipient of the experience is rarely in control. The ESS is thus increasing its dominance; biological memory is, more and more, unable to draw on its own experiences without reference to the ESS.

But books and visual symbols are dead things which, left to themselves, sit on shelves. The ESS is not the field of action; the process of thought occurs in the individual biological mind. Traditional models of the thought process place it in an arena sometimes referred to as consciousness or working memory, a short-term, labile part of the mind that seems an unlikely locus for the extraordinary theoretic products that humans can produce. This issue is not simple and requires a reexamination of the traditional psychological ideas of both consciousness and working memory.

A Revised Concept of Working Memory

The concept of working memory has been conceptualized as a temporary holding system for various cognitive operations (Baddeley, 1986). Numerous other proposals about human cognitive architecture have made essentially the same claim, and many earlier theorists called the temporary holding system "short-term memory" (for example, Broadbent, 1958; Atkinson and Shiffrin, 1968). Although there is little agreement on the structure of working memory, psychologists have generally used the term to refer to a biological memory system, related to the idea of a short-term store, which supports problem solving, learning, and thought. Even the process of solution itself is sometimes thought to take place in working memory; if longer-term memory material has to be brought into the process of solution, it has, supposedly, to be brought in via the working-memory system.

Baddeley (1986) and his coworkers have systematically examined whether working memory consists of a single nonspecific storage system or a multitude of modality-specific systems. He has studied the patterns of interference between various simultaneous mental tasks and concluded that both possibilities are partially correct. On the one hand, there is some evidence to support the idea of a general working memory capacity, which overloads whenever the mind is burdened with a difficult task. This overloading is often evident in the breakdown of performance of a second task, attributed to interference. On the other hand, the magnitude of interference effects is much greater for some combinations of tasks than for others; certain combinations allow human subjects to maintain very good performance on one task while performing another just as well. Hitch and Baddeley (1976) showed that subjects were able to solve a difficult visual sentence-parsing test while continuously performing a digit-recall task set at the upper limits of their ability; they slowed their rate of performance slightly, but maintained 95 percent accuracy on the sentence task while maintaining their digit-repetition accuracy. If there had been only a single working memory system receiving inputs from all sensory modalities, loading it to capacity in one modality should have wiped it out in all other modalities. But this did not happen, and Baddeley was forced to conclude that, at the very least, visual and acoustic working memories were somewhat independent of one another.

Baddeley has attributed generalized interference effects to a "centralized executive system," closely related to the system proposed by Norman and Shallice (1986). Baddeley has explained specific interference effects by proposing two slave systems, the "articulatory loop" or auditory buffer system (already discussed in Chapter 7) and the "visuo-spatial scratchpad," or simply scratchpad memory. Baddeley acknowledges that there may be other slave systems, but he has documented these two in particular, perhaps because they are the most salient ones. His conceptualization of the architecture of working memory fits well with many other lines of investigation that have demonstrated the relative independence of visual thinking and spoken language.

The ESS has been, until recently, almost exclusively visual; the arena of the great new human cognitive skill was thus somehow visual. But there is no evidence that human visual powers per se have changed, in the biological sense, over the past 4000 years. Could visual scratchpad memory, as traditionally conceived, have sufficed as the arena of complex visual thought? This is highly unlikely; a list of the paradigms used by Baddeley suggests that the scratchpad cannot function well except in the *absence* of competing visual stimulation. Some of the visual tasks he used to study scratchpad memory include: (1) image generation following the presentation of high-imagery phrases such as "nudist devouring bird"; (2) tracking a mental image of a familiar object, for instance the letter F, and then answering successive imagery-related questions about it that test detailed recall; (3) Brooks' (1967) sketchpad paradigm, which involves mentally filling a four-by-four matrix with numbers, which are presented verbally; (4) abacus-derived mental calculations, including an imagery-based digit-span test; (5) memory for briefly exposed visual patterns; (6) retention of sequences of briefly exposed visual patterns; (7) visual memory for briefly exposed letters; and (8) memory for visually presented words, or what Baddeley calls "lexigens."

All of Baddeley's paradigms address unaided visual imagery, and serve to emphasize the delicate and transient nature of scratchpad memory. As the locus of visuospatial imagery, the unaided scratchpad is an especially unlikely field for elaborate mental processing of visual material of any kind, for the very good reason that such imagery is so prone to interference, and that the visual field is usually full of competing stimulation that would simply erase mental imagery. Since the articulatory loop, although undoubtedly important in listening to speech, is also too transient and vulnerable to interference to make it a likely candidate for long-drawn-out thought processes, neither of Baddeley's slave memory systems are good candidates for the kind of iterative, interactive thought process proposed here. Assuming the need for some arrangement between working memory and the ESS, we seem to be forced to return, once again, to the ever-present "central executive" to explain how biological working memory interacts with the ESS. But is this a satisfactory resolution?

The "central executive" is a hypothetical entity that sits atop the mountain of working memory and attention like some gigantic cognitive Buddha, an inscrutable, immaterial, omnipresent homunculus, at whose busy desk the buck stops every time memory and attention theorists run out of alternatives. Baddeley (1986) acknowledged the conceptual fragility of the central executive, entitling a section of his book, with admirable, if somewhat despairing humor, "The central executive: ragbag or supervisor?" The idea of a central executive process seems to represent a return to the unitary model of mind rejected earlier in this book.

Animals, as well as humans, are supposed to employ working memory to solve problems; the anatomy and structure of working memory in animals (see O'Keefe and Nadel, 1977; Olton, 1977) dovetails nicely with the human neuropsychological data (Milner, 1966; Squire, 1984, 1987), and there is evidence of evolutionary continuity in the emergence of working memory (Sherry and Schacter, 1987). Entry into working memory could mean that certain memory files are in a state that could be called "open" or "active," while others remain closed; or it could mean that certain traces are literally moved into a physical area of memory reserved for working memory. Whatever its physical basis, working memory material is more relevant, and more "available," to conscious thought than material that is in longer-term storage and thus supposedly not as easily available for processing.

Thus, when rats are trying to solve a radial-arm maze, they must carry, in working memory, a map of the maze, perceived in reference to the room surrounding it, and the objects, if any, contained within it. This map serves as a working memory, a reference system for the task while they work at solving the maze. Similarly, if humans are asked to solve a maze or a puzzle, they supposedly construct a map of it in working memory, which they use to guide their decisions. The same applies to language; if asked to remember and paraphrase the contents of a news broadcast, they store a digested version in working memory for ongoing semantic reference. And if they are required to recall and reconstruct an episode from long-term memory, they do this by selectively retrieving items from longer-term storage and holding them in working memory for analysis.

To a degree, the idea works. In animal cognition, and in simple stimulus-driven problem solving of the kind typically tested in the laboratory, there may be some value in the concept, although it is not proven that working memory exists as a definable *biological* entity. The problem becomes acute when trying to extend the idea to include more elaborate human thought projects, including what Darwin called "longer trains of thought." Many current theorists seem to believe that humans must think with essentially the same array of mechanisms that animals use, with the important addition of language: thus, short-term memory, perhaps supported by certain mysterious uncharted "unconscious" thinking mechanisms, and supplemented in humans by spoken language and generative visual imagery, constitutes the field—that is, functional location—of thought.

However, there are serious problems with this idea. By their own experimental evidence, memory theorists have shown the tremendous limitations on human awareness and short-term memory. The articulatory loop endures, without further rehearsal, for less than two seconds (Baddeley, 1986); memory for lists of words seems to persist slightly longer, but no more than fifteen seconds (Peterson and Peterson, 1959). Humans can take only five to seven items into their temporary store at the same time (Miller, 1956) and are extremely vulnerable to interference from irrelevant environmental events (Broadbent, 1958). Moreover, the temporary store is highly unreliable; brief glimpses of scenes lead to highly unreliable accounts of what transpired, to the point where eyewitness testimony, even when tested immediately after the event, has been thrown into question (Buckout, 1982). Visual working memory is especially vulnerable to interference (Baddeley, 1986).

Thus, although adult conscious experience, in the sense of that which is immediately present to the mind, may appear to be somewhat clearer than the "buzzing, blooming confusion" attributed to infants by William James, it is nevertheless very limited, a small window of awareness staggering through the episodic stream, battered about by every wave and wind that happens along. And that is under the best of circumstances; as soon as the company of others is removed for a long time, our limitations become even more striking. In complete sensory isolation, the human mind rapidly becomes disorganized and psychotic (Heron, 1967). How could such a delicate and unreliable instrument be the cognitive field where novels are written and treaties negotiated?

The answer has to be that the idea was, and is, wrong. Humans do not think complex thoughts exclusively in working memory, at least not in working memory as traditionally defined; it is far too limited and unstable. In modern human culture, people engaged in a major thought project virtually *always* employ external symbolic material, displayed in the EXMF, as their true "working memory." They use their biological working memory system, along with their perceptual apparatus, more as an iterative data-crunching device, or a processor of visual analog images. In this rearrangement, biological memory (and not necessarily only its working memory aspect) becomes the loop in the thought process that performs transformations and analyses on the database provided by external symbols, while *the EXMF* becomes the *real* working memory, or temporary holding tank. Simultaneously, of course, relevant retrieval routes within ESS systems become an integral part of the thinking process, as working memory must be part of any complex cognitive operation.

This arrangement can be visualized (Figure 8.6) as a temporary cognitive structure in which the mind uses the external symbolic system for a limited time both to hold certain items and to organize the

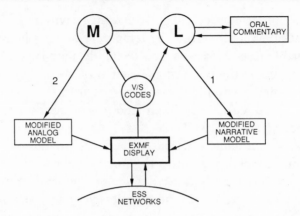

Figure 8.6 Schematic of two external working memory arrangements in common use. Path 1 involves the development and refinement of written narrative models by means of successive iterations through the linguistic controller (L) and the external memory field (EXMF). Path 2 involves a parallel refinement of visual analog models, primarily through the mimetic controller (M) and the EXMF. Both loops also profit from the generation of concurrent narrative commentaries on their progress. V/S codes: visuosymbolic codes.

memory material in certain ways. Advanced thinking virtually always depends on this arrangement: the thinker selects and organizes ESS items according to some objective and uses the power of both the oral–narrative system and the visual symbolic systems to examine, process, rearrange, and invent new ESS entries. Each iteration is usually a short one; the limitations of consciousness are so great that, in any truly original creative thought, the thinker has to revolve around the ESS database, verbalizing, sketching ideas (a part of the ESS usually serves as an *external* visuospatial sketchpad), producing new outputs, in an iterative loop, until a satisfactory resolution is reached. With highly familiar (overpracticed) material, the thinker may be able to continue the thinking process to a degree without the ESS being immediately present, but it is the essential reference point to which the thinker must return. The field where these symbolic interactions takes place is thus appropriately labeled the external working memory field.

This proposal is not easily reconciled with the traditional version of working (biological) memory as the primary field for complex mental processing. And yet the latter could not possibly have supported the type of theoretic development that humans have come through during the past four millennia. Working memory is too transient, too vulnerable to distraction, and too limited in capacity to manage a major cognitive project that may eventually result in theoretic products. In this case, the EXMF is the crucial working memory locus (displayed by means of the visual system and interpreted by the visuosymbolic encoding machinery). This symbiosis of human working memory and the EXMF is basic to modern thought. It also might help explain why the trained introspectionists of the Würzburg School were unable to find much evidence of conscious thought in their own minds; for the most part, they were looking in the wrong place. The EXMF is an essentially external loop for organized thought process, and there is no substitute for it in biological memory.

How can working memory be usefully redefined to incorporate the external field of complex thought? Assume for the moment that working memory, as traditionally conceived, probably serves fairly adequately to describe the role of human short-term memory before the third transition—that is, as working memory was, and is, in oral–mythic cultures. The central executive controlled memory traffic, while the articulatory loop and visual scratchpad memory supported specialized uses of visual and auditory recall—a pretty limited system for the purpose of thinking. The addition of the EXMF to this conceptual diagram both offers a structural explanation for the radical changes that have taken place and suggests a different role for traditional memory systems, especially for scratchpad memory. The central executive and articulatory loop probably retained their primary function, *to support spoken language;* but the visuospatial scratchpad was relegated to a very secondary role.

The thought process itself could not take place in any one of these traditional working memory areas. The algorithms of thought, like the subroutine and assembly-language commands underlying higher-level computer programs, involve various separate functions, carried out in a sequence, and each stage requires its own appropriate mental arena. The specific functions supporting a particular type of thought vary according to the task. Thus, if a symbolically trained mind wishes to take notes from a book and formulate a theory, the surface procedure would be, roughly: (1) scan and enter into the mind some ESS item, drawn from the EXMF; (2) decode the item; (3) reduce the semantic content of the item to a few essentials; (4) write down the reduced content, that is, reenter it into another EXMF location, for later retrieval; and (5) continue this iterative process until some criterion of completion is reached. The underlying functional sequence undoubtedly involves biological working memory systems, but in what role? They do not perform the actual scanning; they do not decode; and they do not control the act of writing. Do they achieve the semantic reduction? This is doubtful; the language systems of the brain presumably do that. Then what do they do? One remaining role might be attention control; but attention control in this sort of task, as mentioned earlier, is vested largely in pointers and flags, procedural items that would not be stored in working memory, at least not as it is usually defined. Probably the principal remaining function of biological working memory is supporting the construction of an oral-narrative commentary on the behavioral process of taking notes and theorizing.

What about searching and scanning the ESS? This might well be another residual function of biological working memory. But scanning and searching also depend heavily on the EXMF. Cognitive maps of the ESS, and the relevant search strategy information (for instance, category labels and hierarchies, and their relevant locations on the ESS map), might be retrieved and held in linguistic working memory, but ultimately the search depends on what is passing through the EXMF. The EXMF contents indicate the current location and direction of movement through the ESS network and ultimately display the sought item. While the biological working memory might direct the search, ultimately even this operation would be impossible without the EXMF.

The truth is, the most useful working memory in an ESS-based thought loop is the EXMF itself. Skinner (1957) remarked on the importance of external stimuli in self-regulation, even in speech. In this, as in many things about language, he was correct, but he understated the case. External symbolic stimuli not only drive the thought process; they serve as the brain's holding tank while its various systems go about the business of processing and altering the symbolic environment. The brain's role is paramount, of course; but the role of biological "working memory" is not at all clear once the ESS is built into the picture. At best it is probably a reservoir of immediately relevant ESS-related scanning and searching strategies, and current semantic content.

The Emergence of Theoretic Culture

Visuosymbolic invention and external memory are highly important cultural innovations, but they should be regarded as the surface reflections of a more fundamental change in human cognitive evolution, and they did not arrive in isolation. The momentum for visual symbolic invention, and for extensive use of external storage, must have come from the modeling process itself. Just as in oral linguistic invention, conceptual development was at the core of the drive toward extensive ESS use. The emerging new level of conceptual development may be called theoretic, judging by its ultimate product; but theoretic development was gradual and took place in several parallel paths.

The impression may be developing that I attribute all systematic thought to visual symbols, the EXMF, and the ESS; but this is not so. Conceptual development often led the way, historically preceding the invention of the EXMF and ESS. This is particularly evident in the history of writing. The idea that writing produced scientific and technological development probably reverses the real order of things; it is just as likely that the invention of writing and other notational devices was driven by the conceptual needs of emerging theoretic culture. Writing, and graphic symbolism in general, arrived on the scene long after a number of major conceptual developments had already taken place. The historical order is important; if all theoretic development had followed writing, and especially if it had followed the invention of the alphabet, an argument could be made that it somehow depended on the latter. But if a certain level of implicit theoretic development preceded writing, then other factors have to be considered.

A long list of technological and protoscientific inventions preceded writing. Fired ceramics appeared in what is now Czechoslovakia about 25,000 B.C. Cave paintings in France strongly suggest the presence of percussive musical instruments about the same time. The first boomerangs appeared about 15,000 B.C., along with the sewing needle, tailored clothing, the bow and arrow, the spear thrower, the first lunar records (notched bones), the first simple maps, and the first uses of rope. Sun-dried bricks were developed about 10,000 B.C.; about the same time, dogs were domesticated in Mesopotamia, followed by sheep and goats in Persia, pigs and water buffalo in eastern Asia, and chickens in southern Asia (Hellmans and Bunch, 1988).

Wheat and barley were cultivated about 9000 B.C.; potatoes were cultivated in Peru, and rice in Indochina, around 8000 B.C.; sugar cane was grown in New Guinea, yams and bananas in Indonesia, flax in southwestern Asia, and maize in Mexico around 7,000 B.C. Mortar was introduced in Jericho, woven mats in Jordan, and woven cloth in Turkey around 6000 B.C. Domesticated bread wheat arrived around 5000 B.C., along with lentils, citrus fruits, peaches, avocados, date palms, cotton, millet, and squash. Irrigation was in use in Mesopotamia by 5000 B.C.; techniques of mummification were also known by this time, as was the domesticated horse, first seen in the Ukraine. Sailing ships were also invented around 5000 B.C., in Mesopotamia. At the same time the Egyptians developed extensive copper mines and smelters. By 4000 B.C., when the first rudimentary writing systems were invented, humans already had invented high-temperature kilns for firing bricks and pottery; they used plows pulled by domesticated cattle; they had the beginnings of copper, silver, and gold metal technology; and they already made wine and beer.

This very rough historical sequence emphasizes the importance of looking at the whole cultural context of symbolic invention; visuographic invention did not take place in a vacuum, and the ESS did not spring fully armed from the bosom of simple aboriginal cultures. By the time writing and monumental art came into widespread social use, human culture had already showed an accelerating rate of change and had produced a proliferation of nonsymbolic inventions, many of which undoubtedly eclipsed early writing in complexity and sophistication.

The complex technological and social developments that preceded writing might suggest the existence of some apparently analytic thought skills that contained germinal elements leading to later theoretic development. However, early inventions were pragmatic and generally not far removed from nature: for example, the domestication of animals and plants would not have required more than a recognition, transmitted over time, that certain species were desirable and domesticable for human use. Complex constructional products, such as brick structures and sailing vessels, might be seen as grand elaborations on the ancient toolmaking skills of humans. The social organization of the first towns and cities presumably borrowed heavily from existing family and tribal structures. These pragmatic developments, impressive as they were, lacked the essentially reflective and representational nature of theory.

But as technology and social organization depended more and more on some form of record keeping (that is, external memory devices), visual symbolic devices emerged in increasing numbers. This probably did not have an immediate radical effect on human society, other than to enable the formation of larger trading patterns and social entities. But the proliferation of the ESS eventually created the intellectual climate for fundamental change: the human mind began to reflect upon the contents of its own representations, to modify and refine them. The shift was away from immediate, pragmatic problem solving and reasoning, toward the application of these skills to the permanent symbolic representations contained in external memory sources.

Early Analog Models: Time and Space

The earliest move in this direction was apparently in a form of visual thinking, especially evident in the construction of analog models of time and space. As in modern science, the theoretical products—analog models—were not totally the result of any single symbolic invention. The earliest evidence of an elementary form of theory formation is found in ancient astronomy. Astronomical knowledge, like writing, was a powerful device of social control; the measurement of time in terms of astronomical cycles was probably the ultimate controlling activity in early agricultural societies, setting dates for planting, harvesting, storage, and distribution of grain and for religious observations, as well as a number of cyclical social functions. Practical astronomic knowledge also contributed to improvements in navigation.

Quite early in the history of visuographic symbolism, analog devices were invented that served both a measurement and a predictive function in representing time. These devices eventually allowed humans to track celestial events, construct accurate calendars, and keep time on a daily basis. They ranged from primitive observatories to small portable calendars for the use of individuals, like the Sumatran

string calendars, which indicated the days of the lunar month by
means of a string looped through a series of 30 holes; the string was
advanced one hole for each day of the lunar cycle. Smaller intervals
of time, closer to the hour in length, were often estimated by using
noon as a temporal anchor in daylight; for example, the Egyptians
used a right-angled "time-stick" to judge how many "hours" it was
before or after noon; this was indicated by the angle of the sun's
shadow, which struck different marks engraved on the time-stick as
the sun followed its apparent course across the sky. The Chinese used
a similar device about 2500 B.C. Later there were also water clocks,
sand clocks, and candle clocks, all of which operated by a similar comparative principle and none of which depended upon writing or formal
numeration for their effectiveness.

Space was also represented in early visuographic symbolism. Some
of the earliest analog symbolic devices were primarily geometric and
reflected an abstract understanding of spatial relationships. A basic
form of analog visual modeling, sometimes called iconography, is still
used in communicative rituals in some preliterate societies. An example is found in the sand drawings and sand stories of the Walbiri,
a people found in western Australia (Munn, 1973). Their iconography
seems to be an extension of metaphoric and iconic gestures, except
that the images are given external form in sand, body paints, or some
other medium. They draw graphs in the sand during formal storytelling, as a running notation to illustrate narrative meanings. Some
simple design elements such as circles and lines are fixed, but the
essential message of a drawing is usually conveyed by the spatial juxtaposition of the elements.

There are layers of meaning in these simple illustrations. For instance, on the most straightforward level of interpretation, to describe
someone's travels one might show different places by circles, and connect then with lines to indicate their path; this overall arrangement
might be regarded as a kind of prototypal map. But the same type of
symbol arrangement might also be used to illustrate a more abstract
mythic voyage, such as the life-cycle of an ancestor or a myth about
the replenishment of life. In this case the circle might represent a
"campsite" which in turn represents a phase of life; lines might represent not only a path but also a motion, recurrence, or transformation.

Simple design elements can therefore be used in various ways to create a variety of quite expressive images. For instance, Walbiri convention might represent a tree from above as two concentric circles: one small circle for the trunk, and the larger for the circumference of the leafy area above. Within the larger circle there might also be several semicircles, each representing a sitting man. The image thus conveys a complex geometric arrangement of people and things. But the sitting men may also be seen to have a symbolic relationship with the roots of the tree, and the tree might stand for society itself, or the cycle of life. Thus, stories can be intertwined with these sand images, by attributing secondary symbolic value to the elements of the drawing, and further by combining other symbols such as painted wooden objects with the sand drawings.

Primitive maps are closely related to this kind of geometric juxtaposition of symbols. A number of preliterate societies have been shown to possess map-drawing skills (Harvey, 1980; Lewis, 1987; Smith, 1987). American natives drew rough maps in mud or sand, and these maps were of use to early European explorers. Harvey (1980) reproduced a simple map drawn in sand in 1884 by a member of the Suya tribe, a central Brazilian people, and recorded by a German explorer. In principle it resembled some of the sand drawings of the Walbiri: lines traced the path of the Xingu river and its tributaries, while cross-strokes indicated the location of each village. Techniques of primitive mapmaking seem to employ universal elements such as lines and dots and circles, in geometric arrangements. In its earliest forms mapmaking was thus a close relative of iconography.

Some of these early symbolic devices were analog in nature; that is, they measured one dimension of reality in terms of another. They were the precursors of modern scientific measurement, which still relies heavily on analog devices: dials, meters, oscilloscopes, polygraphs, thermistors, and so on. Temperature is still measured by means of various analog devices (for example, inches of mercury), and so is weight. These are now supplemented by exact numeration of the measurements; but this was not always the case.

The origins of analog measurements go back far in time, as testified by their common use in aboriginal societies. Simple analog indices of measurement can also be obtained through the use of relative or comparative linguistic statements: for instance, distances can be measured in terms of time (a day's walk), or volume in terms of various carrying vehicles (a handful, pocketful, mouthful, armload). The distance from such verbal metaphors to systematic analog measurement is not very great: if "late in the day" means a longer shadow, it is not a great leap to create a standard device for marking that shadow.

Once visual analog symbols had been created, however, the opportunity arose to use them as memory storage media and as computational devices. The recurrence of midsummer moonrise at the same point on the horizon could be verified by a system of stone markers. Other highly visible, predictable astronomical events could also be marked and verified, and over a period of time, the range of variation—that is, the maximum deviations from the general trend—could be marked. This seems to be the way the observatory at Stonehenge evolved (see Hawkins, 1963, 1964), between 2000 and 1600 B.C. The positioning of the stones allowed the prediction of the summer and winter solstices, the two times of the year when the progressive lengthening or shortening of daylight reverses. The alignment of the "heel" stone of Stonehenge, relative to its center, was oriented precisely toward the midsummer sunrise, testifying to centuries of experience in making such predictions. By means of the 56 "Aubrey holes" on the circumference, and a system of 6 moveable stones, it also could have been used to keep track of the 56-year cycle of eclipses and winter moonrises and thus apparently to predict eclipses and a variety of other astronomical events (Hawkins, 1964).

The megalithic observatory was thus really a gigantic visual machine, a modeling device. By means of the stone markers added and altered to register events, data impressed themselves on the machine, and the carefully positioned stones became external memory traces of geometric impressions left on Earth by the light emitting from astronomical events, more like a time-lapse photograph than a system of writing. But it was nonetheless a symbolic invention: it allowed the storage and analysis, in an external device, of complex information. The entire megalithic complex became a large-scale, and very abstract, model of astronomical motion. Its later and more sophisticated relative, the astrolabe, employed digital markings but remained essentially an analog modeling device.

The construction of analog visual devices involved a different ESS loop than oral linguistic invention. Before writing, the earliest analog models must have been created through an interplay of mimetic visual modeling and oral commentaries on the modeling process. As pointed out in Chapter 7, the mimetic system would have provided the basis for simple metaphoric gesture; and oral narrative ability would have provided a device for "commenting upon" such visual metaphors, that is, for assigning verbal symbols to the analog visual process. This would have afforded a degree of control over the analog process itself to the linguistic system. Thus, the earliest analog models were probably joint products of the two systems in interaction.

In the case of megalithic astronomy, the storage devices themselves constituted an analog model. A ground-based perspective on their gigantic analog model was undoubtedly not the schematic aerial view shown by authors such as Hawkins; nevertheless, their understanding of the model, for predictive purposes, was similar. Through aligning two specific stones, they could predict the place on the horizon where various events—especially important seasonal moonrises, moonsets, sunrises, and sunsets—would take place. The visual field in which the model was created and refined thus became the EXMF. The V/S codes were complex and, for us, unconventional: the stones had to be "read" according to a complex set of interpretative strategies.

The construction of early calendars always implied some theoretic development, in the sense that implicit models were being built of astronomical events. Many of the elements of modern science were already present in primitive astronomy: systematic and selective observation, and the collection, coding, and eventually the visual storage of data; the analysis of stored data for regularities and cohesive structure; and the formulation of predictions on the basis of those regularities. All early agricultural societies, out of necessity, had calendars based on astronomical science. Therefore, procedurally, the groundwork for scientific observation and prediction had already been constructed 5,000 to 10,000 years ago, not necessarily in written symbols but in a different kind of visuographic invention, which represented an analog *system* of knowledge. Written lists of astronomical observations came much later, in China and Babylon.

This shows how writing followed, rather than preceded, the invention of analog visual symbolic devices, at least in some cases. A great deal of knowledge existed before visual symbols were developed for the denotative encapsulation of that knowledge. The visual recording and modeling of data was necessary to the scientific progress being made; but written words were not always the prime movers in early scientific thinking, although oral commentaries undoubtedly were. And while it is common now to reformulate analog data in written words, the early history of astronomy shows the power of purely analog modeling, even in the absence of writing.

In conclusion, astronomy was probably the earliest example of widespread, socially important theoretic development in human history. Astronomical observations and predictions could not have been achieved without some form of external storage of data and could not have been modeled without some form of computation. By combining a simple token system of counting with various analog measurement and computational devices, humans were able to improve their mental models of time and space in significant ways, while using these calculations to run their growing agricultural society. Analog measurement devices were thus interwoven with the evolution of the theoretic process. The resulting visual "models" reflected the state of theory, as they were a direct product of the theorizing process. Theory had not yet become as reflective and detached as it later would; but the symbolic modeling of a larger universe had begun.

Externalizing the Oral Commentary

Early visual analog models were exceptional; there is little evidence of other theory formation until the first millennium B.C., when writing was used by the Babylonians and Chinese to order the collection of astronomical data, which allowed further refinement of astronomical models and more accurate prediction of eclipses. But writing in this case remained locked into a limited role; for the most part, astronomical record keeping was similar to commercial record keeping and involved mainly the construction of lists of observations.

Until writing could be combined with other visual symbolic media, the *theoretic attitude*, that is, the deliberate analytic use of symbolic thought, had little chance to develop. Ancient Greece, from around 700 B.C., was undoubtedly the birthplace of theoretic civilization. In Greece, human thought suddenly came into its own simultaneously in a number of fields: philosophy, mathematics, geometry, biology, and geography, to name a few. From that time on, there was a steady progression of theoretic development, one that has been well documented by historians and philosophers.

Our concern here is not so much with the history of science or philosophy per se as with the cognitive framework that enabled such accelerated change. How had the *structure* of the human thought process changed? The answer appears to be at least partly that, in the ancient Greeks, all of the essential symbolic inventions were in place for the first time. The evolution of writing was complete; the Greeks had the first truly effective phonetic system of writing, so successful that it has not really been improved since. They also possessed advanced systems of numeration and geometric graphing. Astronomy had advanced considerably under the Babylonians, and the Greeks were aware of that body of knowledge, as they were of Egyptian mathematics. Moreover, their society was open, intensely competitive, and sufficiently wealthy that education went beyond the immediately pragmatic.

Logan (1986) has suggested that the greatest symbolic invention of the Greeks, and the key to their progress, was the alphabet. Indeed, this was their most original invention on this level, and the alphabet was a more efficient script than its predecessors. But it is difficult to see how the mere possession of the alphabet could have triggered such a creative explosion; phonetic scripts, in the form of various syllabaries, had been in existence for almost 2,000 years, and the phonetic transcription of speech had long since been achieved. The alphabet made such transcriptions much easier and more accurate, and this allowed the spread of literacy and facilitated the writing of books, so that it cannot be discounted as a factor; but it is difficult to assign it a determining role in the Greeks' extraordinary achievements.

The key discovery that the Greeks made seems to have been a combinatorial strategy, a specific approach to thought that might be called the theoretic attitude. The Greeks collectively, as a society, went beyond pragmatic or opportunistic science and had respect for speculative philosophy, that is, reflection for its own sake. In the brief flowering of the Greek city-states, they founded abstract geometry and the idea of formal mathematical proofs, and the first systematic taxonomy and embryology of living species from all over the known world. They had the first theory-based system of cosmology and achieved great advances in theater, sculpture, public administration, and architecture. In all of these areas, they employed a strategy that combined all the previous isolated cognitive advances we have noted. In effect, the Greeks were the first to fully exploit the new cognitive architecture that had been made possible by visual symbolism.

A number of factors facilitated this process. The process of de-mythologization was well advanced in Greece, perhaps more so than in any society until two millennia later, during the European Enlightenment. The possibility of disengagement from mythic thought was a feature of Greek city life, which lacked a single overriding religion. A good example of the difficult process of separation from mythic thought may be found in the case of Hippocrates, the founder of professional medicine, who fought for the separation of medicine from religion at his school in Cos, in the fourth century B.C. This might not, in itself, have led to anything. But the timing was perfect; symbolic invention had progressed to the point where most of the basics of the modern repertoire of symbols were available, and the secularization of attitudes left the Greeks free to explore the new possibilities offered by their new mental tools.

The critical innovation was the simple habit of recording speculative ideas—that is, of *externalizing the process* of oral commentary on events. Undoubtedly, the Greeks had brilliant forebears in Mesopotamia, China, and Egypt; but none of these civilizations developed the habit of recording the verbalizations and speculations, the oral discourses revealing the *process* itself in action. The great discovery here was that, by entering ideas, even incomplete ideas, into the public record, they could later be improved and refined. Written literature for the first time contained long tracts of speculation—often very loose speculation—on a variety of fundamental questions. The very existence of these books meant that ideas were being stored and transmitted in a more robust, permanent form than was possible in an oral tradition. Ideas on every subject, from law and morality to the structure of the universe, were written down, studied by generations of students, and debated, refined, and modified. A collective process of examination, creation, and verification was founded. The process was taken out of biological memory and placed in the public arena, *out there* in the media and structures of the ESS. What the Greeks created was much more than a symbolic invention, like the alphabet, or a specific external memory medium, such as improved paper or printing. They founded the *process* of externally encoded cognitive exchange and discovery.

From the first, Greek thinkers were employing external memory devices to their fullest effect, in a way that was totally new. Although phonetic writing had been invented by the Egyptians and later refined by many cultures, it had never been used to record the thought process itself. Both spoken narrative thought and visual thought were symbolizable, that is, representable in graphic form. This habit of externalizing the process was to be the key to all further development in the arts and sciences. The Greeks seemed to have realized this early in their theoretic development, because they invented the written dialectic, a formal method of presenting ideas that had never been tried before. The dialectic involved an exchange of views, rather than a single manufactured "official" viewpoint. Doubts were raised, contradictions were resolved, the flaws in argumentation were examined and reexamined.

The exchange of views and exposure to public debate was always a characteristic of spoken narrative thought, which by nature is a public process. But spoken narrative thought was quite limited, when it had no other resources to draw on, in the construction of theory; this was due to the limitations of auditory working memory. In order to circumvent this limitation—for instance, by rote oral memorization—thoughts tend to be repeated literally. Thus, the natural tendency of spoken thought, stripped of *any* ESS input, is toward fairly loose narration of events, metaphoric fantasy, and storytelling. Fine-grained analysis of the thought process itself is difficult, because the memory trace of an oral narration is so ephemeral. There are ways around this, and the power of spoken thought was later greatly increased, but only *after* the ESS was brought into play. When the Greeks began their great intellectual adventure, widespread use of the ESS as a field for thought was still unknown.

The Greeks thus invented a new ESS loop for the iterative refinement of linguistic productions. Various written discourses were entered into the ESS and selectively displayed in the EXMF. Any sample of discourse in EXMF can be subjected to oral commentary and analysis and modified in the usual iterative manner. In this case, discourse-level metalinguistic skills are necessary for making modifications to the ESS entry. Further metalinguistic evolution was trig-

gered by this development, as will be discussed in the next two sections.

To summarize, the Greeks possessed all of the basic visuographic
inventions: they had advanced pictorial skills; they knew the ideographic and analog uses of visual symbols, especially from mathematics and astronomy; they had the best system of phonetic writing
ever invented; and they possessed all the media needed to record their
ideas. Moreover, their political–religious climate encouraged and tol-
erated (with some notable exceptions) free speculation. Early in their
history, they started the habit of recording not only a few observations, or lists of names and events, but the very process of dialectical
exchange. The result was that, for the first time in history, complex
ideas were placed in the public arena, in an external medium, where
they could undergo refinement over the longer term, that is, well
beyond the life-span of single individuals. This meant that the EXMF
could be fully exploited for the first time; where its use had been
restricted to analog models, lists, and a few simple narratives, it was
now the field of more elaborate symbolic structures. The representation of narrative thought skills in the EXMF allowed the process of
reasoning and verification to be laid out in hard storage, as in the
Socratic dialogues. And the combination of analog models and narrative commentary led to more elaborate philosophical and mathematical ESS entries.

Pedagogy and Metalinguistic Evolution

Thought habits, particularly symbolically based ones, are learned.
One might speculate that the symbolic skills of educated post-
Enlightenment Europeans were not ingrained, judging from the behavior of earlier Celtic and Germanic civilizations. The process of
collective skill acquisition by a society must be transmitted and accumulated across generations, and symbolic skills are so complex that
they require extensive formal education. The latter has come to account for many of the programmed intellectual skills and habits of
adults, particularly in the most highly educated sectors of society.
Students are endlessly rehearsed and trained in finding ESS entries,
analysis and reduction of these entries, and the creation of new ones.
Therefore, major cultural changes in the balance of cognitive skills

might be tracked by tracing the history of education over the two and a half millennia since the pattern of Western education was established. The curriculum of education, in other words, constitutes a valuable cultural artifact in itself and is a particularly useful source of evidence, one that testifies directly to the pattern of early cognitive training imposed by society. The curriculum reflects not only the values held to be important but also the socially programmed structure of thought skills. It follows that the history of pedagogy might reflect the process of metalinguistic evolution.

Metalinguistic processes may be defined as superordinate cognitive structures that regulate the uses to which words and sentences are put. The need for such processes seems to have been overlooked by psycholinguists until the advent of computational linguistics. Metalinguistic processes are essential for spoken discourse; for instance, Levelt (1989) refers to the construction of shared "discourse models" by participants in a conversation; these are more or less equivalent to Johnson-Laird's mental models, and they serve as a common conceptual context in which to interpret what is being said. These models control the highest aspects of speech production in conversation: themes, topics, roles, and interpretations.

Levelt describes the process of deciding what information to express, and how to order it, as macroplanning. Both discourse models and macroplans are metalinguistic processes inasmuch as they dominate the relatively "front-end" operations of sentence construction, lexical selection, and phonology. In a similar vein, Kintsch (1988) refers to the need for "discourse representations," that is, a knowledge structure that is constructed of concepts and propositions. Without such metalinguistic representations, the interpretation and production of meaningful discourse is very limited.

Discourse representations and macroplans, however, control only the interpretation and production of speech over the relatively short term; for instance, Kintsch (1988) used a simple three-line mathematical problem as an example of a problem needing a knowledge-based discourse model of the contents of the problem statement. Levelt's examples are a little more broadly representative of real discourse, ranging over a variety of examples from simple statements to longer conversations. However, his examples do not address the level of discourse organization and macroplanning evident in, for instance, a long speech or a complex, intricately argued article or book, let alone a comprehensive scientific theory.

But it is the latter that have continually evolved in the history of Western education; a consistent theme, starting with the ancient Greeks and continuing until recently, has been the systematic training of generations of students in what may be called the broader meta-linguistic skills. The skills that were central to the curriculum focused initially on the art of constructing spoken discourse and engaging in dialectics, and then generalized to the art of written composition, evolving gradually into what amounts to a systematic curriculum for the training of thought skills and habits. Such habits are metalinguistic; they control the uses to which words are put and the interpretations that are placed on what is said. They are evidently not innate, since they have developed fairly recently and have changed in a culturally cumulative manner over several millennia. In fact, the thought habits of postindustrial civilization are a very recent invention, the latest in our long history of symbolic invention.

Educated persons were a relatively small percentage of the population of the ancient world, but they wrote all the books, kept all the records, ran the schools, academies, and universities, controlled religious observances, wrote the history of the period, interpreted the past, determined fiscal and political policies, regulated trade, produced most of the art and all of the literature, and increasingly dominated the social order as a class. The forms of culture, a good deal of the technology, and the techniques of information storage were products of the educated classes. Virtually all major cultural change originated in this class of people, and therefore their thought habits and skills were the engine of cognitive evolution in Western society.

Of course this subject could lead in a number of directions that would prove irrelevant to this inquiry. The details of the curriculum, on one level, are completely tangential to the question of cognitive evolution; it does not matter what versions of history are taught, or which languages, or which cosmological or theological theories. What matters to this inquiry is which cognitive *skills* are formulated, emphasized, tested in the marketplace, and hammered in year after year through repetition and practice. Fortunately for our purposes, the curricular skeleton of Western education was fairly constant for a very long time. It is well-documented and is easily traceable; and at the core of Western education, there has been a strong tradition of training specific metalinguistic skills. The training of these skills has, in fact, been at the heart of the competitive societies of the West and has in itself been an agonistic, or competitive, process.

Before starting into this subject, we must remind ourselves that, long after the invention of powerful writing systems, and long after the role of writing in the government and control of human affairs was established, the uses of writing (and visuographic skill in general) remained subordinate to mythic thought and narrative skills. This was evident in the first (mythic) uses of pictorial symbols on cave walls; it was also evident in early written records such as the Homeric myths, which may have been recorded in writing but originated in oral tradition; and it remained evident in the later use of written instructions to refine oral narrative skills.

For a very long time, the skills acquired in building the ESS were used as a supplement to the education of oral and social skills, which were still considered the central assets of the ruling classes. Two types of training epitomized this attitude: ancient rhetoric and the training of memory, or formal mnemonics. Formal mnemonics once served a valuable role but has long ceased to be relevant; but ancient rhetoric subsequently differentiated into a structured curriculum that formed the basis for the entry of many generations of Western adults into the highest reaches of their cognitive culture.

Barthes (1985; translated in 1988) has written a useful short history of ancient rhetoric and traced its influence on the later forms of Western education. Rhetoric is what Barthes calls "metalanguage"; it is a set of skills that controls language use on the level of discourse. From the fifth century B.C. to the nineteenth century A.D., it was one of the major subjects of Western education; in fact, it was the original subject around which much of the curriculum of higher education was built. Rhetoric had its origins in town life in ancient Greece and was central to the uses of public debate in politics and law. It rapidly became a school subject and then differentiated into several subjects, with texts, lessons, exercises, and tests. It included rules and practices that served to regulate the use of language on a metalinguistic level; that is, it included training in the art of extended speaking and persuasion, the construction of phrases and orations, and the attainment of excellence in both a communicative and an aesthetic sense. It was, in a very real sense, the linguistic equivalent of coaching in organized athletics. Students were diagnosed, trained, evaluated, and placed by their rhetors (teachers of rhetoric); they entered in competitions to hone their skills, and, on graduating, these linguistic athletes entered the competitive Greek political fray, all according to precepts laid down in writing and discussed publicly. Some rhetors were paid to deliver important public addresses—for instance, a funeral panegyric or an ode to a tyrant. They also served as lawyers, pleading legal cases publicly before juries.

Early in the development of rhetoric as a subject, it fragmented into schools of thought. Students of the Syracuse school of Corax around 450 B.C. (the first professional, that is, paid, rhetor) were taught to follow a fixed order in constructing an argument: (1) the exordium or exhortation to the audience; (2) the narration of relevant facts; (3) the argument proper; (4) a digression; and (5) an epilogue. They were trained in the techniques necessary for constructing each stage of the oration. Gorgias of Leonthium, around 420 B.C., emphasized the elocutionary aspects of oratory: the symmetry of sentences, the contrast of thesis and antithesis, and the uses of metaphor and alliteration.

Thus, from the start, rhetoric emphasized the large-scale, on-line structuring of linguistic thought products. This fits the definition of a very high-level metalinguistic skill and was already a considerable step away from simple, linear narratives and unconstrained imaginative myth. The art of discovering the metalinguistic structure of ideas gradually became the focus of training. The logic of argumentation was also starting to emerge as a *trainable* skill. In effect, the early growth of rhetoric reflected the refinement and formalization of thought strategies and criteria for evaluating and crafting an effective argument. The Sophists later emphasized the techniques of persuasion without regard to content, an attitude that incurred the wrath of Plato, who held the Sophist school in contempt.

In fourth-century B.C. Athens, Plato emphasized that training in rhetoric aided the formation of the whole psyche, and he taught the use of dialectic. But it was in the writings of Aristotle that rhetoric became a firmly established subject. Aristotle's basic format for teaching rhetoric endured in Western education until the nineteenth century. Aristotle's emphasis was on the process of reasoning, composition, and style. He wrote a three-volume work on rhetoric. The first book was on the formation of arguments by the orator, and amounted to a treatise on the populist psychology of persuasion. The second taught how to receive and evaluate the arguments of others; and the third taught proper elocution and the ordering of elements of an oration. He also dwelt on the use of mnemonics to help auditory–oral memory (note that his advice, even on this matter, was written down). His pedagogy was adopted and modified by the Romans, notably by Cicero and Quintilian; importantly, the Romans applied the principles of rhetoric to writing as well as speaking, and worked out an extensive educational program that extended from childhood to the young adult. Children were exercised extensively in language skills, including grammar. Then, at age fourteen they were introduced to the metalinguistic realm of rhetoric, learning both the art of argument and imaginative improvization.

This programmatic approach to language and thought skills extended into the Middle Ages through the efforts of several major figures in the history of Western education, notably Augustine, the Venerable Bede, and Charlemagne. The curriculum of the Middle Ages had a structure—derived from ancient rhetoric but much more elaborated—that served as the basis of Western education until less than a hundred years ago. This structure is of great psychological interest, because it emphasized the training of mental skills; it amounted to a programmatic device for structuring thought in the individual mind, and was increasingly geared toward literacy and the production of ESS entries, rather than the vocal–auditory emphasis of traditional rhetoric. Most of the minds that produced the Renaissance, the Enlightenment, and the industrial revolution were the products of this curriculum.

The Trivium: Rhetoric, Logic, and Grammar

The Greek competitive approach to public debate bridged into the Middle Ages; competitions were built into the training of scholastics. Scholars were trained to be effective advocates—academic warriors representing a side or a school of thought—and their training emphasized generalized cognitive skills rather than specialized knowledge. The educational curriculum at that time evolved into a seven-part program consisting of three skill-related subjects (the Trivium) and

four substantive academic subjects (the Quadrivium). The Trivium
was really about the use of language and thought as weapons in debate, and the Quadrivium was about specific knowledge in fields of
the arts and sciences, especially music, astronomy, and mathematics.
One interesting feature of this curriculum was that the general training of cognitive skill received at least as much emphasis as the teaching of specific subjects. In fact, the training of general cognitive skill
was the dominant objective of higher education at that time.

The Trivium consisted of three subjects: *Rhetorica*, *Grammatica*,
and *Logica* (or *Dialectica*). Rhetoric had by this time become the least
of the three subjects, even though in its origins it had encompassed
all of them; as logic and grammar became subjects in their own right,
rhetoric was left only with decorative, persuasive, and structural skills
related to presenting an argument in oral or written form. Rhetoric
was originally defined so broadly that it included even what we would
call poetic and literary skill; but as these areas differentiated from the
parent concept, rhetoric came to refer only to the secondary aspects
of metalinguistic mastery. In Barthes' judgment, it was surpassed in
importance by grammar and logic as early as the eighth century A.D.
Nevertheless, it persisted in the curriculum until the twentieth century.

Grammatica grew in the early Middle Ages to a position of prom-
inence in the Trivium. Influential Latin grammars were produced in
the fourth century, setting down normative standards for language
use. Grammar was initially regarded as an authoritative set of rules,
virtually a science of correct language use—everything from the rules
of spelling and sentence construction to the uses of punctuation
and rhythms. But in the twelfth century, "speculative" grammar
emerged, dealing with morphology and syntax; at this stage, for the
first time, the idea was proposed that a universal grammar, equally
applicable to all languages, might exist. Where rhetoric had started as
a very general training in language, grammar went into the specifics
to a far greater degree; hence, it can be seen as a specialization within
the global concept of ancient rhetoric.

Logica eventually emerged, however, as the controlling subject of
the Trivium, pushing grammar and rhetoric into the background in
the High Middle Ages, the age of the Scholastic philosophers. Logic
was the ruling skill that employed grammar and rhetoric to its ends.

The basis of logic was the dialectic, the systematic opposition of contradictory points of view, in a kind of syllogistic duel. Its origins were in Aristotle, but the structure of the medieval disputation was more formal, more like a physical duel; the modern "debate" finds its origins in the *disputatio*. A thesis is posited, attacked by an opponent, and finally defended again by the protagonist; a master, or chairman, concludes the exchange with a commentary. The argumentation is harder edged, more either–or in its contrasts, than a Socratic dialogue. Even longer written philosophical works were modeled on the *disputatio*.

Thomas Aquinas's *Summa Theologica* (1272) is a good example of this rigid dialectic structure. Each of the many articles comprising the *Summa* begins with a specific question, followed by statements of the principal "objections" (negative answers) to the question. The author then cites the existing literature and, following this review, lays out his own reasons for answering the question in the affirmative. Following the exposition of his position, he systematically demolishes each of the objections raised at the beginning of the article. Virtually all of Aquinas's highly influential thought was expressed by means of this agonistic, dialectic structure.

In the world of medieval disputation, philosophical and theological authority carried very high economic, military, and political stakes. Thus, the oppositional structure of thought was partly a function of the political ferment of the time. Practice exercises for students included challenges to argue impossible or unsolvable positions, which was thought to be helpful in honing argumentative skill. The rules and procedures for disputations were written down in great detail, and these confrontations were central to the formation of the medieval mind. Aquinas, who stood dominant at the top of this competitive scholastic world, saw his task in competitive terms; he chose not only to break with established authority (for instance, he was the first to refuse to accept the Arabic philosophers' interpretations of Aristotle), but he took the radical step of becoming a Dominican when he had been raised in the Benedictine camp (roughly equivalent to, say, abandoning General Motors to work for Toyota). His academic life was a series of pre-emptive strikes and sieges against philosophical enemies.

These patterns of curricular change testify to the changes simultaneously occurring in the skill structure underlying the Western mind.

When rhetoric developed in ancient Syracuse and Athens, its emphasis was on the development of oral rhetorical skills; and even though
its rules were written down, the written instructions were destined
mostly for oral–auditory use. Moreover, the practical stakes driving
academic discussion were not so high. By the late Middle Ages, most
of the important *disputationes* were written down, even though there
were many oral debates as well, and the emphasis had shifted deci-
sively toward skill with written prose, written mathematics, and geometry. And because academic and political power were so closely
allied, the competitive atmosphere was heating up. Although this intense competition may not have encouraged the "free play of the
mind" so beloved of the nineteenth century, it produced great pressure to improve thinking skills and to put in place institutions that
could cultivate them. The universities found their origins here, and
the great libraries. Ultimately, this tradition led to Francis Bacon and
William of Occam. The analytic habit—the process of public, iterative verification and improvement of ESS entries—was gradually
eroding the dominant mythic forces atop the cognitive hierarchy. But
it did not happen quickly; metalinguistic revolutions apparently take
some time to become established.

The trend toward the preeminence of logic, and away from the
imaginative and purely persuasive uses of rhetorical skill, continued.
The fine-grained analysis of language taught and debated so passion-
ately in the *Grammatica* had been motivated by a realization of the
power and importance of language and an attempt to understand its
innate rules. In much the same way that the second half of the twentieth century has been obsessed with the properties of computational
devices, medieval scholars were entranced with their logic-machine;
it promised to reveal the secrets of the universe.

Language, of course, was identified with thought and with its products in literature, philosophy, and poetry. We might prefer to distinguish between the "purely" linguistic aspects of language and the
process of thinking with the help of language, but in fact most of the
topics of medieval disputations were impossible to conceive through
any other medium. Insubstantiality, the existence of will in animals,
the transmigration of souls, the resolution of the contradictions inherent in a monotheistic Trinity—these were problems that existed
only in the realm of language. They were abstract, they were some-

times intricate beyond belief, and they were taken with deadly seriousness, because language and logic were still seen as having an inherent bond to absolute truth. The Scholastics were unlike the adherents of any modern discipline in their sheer love of language as an end in itself.

The point is, in their training patterns, in their competitive disputations, and in their search after linguistic truth, they were forging the future shape of cognition. They, like the Greeks and Romans before them, were actually inventing new cognitive tools, algorithms for thought, which later served as the basis of modern scientific thought. Symbolic invention was a force in every stage of human evolution, and this period was no exception. The symbolic inventions at this stage, however, went well beyond the limits of the individual psyche; the new inventions were elaborate, socialized metalinguistic procedures that were to serve as the superordinate controlling aspects of the thinking process. These were being formulated, tested in the field, and introduced into education and written literature. Thus, what we see in the pages of the growing written records of human affairs and ideas is the evolution of strategies for handling, cultivating, and institutionalizing the human capacity for thought. The whole apparatus—rhetoric, grammar, logic, disputations, and their many descendants—was as much an invention as more specific symbolic devices like the alphabet itself.

Figure 8.7 shows the configuration of systems in the types of ESS loop that supported early metalinguistic development. Rhetoric was really a complex of skills related to the art of oral and written persuasion; thus, complex social skills, such as the judging of audience expectations, were involved. Standardized grammars were a means of guaranteeing precision in the assignment of reference and the denotation of relations. Grammars were linguistic inventions, like words; and new grammars continue to be invented, especially in computational science. Grammatical conventions, like rhetorical skills, were developed and stored through an ESS loop.

Logic evolved out of an external, formalized process of verifying the "truth" of propositions. The "rules" of logic are themselves a working model of the verification process. The development and acquisition of purely symbolic—that is, logical—verification has a long and arduous history of symbolic invention; it was the furthest pos-

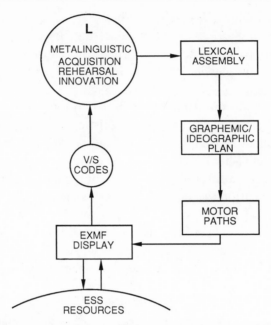

Figure 8.7 Configuration of biological and external memory structures involved in metalinguistic evolution, that is, in the development of discourse-level thought skills. The formalized conventions and rules of rhetoric, grammar, and logic were invented, tested, and elaborated in an iterative loop involving the linguistic controller (L) and the external memory field (EXMF).

sible thing from an innate process. Very few of these metalinguistic skills can be applied without continuous use of the EXMF.

Although logical thinking, rhetoric skill, and formal grammatical conventions may be internalized from the ESS after extensive training and applied (to a limited degree) within the monadic confines of the biological mind, they could not have begun to evolve without the support of the ESS loop. And no major graphemic products—things such as novels, scientific theories, economic forecasts—have an equivalent in purely oral expression. They are products of hybrid minds with extensive ESS linkages.

Long after the High Middle Ages had blended into the Renaissance, and long after thought had become secular, the Trivium served as the basic structural element of Western education. The habits of competitive disputation eventually passed into history, and thought took on a new programmatic structure, especially with the growth of empiricism, positivism, and experimentalism and the concomitant expansion of the natural sciences. This transition was entirely managed by minds whose formation was still largely beholden to the Trivium. However, the accumulation of specialized knowledge and the sheer exponential growth in the size of the external symbolic storage system led to a shift, starting in the nineteenth century, toward more specialized education.

The history of the Trivium, especially its shift from an oral–rhetorical to a written–logical emphasis, dovetails well with the length, style, and organization of books and articles during that long period. The great cognitive difference between postmedieval Europe and all its predecessors was in the gradual insinuation of the ESS into its very fabric. Written political and social works, applications of geometry and mathematics to engineering, and systematic empirical science became an intimate part of the social order in a way that had never been true in the past, not even in Greece and Rome. Mass literacy became possible with the printing press, and although that invention had a very great effect, the society that invented the printing press already possessed tremendous cognitive resources. Print served mainly to disseminate those resources, creating a spiraling involvement of more and more of the population in the utilization of the ESS.

Conclusion: The Hybrid Modern Mind

The modern era, if it can be reduced to any single dimension, is especially characterized by its obsession with symbols and their management. Breakthroughs in logic and mathematics enabled the invention of digital computers and have already changed human life. But ultimately they have the power to transform it, since they represent a potentially irreversible shift in the cognitive balance of power toward complete ESS-based dominance of human cognitive structure. Modern scientific graphics, cinematic special effects, and various other computer-graphic capabilities are direct extensions of what vision can do and what it can be used for. All forms of human representation, from our archaic episodic experiential base, through mimesis and speech, to our most recent visuographic skills, are now refinable and expandable by means of electronic devices. Our modern minds are thus hybridizations, highly plastic combinations of all the previous elements in human cognitive evolution, permuted, combined, and recombined. Now we are mythic, now we are theoretic, and now we harken back to the episodic roots of experience, examining and restructuring the actual episodic memories of events by means of cinematic magic. And at times we slip into the personae of our old narrative selves, pretending that nothing has changed. But everything has changed.

The growth of the external memory system has now so far outpaced biological memory that it is no exaggeration to say that we are permanently wedded to our great invention, in a cognitive symbiosis unique in nature. External memory is the well of knowledge at which we draw sustenance, the driving force behind our ceaseless invention and change, the fount of inspiration in which succeeding generations find purpose and direction and into which we place our own hard-won cognitive treasures. Paradoxically (since any symbiosis with external elements seems alien to human nature), the individuation of humans has greatly increased with the growth of the ESS, perhaps because it holds a much larger reservoir of alternatives for individuals to choose from, and because it challenges the tradition-bound mythic elements of society to find significance in individual life rather than in the group.

The central point deriving from the history of the third transition, as it moved from visuographic invention to the management of external memory devices to the development and training of metalinguistic skill, is that it was *not* a given of human nature but rather a structure dependent upon both symbolic invention and technological hardware. The hardware may not have been biological, but from the viewpoint of a natural history of cognition this does not matter; the ultimate result was an evolutionary transition just as fundamental as those that preceded it. Once the devices of external memory were in place, and once the new cognitive architecture included an infinitely expandable, refinable external memory loop, the die was cast for the emergence of theoretic structures. A corollary must therefore be that no account of human thinking skill that ignores the symbiosis of biological and external memory can be considered satisfactory. Nor can any account be accepted that could not successfully account for the historical order in which symbolic invention unfolded.

This natural history of human cognitive emergence, and particularly the last part of the scenario, started off as a highly speculative enterprise. But in fact there have been fewer degrees of freedom in constructing an evolutionary account than one would have expected. Each of the three transitions has involved the construction of an entirely novel, relatively self-contained representational adaptation—that is, a way of representing the human world that could support a certain level of culture and a survival strategy for the human race. Each style of representation acquired along the way has been retained, in an increasingly larger circle of representational thought. The result is, quite literally, a system of parallel representational channels of mind that can process the world concurrently.

This concurrence may be illustrated in common modern theoretic institutions, such as symposia or public lectures. All the elements of human cognitive evolution come into play. The symposium or lecture will take place in an established mimetic framework; the audience, lecturer, and outside world all understand the unwritten mimetic rules of behavior governing their roles in the event. The lecture proper usually combines a spoken narrative with pictorial, ideographic, and phonological visual devices, using all three visuographic channels described in this chapter. The lecturer is also embedded in a social framework, replete with symbols and signs of status, credibility, and power. The audience and lecturer alternate between various media and cognitive levels. The lecturer says one thing while illustrating another, while the audience registers yet a third message, embedded in a complex fabric of communication surrounding the event. Other players in the symposium may use gestures, ideograms, analog models or logograms; logic might dominate, and then induction. The minds of the participants harness every symbolic device the human race has painstakingly invented over the last 200,000 years, to the common end of modeling, in the external arena, some idea. Working memory becomes a loosely accessible, shared, and highly selective field of the external store. Individuals construct and exchange narrative commentaries and move easily within their shared mimetic framework. And ultimately the refined product of the exchange will be developed, manufactured, and stored externally. The participants in this event are clearly hybrid minds, whose hybrid structure is still in the process of evolving.

Visuographic invention and the resulting growth of external symbolic memory media have altered the nature of working memory and the role of biological memory in humans. The major locus of working memory, for theoretic endeavor, is now external, and the major slave systems of working memory are also external: the true visuospatial sketchpad is the external working memory field, or EXMF, and the written narrative is much more important than the articulatory loop, or inner speech, at least in constructing a finished theoretic product. This is because permanent external devices allow an iterative, interactive thought process to operate repetitively on its own products; and, more importantly, the thought process itself can be largely externalized and institutionalized. Since external memory devices can provide a linkage to all preceding levels of cognitive evolution, they also serve in the construction of an integrative field, where the products of various types of thinking can be juxtaposed and combined.

Metalinguistic skills have been essential to human success with the ESS system and have been the core of educational curricula for two millennia, starting with classical rhetoric and moving through various additional stages. The curricular focus has moved from speech to script; from overall narrative structure to the intricate thought skills embedded in grammar, logic, and induction; and from extended narrative model building to the construction of increasingly specialized theoretic products. The entire theoretic complex of visual symbolic devices, memory management skills, and related metalinguistic skills was completely novel and constituted a new human representational apparatus of unprecedented power. The resulting cognitive architecture is a hybrid structure of great internal and external complexity, very far removed from our closest genetic relatives.

The very recent combination of this new architecture with electronic media and global computer networks has changed the rules of the game even further. Cognitive architecture has again changed, although the degree of that change will not be known for some time. At the very least, the basic ESS loop has been supplemented by a faster, more efficient memory device that has externalized some of the search-and-scan operations used by biological memory (Figure 8.8).

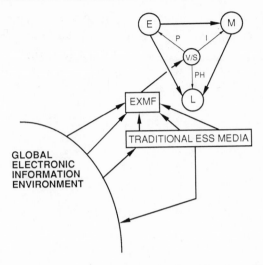

Figure 8.8   Juxtaposition of human representational architecture with the developing global electronic information environment. Many ordering rules and search functions that were entirely internal to biological memory are now resident in external memory systems. A variety of other cognitive operations are now entirely offshore, that is, found only in external memory. Most operations, however, still employ the EXMF loop. EXMF: external memory field; E: episodic memory; M: mimetic controller; L: linguistic controller.

The computer extends human cognitive operations into new realms; computers can carry out operations that were not possible within the confines of the old hybrid arrangement between monads and ESS loops shown in the last few figures. For example, the massive statistical and mathematical models and projections routinely run by governments are simply impossible without computers; so, more ominously, are the synchrony and control of literally millions of monads. Control may still appear to be vested ultimately in the individual, but this may be illusory. In any case, the individual mind has long since ceased to be definable in a meaningful way within its confining biological membrane.

The globalization of electronic media provides cognitive scientists with a great future challenge: to track and describe, in useful ways, what is happening to the individual human mind. The architecture of mind has evolved rapidly when viewed against the background of earlier evolution, and the rate of change seems to be accelerating rather than diminishing. The number of possible temporary configurations between monads and the ESS has increased, and the individual experiences a feeling of greater choice. But the role of the individual mind is changing, not in trivial ways but in its essence. And these changes need watching.

Consciousness and Indeterminacy

Transcending the Episodic Model of Mind

The dominant theoretical model of mental structure in this century has been implicitly a neuropsychological model, and the central figure in that tradition has been D. O. Hebb. Hebb (1949) built a theory of mind that served as a valuable vehicle for integrating the findings of apparently disparate fields, namely Gestalt psychology, behaviorism, and functional neuroanatomy. At the center of Hebb's neuropsychological theory was a model of memory; he recognized that this was a theoretical necessity, since all higher functions rest on the foundation of memory.

Hebb's central memory devices were "cell assemblies," self-propagating loops of neurons formed by synaptic growth during learning. These could be organized into larger "phase sequences" that formed the physical basis of "semi-autonomous" mental processes. These phase sequences, in turn, could be activated by endogenous (internal) as well as exogenous sources of stimulation, allowing the mind a degree of independence from environmental determinants of behavior. This was a major break with the hard-line behavioristic tradition that dominated at the time.

Hebb's ideas had a major impact on cognitive theory. He believed that the cell assemblies that determine a specific behavior pattern could be electrically active in the short-term, while long-term memory storage took the form of a permanent structural change in neural synapses. Thus, Hebb introduced the idea of a labile, neuroelectrical short-term memory store that determined both the contents of consciousness and the specifics of particular behaviors. This idea has remained enormously influential.

Hebb's theory lacked detail, but during the ensuing decades his basic ideas have been expanded and supported by various fields of investigation. The idea of a temporary store has survived in various guises, most recently in the concept of working memory, reviewed in Chapter 8. The synaptic basis of long-term memory storage has been a recurrent theme—for instance, in the epigenetic neuronal selection theory of Changeux (1985) or in Edelman's (1987) concept of the "secondary neuronal repertoire," both of which ultimately place behavioral learning in neuronal matrices or networks formed by altered synaptic patterning. The idea of automatic perceptual learning in neural networks induced through repetitive stimulation of neuronal lattices has survived in both Edelman's neuronal theory and in the recent connectionist theories of McClelland and Rumelhart (1986). Moreover, from Broadbent (1958) to Anderson (1983), cognitive psychology has retained Hebb's implicit architectural model of mind.

Hebb's model of mind, and most of its successors, did not attempt to deal with any of the uniquely human stages of representation that I have proposed in this evolutionary scenario. Basically, Hebb wanted to explain how perceptual learning was possible, how associations might be formed in the nervous system, how general conceptual development might be related to the amount of association cortex, and why intelligence, broadly defined, was difficult to localize. Consciousness was seen as an epiphenomenon, a by-product of reflective short-term memory. In this context, the progress from ape to human appeared to be a seamless and graduated shift, in which the overall power, but not the basic structure of mind, could change. Humans and apes shared the same memory structure, according to Hebb, the same perceptual systems, and the same associationistic mental processes; the differences were of degree. In effect, Skinner's behavioristic agenda was realized in the nervous system with Hebb's theory. At the same time, Gibson's neo-Gestalt perceptual ideas were given a neural foundation. But systems of symbolic representation were left out.

Hebb's theory of mind was thus exclusively a theory of *episodic* mind. Similarly, its descendants in modern connectionism, neural network theory, and perceptual psychology are primarily theories of episodic knowledge and behavior. This is not a problem in traditional physiological psychology, which is built on animal research and therefore somewhat limited in its cognitive objectives. It is a very serious problem, however, in neuropsychology, which is built on human neurological research. And it presents a major obstacle to anyone who wishes to generalize a theory of mind from animals to humans. It is also a problem for any computational theory that purports to simulate uniquely human processes like "reading" or "speaking" with a non-referential solution. For example, the PABLO or TRACE simulation models described in McClelland and Rumelhart (1986) should not be seen as models of reading or speech comprehension but rather as models of basic visual and auditory pattern recognition. Strictly speaking, these are not even episodic models; episodic storage is not (yet) built into them. They are able to attain a degree of categorical recognition, which is no mean achievement, but they should not be misunderstood as models of "reading."

One implication of the theory put forth in this book is that Hebb's conceptual (episodic) nervous system, and its recent derivatives, have been encapsulated in humans by our more recent cognitive acquisitions. Episodic mind may be organized along similar principles in animals and humans, but the locus of attentional control, and the nature of consciousness and thought, are different in humans because of the ascendancy of mimetic and linguistic representation and the externalization of symbolic memory. It follows that *generalizations about the localization and organization of higher mental processes cannot be made from animals, even apes, to humans, except for the purely episodic aspects of behavior.*

By pointing out the limitations of such models, I do not wish to imply that we cannot learn anything from them. On the contrary, the entire enterprise of understanding the human mind would founder if we did not possess the fruits of comparative research, and as Edelman points out (1987), we have to start with a solvable problem. However, it is useful to be as clear-minded as possible about the demarcation lines between the types of models that are direct extrapolations of animal research, which rest upon theory-building habits inherited from the Hebbian tradition, and models that try explicitly to accommodate uniquely human features of mind.

Where Is the Homunculus?

Psychologists working in the human engineering tradition, such as Broadbent (1958), took a position that was roughly similar to that taken by Hebb. They proposed a single central processor at the center of mind whose activity was more or less synonymous with consciousness. The central processor was the locus of working memory, perceptual awareness, thought, and all reflective conscious control of behavior. The central processor was placed at the peak of a monolithic system; it was the ultimate arbiter, the multimodal unencapsulated locus of thought. Even Fodor (1983), viewing the human mind from a more philosophical vantage point, supported the central processor as a necessity, while disavowing monism in subsequent computational work (Fodor and Pylyshyn, 1988).

But there is a not-so-covert homunculus lurking in all such models; the homunculus reads the codes and interprets them, tells the articulatory system whether its outputs are meaningful, invigilates the written scribbles of the hand for good sense, and monitors spoken words for a match to its instructions. It searches through mental dictionaries for word meanings, picks up the grammatical tags Chomskians think words carry, and then matches the "deep structures" of sentences against some criterion of meaning. It reads the events of perception directly, then checks out the narrative descriptions cranked out by the language system for veridicality.

The central processor thus formulates, encodes, and has access to a system of knowledge. It is the semantic daemon of the current version of pan-daemonium; it manages and interprets all the mechanistic systems dreamt up by cognitive scientists, gives them direction and meaning, and structures their memory systems as a bonus. Without the knowledgeable central processor (the homunculus), the mind appears to be as empty as any computer.

But do such inherently homuncular models have any advantages over prior, supposedly less sophisticated, ones, such as medieval ideas about agent intellects? And what about clockwork Cartesian formalisms, like transformational grammars, that allude vaguely to the all-powerful "semantic systems"; are they not also inherently homuncular? Are they not really avoiding the central question of meaning? Of course they are; implicit homunculi are just as suspect as the explicit homunculi of old-fashioned dualists.

And computational science isn't any less homuncular in its approach than cognitive psychology. While arguing against the need for a central processor, Minsky (1985) has attempted to replace it with myriads of subroutines and networks called "agents." Minsky holds that the unity of behavior and consciousness is an illusion; mental processes are a product of competition and cooperation between thousands of independent subroutines that constitute what he calls the "society of mind." The subroutine temporarily in command unifies behavior and creates the illusion of a central processor. What Minsky is really suggesting is that we replace the traditional homunculus with platoons, indeed entire armies, of homunculi; it is difficult to accept that this is a significant improvement in the state of our metaphysics.

Fear of the homunculus begets irrational behavior in cognitive scientists. They dread the truth: in a tiny slab of brain there resides a consciousness capable of all we have achieved and experienced; and obviously, on one level, *there is a homunculus.* The homunculus is synonymous with the reflective, conscious mind, and somehow, somewhere in the protean parenchyma of mind, it must reside. *It cannot be explained away as an epiphenomenon, "reduced" to algorithms or neuronal nets, or simply denied existence.* It is *the* mainstream problem, the principal phenomenon under investigation. It is no help to call it a central processor, when the central processor possesses virtually the same properties as the homunculus.

Simulating Cognitive Evolution

Of course, to admit this exposes the full extent of the challenge facing cognitive science. The modern human mind is not a simple medieval clock; it is not a radio or telephone switchboard; it is not a system of clever software; and it is most definitely not a general-purpose computing device like a Turing machine. These things are its inventions, products of its culture. They are no more than the metaphors it currently uses when contemplating itself. We do not even know whether the neural metaphor, which at least is a product of direct investigation, is apt.

A cognitive–evolutionary approach, such as the one developed in this book, does not "solve" the problem of the homunculus either, in the sense of reducing it to a mechanistic or computational solution. But an architectural scheme can be valid and verifiable even if many of its observations remain, for the moment, outside the reach of a reductionistic explanation. And at least this approach does not ignore the problem or sweep it under the carpet: representational systems are acknowledged as the driving force behind, and an integral part of, each new cognitive transition. In fact, *the essential cognitive adaptation underlying each of the three great cognitive transitions in human evolution is a new system of memory representation.*

Churchland (1986) has extensively considered the problem of representation in the nervous system, and she believes in the long-term promise of nonreferential connectionist models of cognition, such as those offered by McClelland and Rumelhart (1986). Connectionism has become a virtual movement or school in modern psychology. It started with simulations of neural nets, and then extended the same principles to human memory and learning, recapitulating on computers many of the cybernetic and neurophysiological debates of the fifties and sixties. But the leap from the current level at which connectionist models are pitched to simulations of symbolic reference and linguistic invention will not be an easy one. In fact, it would appear that computational science must be faced eventually with having to retrace the steps of cognitive evolution. Retracing cognitive evolution appears to be out of the reach of current technology and knowledge, but I readily concede the possibility (without trying to guess the probability) that connectionism might someday successfully model the processes underlying the perception of objects, three-dimensional space, and the image of one's own body in space.

The reason connectionist models are attractive is that they try to model the brain and mind with a *nonsymbolic* (sometimes called nonrepresentational) strategy. Like a primitive nervous system, a connectionist network constructs its own perceptual version of the world, without relying on a symbolic system given to it by a human operator. Such models are very rudimentary at present, but in principle they could be made much more powerful. Perhaps these types of models might eventually encompass something as complex as social event perception, the highest achievement of episodic mind, where objects in juxtaposition are resolved into a perceived "situation."

However, it is not obvious how a connectionist network could handle the level of abstraction and variability demanded by social event perception, and therefore I see this eventuality as very far off.

If connectionism travels that far, it will approach the rim of the episodic world, and at this point it might be possible to assess whether something even more complex, such as mimetic representation, might be tackled. The mimetic mind is intelligent—very intelligent indeed, when compared with any of its predecessors—and yet it does not possess words or their equivalents. As shown in Chapter 6, it is a level of mind that is capable of reference, generative representation, recursion, and so on; and yet it remains prelinguistic in its achievements, concrete, and incapable of reflection. The first formal model of truly human representation, far in the future, will have to cope with the problem of mimesis before it considers the problem of linguistic invention. This will undoubtedly prove very difficult; but there is no a priori reason for complete despair. If event representation can be solved in a connectionist framework, mimetic representation might also prove solvable in a similar way. Of course, following the scenario outlined in this book, a successful connectionist model of mimetic skill would ultimately have to be able to represent its perceptions of complex events in terms of its own outputs, *spontaneously and intentionally*—a tall order indeed. To consider this as a realistic possibility requires a great leap of faith (which many adherents to the connectionist dream have evidently already taken).

Having come this far, future modelers might redefine the problem of language in ways that are not obvious to this generation. At present, we are left with models of language that remain locked into what Churchland (1986) calls the "sentential paradigm"; a sentential theory holds that semantic systems are built upon propositions that can be formed into sentences ruled by logic. Such theories assume the pre-existence of symbols and minds adept at their rule-governed use. Thought thus becomes, in Churchland's words, sentence-crunching. The main difficulty with sentence-crunching models is their a priori assumption of a symbolizing mind: some intelligent animals (and some deaf-mute humans) are able to think but do not appear to possess any form of language. If thought is essentially sentential, how could this be possible? Fodor (1975) suggested that the mind might, in some cases, use a "language of thought" that represents propositions, even though it lacks the production systems for speech. Unfortunately, this position leads to an infinite regress; how does the language of thought validate its referential material? How does it differ from language itself? How does it help?

The progression from simulations of mimetic representation to models of oral language and narrative skill will be even more difficult. The simulation of mind, having finally achieved the spontaneous mimetic modeling of its world, will have to progress to symbolic invention, and then to some form of rule-governed language use based on the nature of its own modeling objectives. To be an accurate simulation, it will have to cross the abyss from nonsymbolic to symbolic representation, and spontaneously develop narrative models, inventing and assigning symbols for a variety of purposes. There is no particular reason why such a powerful simulation should prefer sound rather than visual displays when communicating with humans (or perhaps it would no longer need or want, at this stage, to communicate with humans), but linguistic modeling would be an essential skill, if it were to be able to reproduce our peculiar cognitive profile. At present, I see no basis for such a breakthrough, but science has always flourished on optimism and faith in the future, and I have no reason to doubt that such a push forward might someday become possible. However, optimism must be tempered with realism; anyone who ignores the problem of reference is doomed to failure. It cannot be ignored; nor can it be deferred, like the spiraling national debt of the eighties, in the hope that no one will notice.

Consciousness in a Hybrid System

Within the context of the hybrid mental architecture proposed in this book, consciousness can take many forms. The subjective nature of consciousness depends entirely on the momentary locus of control in the central representational systems. Consciousness has many definitions, but it is mostly about control and reflection. In an episodic system, consciousness is episodic: situation-bound, and concrete. This is because control resides at the peak of the episodic system, in a device that undoubtedly resembles the central processors of Hebbians and their descendants.

Broadly speaking, the mimetic and oral–narrative systems normally dominate human consciousness. This is because the locus of control in humans has shifted to the highest, most recently evolved cognitive structures. Moreover, whenever external memory devices enter into the temporary architecture of mind, they tend to become dominant. They can also serve to reconfigure the mimetic and oral–narrative regions of mind in ways that would not have been possible before the third transition. In effect, all three of the distinctly human representational systems can serve as the temporary center of control, and since they can also function in parallel and be programmed to alternate with one another in various systematic ways, the subjective quality of consciousness may be extremely rich and varied.

In the hybrid scheme proposed here, the functional locus of "consciousness" can shift, depending upon the representational system currently in command. What we experience as basic, unreflective awareness probably corresponds somewhat to direct episodic experience, uninterpreted by any of our representational systems. Such unreflective states are probably as close as modern humans can get to the episodic cultures of higher mammals. In such a state, the *absence* of mimetic or linguistic representation concedes control to episodic cognitive structures, by default.

Predominantly mimetic states of awareness tend to be event-oriented, action-based, and usually socially interactive. Above all, and in contrast with unreflective episodic experience, mimetic states take an active, modeling approach to experience. The invention and practice of sport, games, dance, ritual, and craft without the engagement of verbal thought are typical of such states; and any extended, intentional nonlinguistic expressive exchange would reflect a predominantly mimetic state of awareness. Such exchanges are more common than most modern humans think. Subjectively we have the impression that we verbalize continuously to ourselves, if not out loud. But it is common, especially in large, anonymous crowds of people, to behave in a relatively pure mimetic manner. Thus, in church, in stadiums, in large demonstrations, in civic celebrations, and in various collective rituals humans behave in ways that belie an unreflective understanding of the social order, and in ways that are collectively representational, but not linguistically based. In such situations, language is often relegated to a support function, and the entire framework of social exchange is mimetic.

When awareness is predominantly oral–linguistic, as it is most of the time in modern humans, the picture is much more complex. However, if we do not confuse the meaningfulness, or semantic content, of such awareness with the state of being linguistically aware, the picture is somewhat simplified. A focused conversation or debate might approach a pure state of oral–linguistic awareness. It has a phonological character, dominated by what Baddeley (1986) calls the "inner voice." It is also capable of being more abstract and discourse-oriented, engaging metalinguistic skills, as it does in myth. But oral language ultimately depends on the incessant rehearsal and recycling of common sentential utterances and remains very heavily dependent on the auditory forms of its material. A purely oral–narrative consciousness would have great difficulty overcoming this limitation.

In all of these cases, the locus of attentional control is determined, to a degree, by default. In the hierarchy of central representational systems described in the first part of Chapter 8, the linguistic controller is at the cognitive peak, and therefore in theory it can encapsulate the other systems by modeling their outputs. But if it is not engaged, control defaults to the next representational level, the mimetic. And so on; control is flexible, and allocated on the basis of the *structure of representation* at any given moment. In this model there is no need for a unified central "supervisory system" like that of Shallice and Norman (Chapter 3), or any other kind of unitary central processor. The locus of conscious control resides in the representational systems and depends entirely on their momentary patterns of activity. Baddeley (1986) has acknowledged that, in principle, this type of sharing arrangement is a viable alternative to the notion of a single central executive system.

Since the various human representational systems can all become active in varying degrees at the same time, we can experience quite subtle and complex states of consciousness. Theatre and cinema, and other multimedia art forms such as opera, use this capability of parallel representation to weave the human mind into such "compound" states. A well-made film in particular can tease the brain simultaneously on the episodic, mimetic, and linguistic levels, sometimes conducting a different theme on each level. Thus, a scene in a horror film might contain an apparently innocent conversation with one or two different levels of linguistic content, while the overall unfolding mimetic scenario elicits growing terror, and the actual individual episodic events portrayed contain a series of disconnected shocks, objects, and events that alternately distract, contradict or support various expectations in the minds of the audience.

The presence of explicit visual representational symbolism changes the picture further. Awareness during a conscious state dominated by the external symbolic storage system is harder to pin down, because control can reside in so many loci. It can have a primarily visual nature, but even within this framework it can be pictorial, ideographic/ analog, or linguistic in content, and visuosymbolic control of consciousness can therefore have a variety of manifestations. Nonvisual ESS-interactive states are possible, especially through the medium of sound, but also through touch, as in Braille reading. The point is, awareness "resides" in, or has the flavor of, the dominant representational system of the moment.

Without doubt the more complex forms of writing, such as that found in advanced ideographic languages, depend on more than one representational system for their operation. The combination of alphabetic print, graphics, and ideographs in various media "plays" the mind's representational systems in parallel. This can be carried to another level of abstraction: modern novels have progressed in their format from the straightforward early narratives that were largely dependent on linguistic modes of awareness, to subtler devices that can indirectly evoke a kind of multimedia effect in the brain, even though they are restricted to print. Thus, a story like Poe's "Masque of the Red Death" elicits all the conflicting emotions and parallel mimetic and linguistic scenarios that a film might evoke. This is because, even when reading a sequentially strung-out series of visual symbols, the reader's brain is able to track the story mimetically. Evidently the reader is able to construct, from printed alphabetic and ideographic symbols, a mimetic scenario. Thus the linguistic controller can not only interpret and model mimetic outputs but can configure and reelicit such outputs at will.

The focus of consciousness can shift quickly, switching the viewer from one locus of control to another with great ease. This can be done voluntarily; many rituals, including forms of meditation, are an attempt to temporarily reduce the dominance of linguistic, and especially analytic, thought. But alternations in the locus of conscious control are fairly routine; events may flip the mind of the viewer rapidly through a series of states in which control resides alternately in imagery, sound, narration, logic, mimetic expression, analog models, ideograms, or a combination of these.

In controlled media like television, where viewing time is monitored in mere seconds, the consciousness of the viewer is seldom allowed to coast or slow down. It is driven from one state to another on a tightly programmed roller-coaster that remains entirely under external control. Humans have very few means of self-defense against such devices. Looked at purely in terms of traditional structure, it is highly probable that extensive multimedia experience of this sort throughout childhood, completely under external control, might do much more than "open up" the mind to new experiences. It could also have an extremely disorienting and scrambling effect on traditional human mental architecture. Alternatively, it might also enable and evoke some interesting new kinds of knowledge structures; but this question would need much more consideration than is possible in this brief chapter.

ESS-interactive states can also serve to reflect upon, and expand, the activity of the oral–narrative system. This is one of the most commonly employed loops through the external memory field; for instance, a speaker might generate an oral–narrative idea that is then transcribed into a purely visual medium, such as a painting. The visual idea then gradually evolves on its own; and as the visual metaphor grows more complex, it outpaces the original verbalized idea, establishing its own symbolic identity and meaning. In such an interaction, the visual and narrative representation are complementary.

Attentional control ultimately rests at the top of the dominant component of the hierarchy. It follows that, in a multicomponent system like the one described here, the locus of attention is highly mobile. Moreover, the locus of attentional control can reside, at least temporarily, in external memory. In the case of television, the viewer yields control to the external system, the screen becoming the external memory field (EXMF). This exporting of control is usually voluntary and reversible, and the degree to which control is truly exported is debatable.

The extent to which a viewer of television, or for that matter a reader of any external medium, yields attentional control to the device is relative to an active, evaluative attitude on the viewer's or read-

er's part. The important difference between active and passive comprehension of external symbolic material is in the primary locus of
control. Active evaluation, and all expressive skills, even those directly dependent on the EXMF, remain under internal attentional
control. Thus, there is nothing inherently uncontrollable about external symbolic storage, provided that the monadic individual mind is in
an expressive or *actively interpretative* role. Those unable to assume
such a role, however, will risk complete domination by external media, to the extent that the media may even structure their experience
at the episodic level.

Localization after the Third Transition

Would consciousness reduce to right- and left-hemisphere modes in
this scheme of things? It would appear not, since mimesis normally
involves both hemispheres, and so does language. Certainly, since
many states of normal awareness evoke all representational systems
in parallel, consciousness cannot be assigned to only one hemisphere.
Some key linguistic functions seem to be left-lateralized, but it would
be a gross exaggeration to claim that verbal thought is an exclusively
left-hemisphere mode of consciousness. "Visual thought" is also an
imprecise term in this context. Some forms of visual thinking are
episodic: for instance, mental rotation, simple visual problem solving,
and probably simple image generation. Other kinds of visual thinking
are representational, either on the mimetic or ideographic-analog
level. There is no evidence that all forms of visual thought are restricted to a single hemisphere.

   Therefore the commonly taught left–right division of function between visual thought in the right cerebrum and verbal thought in the
left paints an oversimplified, misleading, and confused picture. We
have no reason to believe that episodic uses of vision are lateralized at
all. And studies of image generation, which place visual imagery on
the *left*, are hard to read because the source of "image generation"
will always depend on the representational system in control. The
image-generation research reviewed by Farah (1984) that purported
to place image generation in the left hemisphere (see Chapter 3) was
at least partly confounded with language representation. Neurological
patients were asked to retrieve images on the basis of verbal instruc-

tions, or to retrieve images that were essentially tied to verbal representation, such as the letters of the alphabet. Their responses depended upon the language system as well. Thus, they were not being asked simply to generate images but to generate and describe images from the vantage point of the oral–linguistic system. Farah's left-damaged patients, who apparently had difficulty with visual-image generation, might therefore have suffered from a Geschwind-style disconnection syndrome, or some other language-dependent deficit, rather than an isolated deficiency in active visual imagery.

Mimesis is a complex skill system, and as concluded in Chapter 6, there is no reason to believe the mimetic system is lateralized exclusively to one hemisphere. Mimetic representation incorporates sight and sound, body movement, and facial expression and relies on a variety of brain systems. Yet it stands as a hierarchical system and holds together as an integral class of representation. As shown in Chapters 3 and 6, mimesis is also extremely robust after brain injury. It cannot be reduced to some simple dimension like "serial motor control" or "praxis." Kimura's apraxic aphasics are often held up as major evidence in support of the left hemisphere's monopoly on serial praxis and generativity; yet those patients typically retain most of their mimetic skills, including most of the generative nonlinguistic expressive skills described in Chapters 5 and 6. Moreover, right-hemisphere stroke patients suffer a variety of major disturbances in consciousness, but their mimetic skills usually do not appear to be affected. Thus, neither left- nor right-hemisphere injury creates a completely dysmimetic person, even when the injury occurs after puberty, when cerebral lateralization seems to be well established.

Reading and visuographic skill in general present an even more complex picture with regard to brain localization. Reading arrived too recently to depend on a major new neurological adaptation. Yet, as pointed out at some length in Chapter 8, visuographic skill depended on, and encouraged, a drastic reconfiguration of human cognitive architecture and a redeployment of control processes like memory management and attention. Whether ideographic or alphabetic, reading placed unprecedented demands on visual processing and memory. The volume of visuographic encoding demanded by our society, including the learning of specialized languages in the mathematical, computational, and engineering fields must have changed the visual memory load on the brain. In addition there are the subtle visual–semiotic rules of artistic, musical, literary, cinematic, and other nonscientific conventions, and the massive amounts of quantitative information routinely digested by workers in practical fields—for instance, financial traders and managers. These are testimony to a modern reconfiguration of the human visual brain of truly heroic proportions.

But the localization of these expanded visual skills, and particularly of visuographic codes, is problematic. It is not enough to say that the "higher visual areas" might simply have differentiated and redirected their use as the new load on the visual brain was gradually increased. A simple expansion of known visual areas, due to heavy use, might be expected to affect synaptic patterning and tuning in fairly minor ways, perhaps like those observed by Merzenich and colleagues (1987) in the somatosensory regions of monkeys. Increased capacity in a highly specialized brain region seems only to result in a wider repertoire of the same specialized behaviors, but not in a qualitative change of function. This is true even if the increased size has a genetic basis; Nottebohm (1984) has observed that marsh wrens that have larger song regions in their brains simply have a bigger song repertoire than other birds, but not some qualitatively new vocal function.

Expansion of the human brain regions concerned with visual perception would not necessarily support a novel skill like reading. This is because reading is not primarily a visual skill; it is a thought skill. The invention of pictorial, ideographic, and phonetic uses of visual symbols came as part of a novel set of human thought skills, involving not only vision but a whole array of components at the representational level which appear to be distinct from speech. The pictorial and ideographic routes are easiest to understand in this regard, because they stand clearly apart from spoken language. When an artist exhibits a painting, he might try to "explain" it orally; but his message is essentially visual, and the visual "vernacular" by which it may be comprehended cannot usually be reduced to a verbal formula. It is mimesis, but *crafted* mimesis, locked into the external memory field. The same principle applies to ideographs, and in this case a linkage to language is added. The brain needs more than perceptual or episodic skill to interpret what it sees; it needs several sets of encoding rules.

Should we expect there to be any consistencies in the neuronal organization of visuographic skill? Neuropsychological evidence from the study of deep dyslexia proved useful in arguing the case for separate ideographic and phonetic paths in reading; but does it help in the precise localization of these processes in the brain? The short answer would appear to be that, no, it does not. Scrutiny of case histories in this area of research provides evidence against precise localization of reading paths in a fixed brain region.

Deep dyslexia often occurs as part of a complex neurological syndrome that includes other symptoms, notably Broca's aphasia, agrammatism, impaired writing and spelling, and a comprehension deficit. Coltheart, Patterson, and Marshall (1980) reviewed the brain lesions underlying five well-documented cases of deep dyslexia and found that all of them had extensive left-hemisphere damage in the speech regions, including the inferior frontal area (Broca's region), the insular region, and the superior temporal (Wernicke's region) and supramarginal gyrus: in other words, they had usually sustained injuries to the entire perisylvian region. The "semantic" errors made by these deep dyslexic patients, who read "flask" for thermos, "wait" for pause, or "eyes" for glasses, and their near-guesses showing that they were close to finding the meaning of a word, such as "a high person, a queen" for protocol—these errors typically occurred only after the virtual destruction of the classical speech regions.

The remarkable thing about these patients is not that they have serious language difficulties but that they have any residual language function at all. This fact alone argues against fine-grained localization of specific reading functions in the way classical localization theory would have preferred. However, it does allow for a degree of gross localization, in that the left perisylvian area appears to be more critical than the right. Some researchers have invoked the right hemisphere in their explanation of residual reading function in deep dyslexics (Schweiger, Zaidel, Field, and Dobkin, 1989). They reason that if the ideographic reading route is on the right side, deep dyslexics with left-sided injuries to the brain might be relying on it. But if this was the case, why don't the isolated right hemispheres of split-brain patients act like deep dyslexics? And why isn't deep dyslexia more common after left-sided strokes and other injuries?

Other researchers attribute the unusual reading patterns of deep dyslexics to "residual" left-hemisphere reading areas, spared from injury. Roeltgen (1987) described the case of a man who was left with deep dyslexia after a left-sided stroke; the stroke had destroyed Wernicke's area and some adjacent cortex, including the supramarginal gyrus. Three years later, he suffered a second stroke, also on the left, that abolished all his residual reading ability; the second stroke evidently destroyed the remainder of the speech regions on the left side. The author concluded that his patient's residual reading capacity during the three years between his first and second strokes must have depended on the areas destroyed in the second stroke, which were on the *left* side. The right-hemisphere hypothesis was obviously not supported in this patient. The odd thing about Roeltgen's theory is that the residual left-sided cortical regions that seemed to support a deep dyslexic reading pattern between the first and second strokes are not normally included in the classical model of reading. His patient's second stroke was in the Rolandic face area, a motor region, and the middle part of the superior temporal gyrus, which are normally thought to be more important in speaking and listening. One looks in vain for a role for these areas in the many anatomical proposals for the human reading pathways.

One way to resolve some of these anomalies (and they are not atypical) is to abandon the idea of rigidly fixed localization for reading. Neuronal input pathways are genetically fixed, and therefore the initial access routes to the visuographic systems are anatomically constrained. This might explain why some data are in line with the classical model; for instance, in some cases of developmental dyslexia, there is evidence (Galaburda and Kemper, 1979) that the deficiency might be due to abnormal development in Wernicke's region. But there is a broader evidential base for asserting that the entire left perisylvian region, and possibly the right as well, is involved in reading and related linguistic functions. If it could be shown conclusively that the precise pattern of use of the perisylvian region varied among individuals, these cases would no longer seem like anomalies but the inevitable consequence of individual variability in the maturation of the language brain. Such individual variability might be predicted for a skill that is not species-universal and lacks a fixed genetic basis.

But this is not to say that reading skill is without a distinct neuropsychological structure; there is good evidence for the *anatomical* as well as functional independence of the ideographic path from the phonological reading route (their functional independence was established in Chapter 8). In the case of dyslexia in a Japanese man (Hayashi, Ulatowska, and Sasanuma, 1985) known as T.O., who showed preservation of ideographic (*kanji*) reading, with virtually complete loss of phonological (*kana*) reading, pronunciation skill must have been intact, since the ideographic *kanji* symbols could be read correctly. The patient had selectively lost phonological conversion skills without losing the ideographic reading path, confirming the existence of a different, or at least partly different, physical locus for it. The separateness of this path has been confirmed, but it should be emphasized that its exact anatomical location remains unknown and might also be variable from individual to individual.

If reading skills are hard to fix into a common pattern of localization, then the more complex aspects of theoretic culture, which include analytic thought, grammatical invention, various metalinguistic skills, memory-management skills, attentional algorithms, and theoretic knowledge itself, are understandably difficult to place in a neat neurological box. *Yet control must reside in the cortex;* there seems to be no alternative. It is difficult to know what expectations might be realistic with regard to the neuronal localization of the complex human theoretic apparatus.

The most simplistic kinds of neural localization theories, which place each function in some discrete brain structure—in some nucleus, ganglion, or cortical area or, more recently, in mysterious inchoate phantasms called "neurotransmitter systems"—cannot be applied to the changes of the third transition. The theoretical apparatus supporting the third transition arrived very recently, depends heavily on external memory, and is determined by patterns of brain use that appear to be *set by the external memory media themselves.* For example, there is no internal wiring schema to support the kind of synthesis made possible by a scientific diagram; the synthesis is *out there,* in the diagram itself. The theoretician depends heavily on a huge variety of external cognitive props—mathematical notations, curves, plots, histograms, analog measurements, and technical jargon—to arrive at a theory. Without these things, thoughts of this kind would simply not be possible, because the end-state or "conclu-

sion" achieved by the mind is driven directly by the external representation itself. The locus of a process like theoretical synthesis would
thus be difficult to attribute to any single part of the internal–external
network that makes up such a system.

Physical Indeterminacy and Singularity

From the comparative anatomical data reviewed in Chapter 4 it follows that, as the human brain expanded in size, the proportion of
primary and secondary cortex relative to total cortical volume was
reduced, while that of tertiary (association) regions was increased.
Most of the greatly expanded cortical regions in the human brain are
tertiary; that is, they receive great quantities of highly digested inputs from all over the brain. They lack a single genetically entrenched
dominant input path; therefore, in theory, maturational competition
could produce a wide variety of different "wiring" schemes in tertiary
regions, depending on the patterns of early experience. In principle,
this would create an opportunity for considerable individual variability in the functional organization of the adult neocortex.

This appears to be particularly true with regard to laterality. Bilaterally symmetrical tertiary zones would be the most indeterminate
brain regions in their final configuration because competition for synaptic growth would entail lateral redundancies in addition to diverse
intrahemispheric competitive pressures during development. The
consequence of this is probably a degree of physical indeterminacy in
individual localization of function. In other words, while higher cognitive functions must be localized somewhere in each individual, they
might not be localized in the same way from one individual to the
next.

Edelman (1987) has considered this issue in detail and has proposed
a theory of competitive neuronal group selection, which he calls
"neural Darwinism." In his conception of neuronal genesis, there is
room for a great deal of individual variability in neuronal organization; in fact, it is an essential feature of his system. No two animals
will evolve the same nervous system, not even during the earliest
stages of development. The allocation of higher-order neuronal
groups, especially those tertiary structures presumably concerned
with cognition, would thus differ greatly between individuals. Thus,

by Edelman's standards, given the competitive maturational environment in which synaptic growth occurs, there should be a considerable
degree of physical indeterminacy in the final configuration of each
adult human brain. This indeterminacy would extend beyond the cortex to the subcortical projections of tertiary cortical regions, such as
the basal ganglia and thalamus, and presumably even to brain stem
structures.

Edelman did not try to directly address the question of human
higher function. But if his basic principle—epigenetic selection from
variants generated by the brain's populations of neurons—applies to
perception, surely it would apply even more strongly to representational systems further removed from anatomically fixed input and
output paths. And as pointed out elsewhere in this chapter, there is
reason to question *in principle* the idea of consistent patterns of localization of higher function in humans. Classical neuropsychology has
tended to assume that localization of function follows a roughly similar course in each individual, provided gender and handedness are
taken into account. But this principle is almost certainly wrong, given
the competitive maturational environment of neurons at the micro-
biological level and the tremendous culturally programmed changes
in environmental patterning of the brain on the psychological level.

In summary, the degree of consistency across individuals that has
been assumed in neuropsychology may not exist in tertiary cortical
regions at all. This might be expressed as the *principle of singularity:*
the individual human brain develops a unique functional organization
at the representational level. This has serious implications for optimal
research strategies in neuropsychology; at the very least, it is a very
strong argument in favor of the single-case approach. The regions of
the brain that are most characteristically human—especially the great
expanses of the frontal and anterior temporal lobes—are likely to be
the most malleable neurological structures in nature, taking on many
forms. They are configurable and reconfigurable to a remarkable degree, because their resources are allocated on a *competitive* basis to
the many input paths impinging on them. In effect, *the physical
structure of mind has become less and less fixed as neocortical evolution has progressed.* This leaves room not only for the kinds of
radical reconfiguration introduced by literacy but also (presumably)
for larger differences between the brains of individual human beings

than are possible in primates. It also leaves room for further cognitive restructuring, possibly in fundamental ways.

Conclusion: Exuberant Materialism

Mental materialism used to be a bad label. Those who opposed it were called dualists: they believed in the traditional comfortable separateness of mind from the physical world. In most cases, early materialists were naive monists; they supported the unity of matter and mind, as aspects of the same, basically material, universe. This unity usually took the form of a simplistic "reduction" of mind to matter, an easy position to demolish. But discussions of the mind–body problem can often degenerate into semantic quibbles, more vexing than illuminating. More than once I have been led around a conceptual Moebius strip by some well-meaning writer—entering the argument on one side and coming out on the side opposite to the entry point. Enter a dualist, exit a monist, reenter a monist, exit a dualist, and so on, forever.

In past generations, the evidence for a straight material explanation of mind was never seen as particularly convincing, even to the uninitiated, and this has led to visionary compromises and tortured logic. A reading of McDougall's *Mind and Brain* (1911) leaves the reader with a vivid picture of the confusion and obfuscation over dualism that reigned at that time. More recently, Sperry (1966), and Popper and Eccles (1977) tried to cope with the same set of conceptual difficulties. But none of these authors tried to deal directly with the material basis of representation, and their materialism was tentative and uncertain.

But lately mental materialism is back, with a vengeance. It is not only back, but back in an unapologetic, out-of-the-closet, almost exhibitionistic form. This latest incarnation might be called "exuberant materialism." Changeux (1985), Churchland (1986), Edelman (1987), Young (1988), and many others have announced a new neuroscientific apocalypse.

Optimism is basically more productive than pessimism, and exuberant materialists are certainly optimists. Neuroscience is in its adolescence, and the field is drunk with its own dizzying growth; how not to be optimistic? Computational science has outgrown its irrepressible infancy and pulled in its horns but remains determined to move boldly ahead, confronting the Beast of Representation head-on. These are exciting times; our view of humanity is evolving alongside the machinery of our minds. Extending the logic of the last chapter, perhaps we will ultimately solve the problem of mind by constructing a new architecture of mind, one that possesses more powerful representational tools and is capable of comprehending itself.

One thing is certain; if we compare the complex representational architecture of the modern mind with that of the ape, we must conclude that the Darwinian universe is too small to contain humanity. We are a different order. We are not just another family within the superfamily of Hominidae, within the Cattarhine infraorder of the suborder of Anthropoidae, within the order of Primates. Nineteenth-century biology was absurdly, grotesquely wrong in its classification scheme, because it had no adequate vocabulary for assessing the cognitive dimension of human evolution. It was a specialist's error, a classification scheme that glorified anatomy and devalued mind.

Our genes may be largely identical to those of a chimp or gorilla, but our cognitive architecture is not. And having reached a critical point in our cognitive evolution, we are symbol-using, networked creatures, unlike any that went before us. Regardless of whether our current chronological framework is verified or radically changed by future research, this much is not speculation: humans are utterly different. Our minds function on several phylogenetically new representational planes, none of which are available to animals. We act in cognitive collectivities, in symbiosis with external memory systems. As we develop new external symbolic configurations and modalities, we reconfigure our own mental architecture in nontrivial ways. The third transition has led to one of the greatest reconfigurations of cognitive structure in mammalian history, without major genetic change. In principle, this process could continue, and we may not yet have seen the final modular configuration of the modern human mind. Theories of human evolution must be expanded and modified to accommodate this possibility.

<u>References / Acknowledgments / Index</u>

References

Alexander, M. P., M. A. Naeser, and C. L. Palumbo. 1987. Correlations of sub-cortical CT lesion sites and aphasia profiles. *Brain,* 110: 961–991.

Anderson, J. R. 1983. *The architecture of cognition.* Cambridge, Mass: Harvard University Press.

Aquinas, T., St. 1945. *Basic writings of Saint Thomas Aquinas.* New York: Random House.

Armstrong, E. 1981. Mosaic evolution in the primate brain: Differences and similarities in the hominoid thalamus. In E. Armstrong and D. N. Pandya, eds., *Evolution of the primate brain.* New York: Academic Press.

Arnheim, R. 1969. *Visual thinking.* Berkeley: University of California Press.

Atkinson, R. C., and R. M. Shiffrin. 1968. Human memory: A proposed system and its control processes. In K. W. Spence, ed., *The psychology of learning and motivation: Advances in research and theory.* Vol. 2. New York: Academic Press.

Auerbach, E. 1953. *Mimesis.* New York: Doubleday.

Baddeley, A. 1966. Short term memory for word sequences as a function of acoustic, semantic, and formal similarity. *Quarterly Journal of Experimental Psychology,* 18: 362–365.

——— 1986. *Working memory.* Oxford: Clarendon Press.

Badecker, W., and A. Caramazza. 1985. On consideration of method and theory governing the use of clinical categories in neurolinguistics and cognitive neuropsychology: The case against agrammatism. *Cognition,* 20: 97–125.

Barthes, R. 1988. *The semiotic challenge.* Oxford: Basil Blackwell.

Bastian, H. C. 1869. On the various forms of loss of speech in cerebral disease. *British and Foreign Medical-Chirurgical Review,* 43: 209–236, 470–492.

Bates, E., I. Bretherton, C. Shore, and S. McNew. 1983. Names, gestures and objects: Symbolization in infancy and aphasia. In K. E. Nelson, ed., *Children's language*. Vol. 4. Hillsdale, N.J.: Erlbaum.

Bitterman, M. E. 1965. Phyletic differences in learning. *American Psychologist*, 20: 396–410.

Bloomfield, L. 1933. *Language*. New York: Holt.

Boesch, C., and H. Boesch. 1984. Possible causes of sex differentiation in the use of natural hammers by wild chimps. *Journal of Human Evolution*, 13: 415–440.

Bogen, J. E. 1969. The other side of the brain II: An appositional mind. *Bulletin of the Los Angeles Neurological Society*, 34: 135–162.

———— 1985. The callosal syndromes. In K. M. Heilman and E. Valenstein, eds., *Clinical Neuropsychology*. 2nd ed. New York: Oxford University Press.

Bouma, H., and A. H. de Voogd. 1974. On the control of eye saccades in reading. *Vision Research*, 14: 273–284.

Brace, C. L. 1979. Biological parameters and pleistocene hominid life ways. In I. S. Bernstein and E. O. Smith, eds., *Primate ecology and human origins*. New York: Garland Press.

Brain, R. 1965. *Speech disorders*. London: Butterworth.

Broadbent, D. E. 1958. *Perception and communication*. London: Pergamon.

Broadbent, W. H. 1879. A case of peculiar affection of speech, with commentary. *Brain*, 1: 484–503.

Broca, P. 1861. Remarques sur le siège de la faculté de la parole articulée, suivies d'une observation d'aphemie (perte de parole). *Bulletin de la Société d'anatomie* (Paris), 36: 330–357.

Bronowski, J., and U. Bellugi. 1980. Language, name, and concept. In T. A. Sebeok and J. Sebeok, eds., *Speaking of apes*. New York: Plenum Press.

Brooks, L. R. 1967. The suppression of visualisation by reading. *Quarterly Journal of Experimental Psychology*, 19: 289–299.

Brooks, R. A. 1989. Intelligence without representation. Research Report, Artificial Intelligence Laboratory. Massachusetts Institute of Technology, Cambridge, Mass.

Brown, R. 1980. The first sentences of child and chimpanzee. In T. A. Sebeok and J. Sebeok, eds., *Speaking of apes*. New York: Plenum Press.

Bruner, J. 1986. *Actual minds, possible worlds*. Cambridge, Mass.: Harvard University Press.

Bruno, G. 1964. *Cause, principle, and unity: Five dialogues*. Trans. J. Lindsay. New York: International.

Buckout, R. 1982. Eyewitness testimony. In U. Neisser, ed., *Memory observed: Remembering in natural contexts*. New York: W. H. Freeman.

Burr, D. 1976. Neanderthal vocal tract reconstructions: A critical appraisal. *Journal of Human Evolution*, 5: 285–290.

Butterworth, B., and U. Hadar. 1989. Gesture, speech and computational·stages: A reply to McNeill. *Psychological Review*, 96: 168–174.

Campbell, J. 1959. *The masks of God: Primitive mythology*. New York: Viking.

————— 1988. *Historical atlas of world mythology*. New York: Harper and Row.

Caplan, D. 1987. *Neurolinguistics and linguistic aphasiology: An introduction*. Cambridge, Eng.: Cambridge University Press.

Carroll, J. B. 1972. The case for ideographic writing. In J. F. Kavanagh and I. G. Mattingly, eds., *Language by ear and eye*. Cambridge, Mass.: MIT Press.

Changeux, J. 1985. *Neuronal Man*. New York: Pantheon Books.

Chomsky, N. 1959. Review of Skinner's *Verbal Behavior. Language*, 35: 26–58.

————— 1965. *Aspects of the theory of syntax*. Cambridge, Mass: MIT Press.

————— 1980. Human language and other semiotic systems. In T. A. Sebeok and J. Sebeok, eds., *Speaking of apes*. New York: Plenum Press.

————— 1981. Knowledge of language: Its elements and origins. *Philosophical Transactions of the Royal Society of London*, series B, 295: 223–243.

Churchland, P. 1986. *Neurophilosophy: Towards a unified science of the mind/ brain*. Cambridge, Mass.: MIT Press.

Cicone, N., W. Wapner, N. S. Foldi, E. Zurif, and H. Gardner. 1979. The relation between gesture and language in aphasic communication. *Brain and Language*, 8: 324–349.

Clutton-Brook, T. H., and P. H. Harvey. 1980. Primates, brains and ecology. *Journal of Zoology, London*, 190: 309–323.

Cohen, M. 1958. *La grande invention de l'écriture et son évolution*. Paris: Imprimerie Nationale.

Collins, A. M., and M. R. Quillian, 1972. How to make a language user. In E. Tulving and W. Thompson, eds., *Organisation and memory*. New York: Academic Press.

Coltheart, M., K. Patterson, and J. C. Marshall, eds. 1980. *Deep dyslexia*. London: Routledge.

Conrad, R. 1964. Acoustic confusion in immediate memory. *British Journal of Psychology*, 55: 75–84.

————— 1972. Speech and reading. In J. F. Kavanagh and I. G. Mattingly, eds., *Language by eye and ear*. Cambridge, Mass.: MIT Press.

Corballis, M. C. 1989. Laterality and human evolution. *Psychological Review*, 96: 492–505.

————— and L. Beale. 1976. *The psychology of left and right*. Hillsdale, N.J.: Erlbaum.

Dart, R. A. 1925. *Australopithecus africanus:* The man-ape of South Africa. *Nature*, 115: 195–199.

Darwin, C. 1859. *On the Origin of species by means of natural selection*. London: John Murray.

————— 1871. *The descent of man*. London: John Murray.

————— 1874. *The descent of man*. 2nd ed. London: John Murray.

Davies, W. V. 1987. *Egyptian hieroglyphs*. London: British Museum Publications.

Dax, M. 1865. Lésions de la moitié gauche de l'encéphale coincident avec l'oubli des signes de la pensée. Montpellier, 1836. *Gaz Hebdom*, 11: 259–260.

Denenberg, V. H. 1984. Behavioural asymmetry. In N. Geshwind and A. M. Galaburda, eds., *Cerebral dominance: The biological foundations*. Cambridge, Mass.: Harvard University Press.

De Renzi, E., P. Faglioni, and G. Scotti. 1970. Hemispheric contribution to exploration of space through the visual and tactile modality. *Cortex*, 6: 191–203.

Diringer, D. 1948. *The alphabet*. London: Hutchinson.

———— 1962. *Writing*. New York: Praeger.

Dunbar, R. I. M. 1988. *Primate social systems*. London: Croom Helm.

———— 1990. Ecological modelling in an evolutionary context. *Folia Primatologica* (Basel), 53: 235–246.

———— [In preparation.] Co-evolution of cognitive capacity, group size and social grooming in primates: implications for the evolution of language.

Edelman, G. M. 1987. *Neural Darwinism*. New York: Basic Books.

Ekman, P., E. R. Sorenson, and W. V. Friesen. 1969. Pan-cultural elements in facial displays of emotion. *Science* 164: 86–88.

Eibl-Eibesfeldt, I. 1989. *Human ethology*. New York: Aldine de Gruyter.

Eidelberg, D., and A. M. Galaburda. 1984. Inferior parietal lobule: Divergent architectonic asymmetries in the human brain. *Archives of Neurology*, 41: 843–852.

Evarts, E. V. 1973. Brain mechanisms in movement. *Scientific American*, 229: 96–103.

———— 1979. Brain mechanisms of movement. *Scientific American*, 241: 164–179.

———— and I. Tanji. 1976. Reflex and intended responses in motor cortex pyramidal tract neurons of the monkey. *Journal of Neurophysiology*, 39: 1069–1080.

Falk, D. 1976. Comparative study of the larynx in man and the chimpanzee: Implications for language in Neanderthal. *American Journal of Physical Anthropology*, 43: 123–132.

Farah, M. 1984. The neurological basis of mental imagery: A componential analysis. *Cognition*, 18: 245–272.

Fodor, J. A. 1975. *The language of thought*. New York: Thomas A. Crowell.

———— 1983. *The modularity of mind*. Cambridge, Mass.: MIT Press.

————, M. F. Garret, E. C. Walker, and C. H. Parkes. 1980. Against definitions. *Cognition*, 8: 263–267.

———— and Z. Pylyshyn. 1988. Connectionism and cognitive architecture: A critical analysis. *Cognition*, 28: 3–71.

Fodor, J. D., J. A. Fodor, and M. F. Garrett. 1975. The psychological unreality of semantic representations. *Linguistic Inquiry*, 4: 515–531.

Fouts, R. S. 1973. Acquisition and testing of gestural signs in four young chimpanzees. *Science*, 180: 978–980.

Frazer, R. 1890. *The golden bough: A study in magic and religion*. New York: St. Martins.

Frisk, V., and B. Milner. 1990a. The role of the left hippocampal region in the acquisition and retention of story content. *Neuropsychologia,* 28: 349–359.

―――― 1990b. The relationship of working memory to the immediate recall of stories following unilateral temporal or frontal lobectomy. *Neuropsychologia,* 28: 121–135.

Frye, N. 1957. *Anatomy of criticism: Four essays.* Princeton, N.J.: Princeton University Press.

Gainotti, G., C. Caltagirone, and G. Miceli. 1979. Semantic disorders of auditory language comprehension in right brain-damaged patients. *Journal of Psycholinguistic Research,* 8: 13–20.

Galaburda, A. M., and T. L. Kemper. 1979. Cytoarchitectonic abnormalities in developmental dyslexia: A case study. *Annals of Neurology,* 6: 94–100.

――――, A. M., and D. N. Pandya. 1982. Role of architectonics and connections in the study of primate brain evolution. In E. Armstrong and D. Falk, eds., *Primate brain evolution: Methods and concepts.* New York: Plenum Press.

Gallup, G. G. 1970. Chimpanzees: Self recognition. *Science,* 167: 86–87.

―――― 1982. Self awareness and the emergence of mind in primates. *American Journal of Primatology,* 2: 237–248.

Gamkrelidze, T. V., and V. V. Ivanov. 1990. The early history of Indo-European languages. *Scientific American,* March 1990, pp. 110–116.

Gardner, H. 1982. *Art, mind, and brain.* New York: Basic Books.

―――― 1983. *Frames of mind: The theory of multiple intelligences.* New York: Basic Books.

――――, H. Brownell, W. Wapner, and D. Michelow. 1983. Missing the point: The role of the right hemisphere in the processing of complex linguistic materials. In E. Perecman, ed., *Cognitive processes in the right hemisphere.* New York: Academic Press.

Gardner, R. A., and B. T. Gardner. 1969. Teaching sign language to a chimpanzee. *Science,* 165: 664–672.

―――― 1978. Comparative psychology and language acquisition. In K. Salzinger and F. L. Denmark, eds., *Psychology: The state of the art.* New York: New York Academy of Sciences.

Gazzaniga, M. S. 1970. *The bisected brain.* New York: Appleton Century Crofts.

―――― 1983. Right hemisphere language following bisection: A twenty year perspective. *American Psychologist,* 38: 525–537.

Gelb, I. J. 1963. *A study of writing.* Chicago: University of Chicago Press.

Geschwind, N. 1965. Disconnection syndromes in animals and man. *Brain,* 88: 237–324, 585–644.

―――― 1970. The organization of language and the brain. *Science,* 170: 940–944.

―――― 1984. Neural mechanisms, aphasia and theories of language. In D. Caplan, A. R. Lecours, and A. Smith, eds., *Biological perspectives on language.* Cambridge, Mass.: MIT Press.

────── and A. M. Galaburda, eds. 1984. *Cerebral dominance: The biological foundations*. Cambridge, Mass.: Harvard University Press.

Gibson, J. J. 1950. *The perception of the visual world*. Boston: Houghton Mifflin Company.

────── 1979. *The ecological approach to visual perception*. Boston: Houghton Mifflin Company.

Gillan, D. J., D. Premack, and G. Woodruff. 1981. Reasoning in the chimpanzee: 1. Analogical reasoning. *Journal of Experimental Psychology: Animal Behavior Processes*, 7: 1–17.

Goldin-Meadow, S. 1979. Structure in a manual communication system developed without a language model: Language without a helping hand. In H. A. Whitaker, ed., *Studies in neurolinguistics*. Vol. 4, pp. 124–209. New York: Academic Press.

────── and C. Mylander. 1984. Gestural communication in deaf children: The effects and noneffects of parental input on early language development. *Monographs of the Society for Research on Child Development*, 49: 1–120.

Goldman-Racik, P. S. 1982. Organisation of frontal association cortex in normal and experimentally brain injured primates. In M. A. Arbib, D. Caplan, and J. C. Marshall, eds., *Neural models of language processes*. New York: Academic Press.

Goldstein, K. 1948. *Language and language disturbance*. New York: Grune and Stratton.

Goodall, J. 1971. *In the shadow of man*. London: George Weidenfeld and Nicolson.

────── 1986. *The chimpanzees of Gombe: Patterns of behavior*. Cambridge, Mass.: The Belknap Press of Harvard University Press.

Goodman, N. 1968. *Languages of art*. New York: Bobbs-Merrill.

────── 1976. *Languages of art: An approach to a theory of symbols*. Cambridge: Hackett.

────── 1984. *Of mind and other matters*. Cambridge, Mass.: Harvard University Press.

Gould, S. J. 1980. *The panda's thumb*. New York: Norton.

────── and N. Eldredge. 1977. Punctuated equilibria: The tempo and mode of evolution reconsidered. *Paleobiology*, 3: 115–151.

Gracco, V., and J. Abbs. 1985. Dynamic control of the perioral system during speech: Kinematic analysis of autogenic and nonautogenic sensorimotor processes. *Journal of Neurophysiology*, 54: 418–432.

Greenfield, P. M., and E. S. Savage-Rumbaugh. 1990. Grammatical combination in Pan Paniscus: Processes of learning and invention in the evolution and development of language. In S. Taylor-Parker and K. R. Gibson, eds., *Language and intelligence in monkeys and apes*. New York: Cambridge University Press.

Gregory, R. L. 1981. *Mind in science*. London: Penguin.

Hadingham, E. 1979. *Secrets of the ice age: The world of cave artists*. New York: Walker.

Harnad, S., ed. 1987. *Categorical perception: The groundwork of cognition.* New York: Cambridge University Press.

Harris, R. 1986. *The origin of writing.* London: Duckworth.

Harvey, P. D. A. 1980. *The history of topographical maps.* London: Thames and Hudson.

Harvey, P. M., R. D. Martin, and T. H. Clutton-Brook. 1986. Life histories in comparative perspective. In B. Smuts, D. Cheney, R. Seyfarth, R. Wrangham, and T. Strubsaker, eds., *Primate societies.* Chicago: University of Chicago Press.

Hawkins, G. S. 1963. Stonehenge decoded. *Nature,* 200: 306.

——— 1964. Stonehenge: A neolithic computer. *Nature,* 202: 1258.

Hayashi, M., H. Ulatowska, and S. Sasanuma. 1985. Subcortical aphasia with deep dyslexia: A case study of a Japanese patient. *Brain and Language,* 25: 141–147.

Hayes, C. 1951. *The ape in our house.* New York: Harper and Row.

Hayes, K. J., and C. Hayes. 1951. The intellectual development of a home raised chimp. *Proceedings of the American Philosophical Society,* 95: 105–109.

Head, H. 1926. *Aphasia and kindred disorders of speech.* Cambridge: Cambridge University Press.

Hebb, D. O. 1949. *The organization of behavior.* New York: Wiley.

Hécaen, H., and M. L. Albert. 1978. *Human neuropsychology.* New York: Wiley.

———, H., and R. Angelergues. 1963. *La cécité psychique.* Paris: Masson.

Hellmans, A., and B. Bunch. 1988. *The timetables of science.* New York: Simon and Schuster.

Herndon, M., and N. McLeod. 1980. *The ethnography of musical performance.* Norwood, Pa.: Norwood Editions.

Heron, W. 1967. The pathology of boredom. In J. L. McGaugh, N. Weinberger, and R. Whalen, eds., *Psychobiology.* San Francisco: W. H. Freeman.

Hewes, G. W. 1977. Language origin theories. In D. M. Rumbaugh, ed., *Language learning by a chimpanzee: The Lana project.* New York: Academic Press.

Hinde, R. A., and T. E. Rowell. 1962. Communication by postures and facial expressions in the rhesus monkey (*Macaca Mulatta*). *Proceedings of the Zoological Society of London,* 138: 103–113.

Hitch, G. J., and A. D. Baddeley. 1976. Verbal reasoning and working memory. *Quarterly Journal of Experimental Psychology,* 28: 603–621.

Holloway, R. L. 1974. The casts of fossil hominid brains. *Scientific American,* 231: 106–115.

——— and M. C. de la Coste-Lareymondie. 1982. Brain endocast asymmetry in pongids and hominids: Some preliminary findings on the paleontology of cerebral dominance. *American Journal of Physical Anthropology,* 58: 101–110.

Hudson, R. 1984. *Word grammar.* London: Basil Blackwell.

Humphrey, M. E., and O. Zangwill. 1951. Cessation of dreaming after brain injury. *Journal of Neurology, Neurosurgery and Psychiatry,* 14: 322–325.

Isaac, G., and E. McCown, eds. 1979. *Human origins.* Menlo Park, CA: W. A. Benjamin.

Izard, C. E. 1971. *The face of emotion.* New York: Appleton Century Crofts.

Jackendoff, R. 1983. *Semantics and cognition.* Cambridge, Mass.: MIT Press.

Jackson, H. 1868. Notes on the physiology and pathology of the nervous system. In *Medical Times and Gazette,* 2: 526. Reprinted in *Brain,* 38: 65–71.

——— 1874. On the nature of the duality of the brain. In *Medical Press and Circular,* 1: 63. Reprinted in *Brain,* 38: 96–103.

James, W. 1890. *The principles of psychology.* New York: Holt.

Jaynes, J. 1976. Constraints on a solution of when language began. *Annals of the New York Academy of Sciences,* 280: 312–318.

Jerison, H. 1973. *Evolution of the brain and intelligence.* New York: Academic Press.

Johanson, D., and Y. Coppens. 1976. A preliminary anatomical diagnosis of the first Plio-Pleistocene hominid discoveries in the central Afar, Ethiopia. *American Journal of Physical Anthropology,* 45: 217–234.

———, D., and M. Edey. 1981. *Lucy, the beginnings of humankind.* New York: Simon and Schuster.

———, D., and M. Taieb. 1976. Plio-Pleistocene hominid discoveries in Hadar, Ethiopia. *Nature,* 260: 293–297.

———, D., and T. White. 1979. A systematic assessment of early African hominids. *Science,* 203: 321–330.

Johansson, G. 1968. Visual perception of biological motion and a model for its analysis. *Perception and Psychophysics,* 14: 201–211.

Johnson-Laird, P. N. 1983. *Mental Models.* Cambridge, Mass.: Harvard University Press.

Katz, J. J., and J. A. Fodor. 1965. The structure of a semantic theory. *Language,* 39: 170–210.

Kellogg, W. N. 1980. Communication and language in the home raised chimpanzee. In T. A. Sebeok and J. Sebeok, eds., *Speaking of apes.* New York: Plenum Press.

——— and L. A. Kellogg. 1933. *The ape and the child: A study of environmental influence upon early behavior.* New York: Whittlesey House.

Kempen, G., and P. Huijbers. 1983. The lexicalization process in sentence production and naming: Indirect election of words. *Cognition,* 14: 185–209.

Kennedy, A. 1984. *The psychology of reading.* London: Methuen.

Kimura, D. 1976. The neurological basis of language qua gestures. In H. Whitaker and H. A. Whitaker, eds., *Current trends in neurolinguistics.* New York: Academic Press.

——— 1979. Neuromotor mechanisms in the evolution of human communication. In H. D. Steklis and M. J. Raleigh, eds., *Neurobiology of social communication in primates.* New York: Academic Press.

Kinsbourne, M. 1974. Cerebral control and mental evolution. In M. Kinsbourne and W. L. Smith, eds., *Hemisphere disconnection and cerebral function.* Springfield, Ill.: Thomas.

Kintsch, W. 1974. *The representation of meaning in memory.* Hillsdale, N.J.: Erlbaum.

———— 1988. The role of knowledge in discourse comprehension: A construction-integration model. *Psychological Review,* 95: 163–182.

Klima, E., and U. Bellugi. 1979. *The signs of language.* Cambridge, Mass.: Harvard University Press.

Knopman, D. S., O. A. Selnes, N. Niccum, D. Rubens, D. Yock, and D. Larson. 1983. A longitudinal study of speech fluency in aphasia: CT correlates of recovery and persistent nonfluency. *Neurology,* 33: 1170–1178.

Kohler, W. 1925. *The mentality of apes.* New York: Harcourt, Brace.

Kolb, B., and I. Q. Whishaw. 1990. *Fundamentals of human neuropsychology,* 3rd ed. New York: W. H. Freeman.

Kolers, P., M. E. Wrolstead, and H. Bouma, eds. 1979. *Processing of visible language,* vol. 1. New York: Plenum.

Kosslyn, S. M. 1987. Seeing and imagining in the cerebral hemispheres. *Psychological Review,* 94: 148–175.

———— 1988. Aspects of a cognitive neuroscience of imagery. *Science,* 240: 1621–1626.

Kuhl, P. K. 1988. Auditory perception and the evolution of speech. *Human Evolution,* 3: 21–45.

———— and J. D. Miller. 1975. Speech perception by the Chinchilla: Voiced-voiceless distinction in alveolar plosive consonants. *Science,* 190: 69–72.

Kussmaul, A. 1877. Die Storungen der Sprache. *Ziemssens Handbuch der Speciellen Pathologie und Therapie,* 12: 1–300.

Lane, H. 1984. *When the mind hears.* New York: Random House.

Langer, S. K. 1967. *Mind: An essay in human feeling.* Baltimore: Johns Hopkins University Press.

Lashley, K. S. 1929. *Brain mechanisms and intelligence.* Chicago: University of Chicago Press.

———— 1949. Persistent problems in the evolution of mind. *Quarterly Review of Biology,* 24: 28–42.

———— 1950. In search of the engram. In F. Beach, D. O. Hebb, C. Morgan, and H. Nissen, eds., *The neuropsychology of Lashley.* New York: McGraw-Hill.

Leakey, L., and M. Leakey. 1959. A new fossil skull from Olduvai. *Nature,* 184: 491–493.

Leakey, R. E. 1973. Skull 1470. *National Geographic,* June, pp. 819–829.

————, R. E. 1981. *The making of mankind.* London: Michael Joseph.

Lecours, A. R., and Y. Joanette. 1980. Linguistic and other aspects of paroxysmal aphasia. *Brain and Language,* 10: 1–23.

Lee, R. B. and I. De Vore. 1976. *Kalahari hunter-gatherers: Studies of the !Kung Sang and their neighbors.* Cambridge, Mass.: MIT Press.

LeGros Clark, W. E. 1967. *Man-apes or ape-men?* New York: Holt, Rinehart and Winston.

Leibnitz, G. W. 1714. The Monadology. In P. Weiner, ed. 1951. *Leibniz: Selections*. New York: Scribner's.

LeMay, M. 1975. The language capability of Neanderthal man. *American Journal of Physical Anthropology,* 42: 9–14.

———— 1976. Morphological cerebral asymmetries of modern man, fossil man and non-human primate. *Annals of the New York Academy of Sciences,* 280: 349–366.

———— 1984. Radiological, developmental, and fossil asymmetries. In N. Geschwind and A. M. Galaburda, eds., *Cerebral dominance: The biological foundations*. Cambridge, Mass.: Harvard University Press.

Lenneberg, E. 1967. *Biological foundations of language*. New York: Wiley.

———— 1980. Of language knowledge, apes, and brains. In T. A. Sebeok and J. Sebeok, eds., *Speaking of apes*. New York: Plenum Press.

Levelt, W. J. M. 1989. *Speaking: From intention to articulation*. Cambridge, Mass.: MIT Press.

Lewis, C. M. 1987. Indian delimitations of primary geographic regions. In T. E. Ross and T. G. Moore, eds., *A cultural geography of North American Indians*. Boulder, Colorado: Westview Press.

Lichtheim, L. 1885. On aphasia. *Brain,* 7: 433–484.

Lieberman, P. 1973. On the evolution of human language: A unified view. *Cognition,* 2: 59–94.

———— 1975. *On the origins of language: An introduction to the evolution of human speech*. New York: Macmillan.

———— 1984. *The biology and evolution of language*. Cambridge, Mass.: Harvard University Press.

————, P. 1990. Rule-governed processes and brains—speech, syntax and toolmaking. Paper delivered at the Wenner-Gren Symposium for Anthropological Research, Cascais, Portugal, March 16–24.

————, P., J. Friedman, and L. S. Feldman. 1990. Syntax comprehension deficits in Parkinson's disease. *Journal of Nervous and Mental Disease,* 178: 360–365.

Lindsley, D. B. 1958. The reticular system and perceptual discrimination. In H. H. Jasper, ed., *Reticular formation of the brain*. Boston, Mass.: Little Brown.

Logan, R. K. 1986. *The alphabet effect*. New York: William Morrow.

Lorenz, K. 1965. *Evolution and modification of behavior*. Chicago: University of Chicago Press.

Lovejoy, C. O. 1980. Hominid origins: The role of bipedalism. *American Journal of Physical Anthropology,* 52: 250.

Luria, A. R. 1966. *Higher cortical functions in man*. New York: Basic Books.

Lyons, J. 1988. Origins of language. In A. C. Fabian, ed., *Origins: The Darwin College Lectures*. Cambridge: Cambridge University Press.

Mackinnon, J. R. 1974. The behaviour and ecology of wild orang-utans (Pongo Pygmaeus). *Animal Behaviour,* 22: 3–74.

MacNeilage, P. 1980. Emerging concepts in speech control. In G. S. Stelmach and J. Requin, eds., *Tutorials in motor behavior*. Amsterdam: North Holland.

Marie, P. 1906. Révision de la question sur l'aphasie: La troisième convolution frontale gauche ne joue aucun rôle spécial dans le fonction du langage. *Semaine Medicale* (Paris), 21: 241–247.

Marshack, A. 1972. *The roots of civilization: The cognitive beginnings of man's first art, symbol and notation*. New York: McGraw-Hill.

Marshall, J. C., and F. Newcombe. (1973). Patterns of paralexia: A psycholinguistic approach. *Journal of Psycholinguistic Research*, 2: 175–199.

——— 1980. The conceptual status of deep dyslexia: An historical perspective. In M. Coltheart, K. Patterson, and J. Marshall, eds., *Deep dyslexia*. London: Routledge.

Marshall, L. 1962. !Kung bushman religious beliefs. *Africa*, 32: 227.

Martin, R. D. 1984. Body size, brain size and feeding strategies. In D. J. Chivers, B. Wood, and A. Bilsborough, eds., *Food acquisition and processing in primates*. New York: Plenum Press.

Martin, S. E. 1972. Nonalphabetic writing systems: Some observations. In J. F. Kavanagh and I. G. Mattingly, eds., *Language by ear and eye*. Cambridge, Mass.: MIT Press.

McCabe, V., and G. Balzano. 1986. *Event cognition: An ecological perspective*. Hillsdale, N.J.: Erlbaum.

McClelland, J., and D. Rumelhart. 1986. *Parallel distributed processing: Explorations in the microstructure of cognition*. Vol. 2. Cambridge, Mass.: MIT Press.

McDougall, W. 1911. *Body and mind*. London: Methuen.

McLean, P. D. 1973. *A triune concept of the brain and behavior*. Toronto: University of Toronto Press.

McNeill, D. 1985. So you think gestures are nonverbal. *Psychological Review*, 92: 350–371.

——— and E. Levy. 1982. Conceptual representations in language activity and gesture. In R. Jarvella and W. Klein, eds., *Speech, place, and action: Studies in deixis and related topics*. Chichester, England: Wiley.

Merton, P. A. 1964. Human position sense and sense of effort. In *Homeostasis and feedback mechanisms*. Cambridge, Eng.: Cambridge University Press.

Merzenich, M. M., R. J. Nelson, J. H. Kaas, M. P. Stryker, W. M. Jenkins, J. M. Zook, M. S. Cynader, and A. Schoppana. 1987. Variability in hand surface representations in areas 3b and 1 in adult owl and squirrel monkeys. *Journal of Comparative Neurology*, 258: 281–296.

Mewhort, D. J. K. 1974. Accuracy and order of report in tachistoscopic identification. *Canadian Journal of Psychology*, 28: 383–398.

Miceli, G., M. Silveri, G. Villa, and A. Caramazza. 1984. On the basis for the agrammatic's difficulty in producing main verbs. *Cortex*, 20: 207–220.

Miller, G. 1956. The magical number seven plus or minus two: Some limits on our capacity to process information. *Psychological Review*, 63: 81–97.

————, G. Heise, and W. Lichten. 1951. The intelligibility of speech as a function of the context of the test materials. *Journal of Experimental Psychology,* 41: 329–335.

Milner, B. 1965. Brain mechanisms suggested by studies of temporal lobes. In F. L. Darley, ed., *Brain mechanisms underlying speech and language.* New York: Grune and Stratton.

———— 1966. Amnesia following operations on the temporal lobes. In C. Whitty and O. Zangwill, eds., *Amnesia.* London: Butterworth.

———— 1975. Psychological aspects of focal epilepsy and its neurosurgical treatment. In D. O. Purpura, J. K. Penry, and R. D. Walter, eds., *Advances in Neurology.* Vol. 8. New York: Raven Press.

————, C. Branch, and T. Rasmussen. 1964. Observations on cerebral dominance. In A. de Reuck and M. O'Connor, eds., *Disorders of language.* London: Churchill.

Milton, K. 1988. Foraging behaviour and the evolution of primate intelligence. In R. Byrne and A. Whiten, eds., *Machiavellian intelligence.* New York: Plenum Press.

Minsky, M. 1985. *The society of mind.* New York: Simon and Schuster.

Moerck, E. L. 1989. The fuzzy set called "imitations." In G. E. Speidel and K. E. Nelson, eds., *The many faces of imitation in language learning,* pp. 277–303. New York: Springer-Verlag.

Mohr, J. P. 1976. Broca's area and Broca's aphasia. In H. Whitaker, ed., *Studies in neurolinguistics.* Vol. 2. New York: Academic Press.

Morton, J. 1980. The logogen model and orthographic structure. In U. Frith, ed., *Cognitive approaches to spelling.* London: Academic Press.

———— 1981. The status of information processing models of language. *Philosophical Transactions of the Royal Society of London B,* 295: 387–396.

———— 1984. Brain-based and non-brain-based models of language. In D. Caplan, A. R. Lecours, and A. Smith, eds., *Biological perspectives on language.* Cambridge, Mass.: MIT Press.

Mountcastle, V. B., B. C. Motter, M. A. Steinmetz, and C. J. Duffy. 1984. Looking and seeing: Visual functions of the parietal lobe. In G. Edelman, W. Gall, and W. Cowan, eds., *Dynamic aspects of neocortical function.* New York: Wiley.

Munn, N. D. 1973. The spatial presentation of cosmic order in Walbiri iconography. In A. Forge, ed., *Primitive art and society.* London: Oxford University Press.

Munn, N. L. 1971. *The evolution of the human mind.* New York: Houghton Mifflin.

Murray, D. J. 1967. The role of speech responses in short term memory. *Canadian Journal of Psychology,* 21: 263–276.

Myers, R. E. 1976. Comparative neurology of vocalisation and speech: Proof of a dichotomy. *Annals of the New York Academy of Sciences,* 280: 745–760.

———— 1984. Comment: Myers on Gazzaniga. *American Psychologist,* 39: 542–546.

Norman, D. A., and T. Shallice. 1986. Attention to action: Willed and automatic control of behavior. In R. J. Davidson, G. E. Schwartz, and D. Shapiro, eds., *Consciousness and self-regulation.* Vol. 4. New York: Plenum Press.

Nottebohm, F. 1984. Learning, forgetting and brain repair. In N. Geschwind and A. M. Galaburda, eds., *Cerebral dominance: The biological foundations.* Cambridge, Mass.: Harvard University Press.

Oakley, D. A. 1983. The varieties of memory: A phylogenetic approach. In A. Mays, ed., *Memory in animals and humans.* Wokingham: Van Nostrand Reinhold.

——— 1985. Cognition and imagery in animals. In D. Oakley, ed., *Brain and Mind.* London: Methuen.

O'Keefe, J., and L. Nadel. 1977. *The hippocampus as a cognitive map.* New York: Oxford University Press.

Oldfield, R. C. 1963. Individual vocabulary and semantic currency: A preliminary study. *British Journal of Social and Clinical Psychology,* 2: 122–130.

Olton, D. S. 1977. Spatial memory. *Scientific American,* 236: 82–98.

Paradis, M., H. Hagiwara, and N. Hildebrandt. 1985. *Neurolinguistic aspects of the Japanese writing system.* Orlando, Fla: Academic Press.

Parker, S. T., and K. R. Gibson. 1979. A developmental model for the evolution of language and intelligence in early hominids. *Behavioral and Brain Sciences,* 2: 367–408.

Pasnak, R. 1979. Acquisition of prerequisites to conservation by Macaques. *Journal of Experimental Psychology: Animal Behavior Processes,* 5: 37–51.

Passingham, R. E. 1982. *The human primate.* New York: Freeman.

——— and G. Ettlinger. 1974. A comparison of cortical functions. In C. C. Pfieffer and J. R. Smythies, eds., *International Review of Neurobiology.* Vol. 16. New York: Academic Press.

Patterson, F. G. 1980. Innovative uses of language by a gorilla: A case study. In K. Nelson, ed., *Children's language,* pp. 497–561. New York: Gardner Press.

Patterson, K., and Besner, D. 1984. Is the right hemisphere literate? *Cognitive Neuropsychology,* 1: 315–342.

Pedelty, L. 1985. *Gestures in aphasia.* Unpublished Ph.d. Diss., University of Chicago, Chicago.

Penfield, W., and L. Roberts. 1959. *Speech and brain mechanisms.* Princeton, N.J.: Princeton University Press.

Peterson, L. R., and M. J. Peterson. 1959. Short term retention of individual verbal items. *Journal of Experimental Psychology,* 58: 193–198.

Pfungst, O. 1965. *Clever Hans, the horse of Mr. Von Osten.* Translated from the 1911 German edition by C. L. Rahn. New York: Holt, Rinehart and Winston.

Piaget, J. 1980. Language and learning. In M. Piattelli-Palmarini, ed., *Language and learning: The debate between Jean Piaget and Noam Chomsky.* Cambridge, Mass.: Harvard University Press.

Ploog, D. 1981. Neurobiology of primate audio-vocal behavior. *Brain Research Reviews*, 3: 35–61.

Plotkin, H. C. 1982. Evolutionary epistemology and evolutionary theory. In H. C. Plotkin, ed., *Learning, development, and culture: Essays in evolutionary epistemology*. Chichester: Wiley.

────── 1987. Evolutionary epistemology as science. *Biology and Philosophy*, 2: 295–313.

────── 1988. An evolutionary epistemological approach to the evolution of intelligence. In H. J. Jerison, ed., *The evolutionary biology of intelligence*. New York: Springer Verlag.

Poizner, H., E. S. Klima, and V. Bellugi. 1987. *What the hands reveal about the brain*. Cambridge, Mass.: MIT Press.

Pollack, I., and J. M. Pickett. 1964. Intelligibility of excerpts from fluent speech: Auditory vs. structural context. *Journal of Verbal Learning and Verbal Behavior*, 3: 79–84.

Popper, K. S., and J. C. Eccles. 1977. *The self and its brain*. Berlin: Springer.

Premack, D. 1976. *Intelligence in ape and man*. Hillsdale, N.J.: Erlbaum.

────── 1987. *Gavagai*. Cambridge, Mass.: MIT Press.

Pribram, K. 1971. *Languages of the Brain*. Englewood Cliffs, N.J.: Prentice-Hall.

Rayner, K. 1978. Eye movements in reading and information processing. *Psychological Bulletin*, 85: 618–660.

────── 1979. Eye movements in reading: Eye guidance and integration. In P. Kolers, M. Wrolstead, and H. Bouma, eds., *Processing of visible language*. Vol. 1. New York: Plenum Press.

Recorde, R. 1557. *Algebra*.

Renfrew, C. 1987. *Archaeology and language*. London: Jonathan Cape.

Rodman, P. 1973. Population composition and adaptive organization among orang-utans of the Kutai Reserve. In R. P. Michael and J. Crook, eds., *Primate ecology*. London: Academic Press.

Roeltgen, D. P. 1987. Loss of deep dyslexic reading ability from a second left-hemisphere lesion. *Archives of Neurology*, 44: 346–348.

Ross, E. D. 1981. The aprosodias: Functional-anatomical organization of the affective components of language in the right hemisphere. *Archives of Neurology*, 38: 561–569.

────── and M. Mesulam. 1979. Dominant language functions of the right hemisphere? Prosody and emotional gesturing. *Archives of Neurology*, 36: 144–148.

Rumbaugh, D. M. 1974. *Language learning by a chimpanzee: The Lana project*. New York: Academic Press.

────── 1977. Language behavior of apes. In A. M. Schrier, ed., *Behavioral primatology*. Hillsdale, N.J.: Erlbaum.

Sarich, V. M. 1980. Molecular clocks and hominid evolution after 12 years. *American Journal of Physical Anthropology*, 52: 275–276.

Sasanuma, S. 1985. Surface dyslexia and dysgraphia: How are they manifested in Japanese? In K. E. Patterson, M. Coltheart, and J. C. Marshall, eds., *Surface dyslexia*. London: Erlbaum.

Savage-Rumbaugh, E. S. 1980. Linguistically-mediated tool use and exchange by chimpanzees. In T. Sebeok, and J. Sebeok, eds., *Speaking of apes*. New York: Plenum Press.

——— 1986. *Ape language: From conditioned response to a symbol*. New York: Columbia University Press.

———, K. McDonald, R. A. Sevcik, W. Hopkins, and E. Rubert. 1986. Spontaneous symbol acquisition and communicative use by pygmy chimpanzees (*Pan paniscus*). *Journal of Experimental Psychology: General*, 115: 211–235.

———, D. M. Rumbaugh, and S. Boysen. 1978. Linguistically-mediated tool use and exchange by chimpanzees (*Pan troglodytes*). *Behavioral and Brain Sciences*, 4: 539–554.

———, R. A. Sevcik, K. E. Brakke, D. M. Rumbaugh. 1990. Symbols: Their communicative use, comprehension, and combination by bonobos (*Pan paniscus*). *Advances in Infancy Research* 6: 221–278.

Schank, R., and R. Abelson. 1977. *Scripts, plans, goals, and understanding: An inquiry into human knowledge structures*. Hillsdale, N.J.: Erlbaum.

Schmandt-Besserat, D. 1978. The earliest precursor of writing. *Scientific American*, 238: 6.

Schweiger, A., E. Zaidel, T. Field, and B. Dobkin. 1989. Right hemisphere contribution to lexical access in an aphasic with deep dyslexia. *Brain and Language*, 37: 73–89.

Selnes, O. A., D. S. Knopman, N. Niccum, and A. B. Rubens. 1983. Computed tomographic scan correlates of auditory comprehension deficits in aphasia: A prospective recovery study. *Annals of Neurology*, 13: 558–566.

Shallice, T. 1988. *From neuropsychology to mental structure*. Cambridge, Eng.: Cambridge University Press.

——— and E. K. Warrington. 1980. Single and multiple component central dyslexic syndromes. In M. Coltheart, K. Patterson, and J. Marshall, eds., *Deep dyslexia*. London: Routledge.

Sherry, D. F., and D. L. Schacter. 1987. The evolution of multiple memory systems. *Psychological Review*, 94: 439–454.

Skinner, B. F. 1957. *Verbal behavior*. New York: Appleton-Century-Crofts.

Smith, C. D. 1987. Cartography in the Prehistoric Period in the old world: Europe, the middle East, and North Africa. In J. B. Harley and D. Woodward, eds., *History of cartography, Vol. 1: Cartography in prehistoric, ancient, and medieval Europe and the Mediterranean*. Chicago: University of Chicago Press.

Spence, J. D. 1984. *The memory palace of Matteo Ricci*. New York: Penguin.

Sperry, R. 1966. Brain bisection and consciousness. In J. C. Eccles, ed., *Brain and conscious experience*. New York: Springer Verlag.

——— 1968. Mental unity following surgical disconnection of the cerebral hemispheres. *Harvey lectures,* series 62. New York: Academic Press.

Squire, L. 1984. *Neuropsychology of memory.* New York: Guilford Press.

——— 1987. *Memory and brain.* New York: Oxford University Press.

Stephens, D., and K. Tuite. 1983. *The hermeneutics of gesture.* Paper presented at the meeting of the American Anthropological Association, Chicago. Cited in McNeill, 1985.

Steklis, H. D., and S. Harnad. 1976. Critical stages in the evolution of language. *Annals of the New York Academy of Sciences,* 280: 445.

Stringer, C. B., and P. Andrews. 1988. Genetic and fossil evidence for the origin of modern humans. *Science* 239:1263–1268.

Stuart, D., and S. D. Houston. 1989. Maya writing. *Scientific American,* 249: 82–89.

Studdert-Kennedy, M. 1975. Speech perception. In N. J. Lass, ed., *Contemporary issues in experimental phonetics.* Springfield, Ill.: Charles C. Thomas.

Stuss, D., and D. F. Benson. 1986. *The frontal lobes.* New York: Raven Press.

Sutton, D., and U. Jurgens. 1988. Neural control of vocalization. In H. D. Steklis and J. Erwin, eds., *Comparative primatology and biology, Vol. 4.* New York: Liss.

Taylor, I. 1976. *Introduction to psycholinguistics.* New York: Holt.

Terrace, H., and T. G. Bever. 1980. What might be learned from studying language in the chimpanzee? The importance of symbolizing oneself. In T. Sebeok and J. Sebeok, eds., *Speaking of apes.* New York: Plenum.

Trehub, S. E., D. Bull, and B. A. Schneider. 1981. Infant speech and nonspeech perception: A review and reevaluation. In R. L. Scheifenbusch and D. D. Bricker, eds., *Early language: Acquisition and intervention,* pp. 9–50. Baltimore: University Park Press.

Tulving, E. 1983. *Elements of episodic memory.* New York: Oxford University Press.

Twyman, M. 1979. A schema for studying graphic language. In P. Kolers, M. Wrolstad, and H. Bouma, eds., *Processing of visible language.* Vol. 1. New York: Plenum Press.

Van Hooff, J. A. R. A. M. 1967. Facial displays of Catarrhine monkeys and apes. In D. Morris, ed., *Primate ethology.* London: Weidenfeld and Nicolson.

Von Senden, M. 1960. *Space and sight.* London: Methuen.

Vygotsky, L. 1962. *Language and thought.* Cambridge, Mass.: MIT Press.

Walker, C. B. F. 1987. *Cuneiform.* London: British Museum Publications.

Warren, R. M. 1970. Perceptual restoration of missing speech sounds. *Science,* 167: 392–393.

——— 1976. Auditory perception and speech evolution. *Annals of the New York Academy of Sciences,* 280: 708–731.

Weiner, J. S., and K. P. Oakley. 1954. The Piltdown fraud: Available evidence reviewed. *American Journal of Physical Anthropology,* 12: 1–7.

Weiskrantz, L. 1988. *Thought without language.* New York: Oxford University Press.

Wernicke, C. 1874. *Der Aphasische Symptomenkomplex*. Breslau: Cohn and Weigart. Trans. *Boston studies in philosophy of science*, 4: 34–97.

Wind, J. 1976. Phylogeny of the human vocal tract. *Annals of the New York Academy of Sciences*. Vol. 280: 612.

Wittgenstein, L. 1922. *Tractatus Logico-Philosophicus*. Trans. D. F. Pears and B. V. McGuiness. London: Routledge and Kegan Paul, 1961.

Yerkes, R. M. 1943. *Chimpanzees: A laboratory colony*. New Haven, Conn: Yale University Press.

Young, J. Z. 1988. *Philosophy and the brain*. New York: Oxford University Press.

Zaidel, E. 1976. Auditory vocabulary in the right hemisphere following brain bisection and hemidecortication. *Cortex*, 12: 191–221.

———— 1983. Disconnection syndrome as a model for laterality effects in the normal brain. In J. B. Hellige, ed., *Cerebral hemisphere asymmetry: Method, theory, and application*. New York: Praeger.

Acknowledgments

This book has its origins in my undergraduate years, when I encountered the work of two important system-builders in my chosen fields of philosophy and literature—Bernard Lonergan and Northrop Frye. Their ideas piqued my curiosity about what mental structures lay beneath the fantastic cornucopia of theories and myths they had spent their lives exploring. My curiosity was increased by the ideas of Marshall McLuhan about the fundamental cognitive transformations imposed by printing and television: What were humans becoming? How could these cognitive transformations be understood? I decided to study psychology to try to view the problem from a (presumably) more scientific aspect; and in graduate school at McGill University I encountered one of the preeminent neuropsychologists of that time, Donald Hebb. Hebb was also a system-builder, but he worked down in the dark mechanistic depths of the mind. Where Frye had been on the bridge, surveying the passing landscape of human culture, Hebb was in the engine room, tinkering with the mental machinery below. For the past twenty years I have worked down in the engine room, while keeping an eye on the bridge, and this is my attempt to connect what is happening at the cultural surface with what little we know about what is happening below.

I would like to acknowledge the help of the Advisory Research Committee of Queen's University, which provided me with financial support in launching this project, and the Faculty of Arts and Science, which granted me a sabbatical during which the book was conceived and largely written. A productive part of that sabbatical was spent at

University College, London, England, and I gratefully acknowledge
the support of Bob Audley, head of the Department of Psychology of
that institution. The Cognitive Science group at Queen's, which has
included a variety of faculty and graduate students from the depart-
ments of psychology, computing science, biology, and philosophy, has
provided me with an important forum for exchanging ideas. To many
colleagues in the field of human cognitive neuroscience, particularly
those involved in the early EPIC and ICON meetings, I wish to express my appreciation for their company in the search for a modicum
of truth about mind-matters.

The manuscript was read and commented upon, in whole or in part,
by Terry Deacon, Howard Eichenbaum, Patricia Greenfield, Yves Joanette, André Roch Lecours, Philip Lieberman, Dave Oakley, Henry
Plotkin, and Sheldon White. I would like especially to thank Howard
Boyer of Harvard University Press for shepherding this book through
the process of review and negotiation, and Susan Wallace for more
than mere editing; the manuscript was improved in substantive ways
thanks to her rare skill. I had valuable help from Stephen Popiel in
assembling references and material for illustrations, and from
Heather Heintzmann and Najum Rashid in producing the initial
manuscript. Nancy Cutway's expertise sped up the process of prepar-
ing the final mansucript, as did the computer-graphic skills of Julian
Donald and Monica Hurt.

The stanzas from Dylan Thomas's "In the Beginning" are excerpted
from *Dylan Thomas: Collected Poems, 1934–52* (London: J. M. Dent
and Sons, 1952), and reprinted by permission of the publishers. Figures 4.1 and 4.3 are adapted and reprinted by permission of the publishers from *The Human Primate* by R. E. Passingham (New York:
Copyright © 1982 by W. H. Freeman and Company). Figure 4.2 is
adapted and reprinted by permission of the publishers from *The Biology of Language* by Philip Lieberman (Cambridge: Harvard University Press, Copyright © 1984 by the President and Fellows of Harvard
College). Figure 5.1 is reprinted by permission of the publishers from
Brain and Mind by David Oakley (London: Methuen & Co., 1985).
Table 7.1 and Figure 7.1 are adapted and reprinted by permission of
the publishers from *Speaking* by W. Levelt (Cambridge: MIT Press,
Copyright © 1989 by Massachusetts Institute of Technology).

I dedicate this book to the memory of my mother, and also to my
father; to my siblings, Jane and Michael; to my sons, Peter and Julian; and especially to my wife, Thaïs, who has stuck with me through
thick (a great deal of thick) and thin, and who remains after all these
years my closest friend and most respected critic.

*I*ndex